Modern Recording Techniques

Fourth Edition

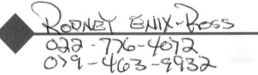

Modern Recording Techniques

Fourth Edition

Focal Press

Boston Oxford Johannesburg Melbourne New Delhi Singapore

Contents

Acknowledgments

First off all, I'd like you to know that the subtle, yet seriously deranged guy on the cover is me…. Now down to biz….

I'd like to thank my partner, Bruce F. Hammerslough, for putting up with the general rantin', ravin', and all 'round craziness that goes into writing a never-ending epic. Same goes for my best buddies: David "Trigger" M. Hines (Seattle), Steve "Stevo" L. Royea (Vancouver) and Wes Bulla (Nashville). Thanks also go to Sean-Michael (freespirit foundation) for his technical assistance and to Shirley Eggerling for constantly dealing with the tons of snail-mail that I received during production. A few bows to the east go to Jordan Gold and Sunthar Visuvalingam at Sams for getting this puppy off the ground and into your hands. Last but not least, I'd like to thank my technical editor, Bruce Bartlett, for his expertise. And special thanks to my editor and recent good buddy, Jodi Jensen, for being a great manuscript editor and an understanding voice over the wires that often kept me from screaming down the streets in a frenzied stupor…. Major time thanx to y'all!

I'd also like to thank the following individuals and companies who assisted in the preparation of this book by providing photographs and technical information:

Ed Simeone, T.C. Electronic; Nan Ferris, Meridian Data, Inc.; Douglas N. Beard, Studer U.S.A.; Chris Good, Sonic Science; Andy Wild, Euphonix; Gregg L. Perry, Young Chang Research and Development Institute; Shirley Beyer, Beyerdynamic; Joyce Greenawalt and Michael MacDonald, Yamaha Corporation of America; Peggy Kennedy, Timeline Vista, Inc.; Barry Fox, QMI; Nick Franks and Steve Harvey, Amek Technology Group, Plc; Mary Conway, Alesis Studio Electronics; Chrissie McDaniel, Aphex Systems; David Kimm, Apogee Electronics Corp.; Fred Balch, Desper Products, Inc.; Carmen David, Sonic Solutions; EveAnna Manley, Manley Laboratories, Inc.; Glen Ilacqua, Nicoll Public Relations; Laura Sordi, Sennheiser Electronic Corporation; Stormy Connor, JLCooper Electronics; Sherri Swingle, Symetrix, Inc.; Art Noxon, Acoustic Sciences Corporation; Carl Tatz, Recording Arts; Danielle Ciardullo, Korg USA, Inc.; Davida Rochman, Shure Brothers, Inc.; David Kilkenny, Twelve Tone Systems; Rick Gentry, Rockford Corp., Haffler Div.; Catherine Moreau, Sony Electronics, Inc.; Deborah Otte, Music Quest; Mark S. Zachmann, The Blue Ribbon Soundworks, Ltd.; Eileen Tuuri, Dolby Labs, Inc.; Wendy DeBernardo, ddrum, Inc.; Bonnie Gardner, Whirlwind Music Dist., Inc.; Lissanne Gillham, Dynatek Automation Systems, Inc.; Andrew Walls, nVision, Inc.; Richard Zimmerman, Night Technologies International; Bernard W. Chlop, Jr., Systems Development Group, Inc.; Jim Giordano, Studiomaster; Micheal Paul Inman, MIDIMAN; Richard J. Mancuso, DIC Digital; Robbie, Popper Stoppers; Bryant, Temporal Acuity Products; Bob Katz Mr. Jitter, Digital Domain; Curt Smith, Sascom Marketing Group; Wynton R. Morrow, Avalon Design; Robert Pursell, Westlake Audio; Jack Kelly, Group One Ltd.; Jim Cooper, Mark of the Unicorn (MOTU); Kristen Wiltse, Giles Communications; Jesse

Walsh Communications; Francine Moran, Sony Electronics Inc.; Jeff Wilson and Al Pickard, Digital Audio Labs; Terri Murphy, Acoustical Solutions, Inc.; Bob Kraft, Versadyne; Sal Greco, Paisley Park Studios; Lenore Zenger, Opcode Systems; Patti Du Fresne, Telex; Eric Bell Top Dog, Howling Dog Systems; H. Hildebrand, Jupiter Systems; John A. Kelly, Nobler; Lesley Cutter, Nelson/Clyne, Inc.; James Goodman, Otari Corporation; Jason Ojalvo, Disc Makers; Jack Knight, Samson; Joanne Darlington, AMS/Neve; Russ Jones and Liz Stasy, Steinberg North America; Jerry Kovarsky, Ensoniq; Kelly Naumann, Russ Berger Design Group, Inc.; Marc Bertrand, TGI North America, Inc.; Georges Arboretum Systems, Inc.; Kathy M. Johnson, Innovative Quality Software; Don Gates, Digidesign, Inc.; James Fowler Minister of Propaganda, Sarah, Mackie; Guy Charbonneau, Le Mobile Remote Recording Studio; Suzi Loritz and Wayne Morris, DOD Electronics Corporation; and Zoraya Mendez-DeCosmis, Music Sales Corporation.

About the Authors

David Miles Huber is widely acclaimed in the recording industry as an author, musician, digital audio consultant, engineer, and guest lecturer. He received his degree in recording techniques (I.M.P.) from Indiana University and also studied in the Tonmeister program at the University of Surrey in Guildford, Surrey, England. Dave has written a number of books, including *The MIDI Manual* (Sams Publishing, 1993) and *Hard Disk Recording for Musicians* (Amsco Publications, 1995). Dave is also an instructor in recording technology at the University of Washington and a contributing editor for *EQ* magazine.

Robert E. Runstein has been associated with all aspects of the recording industry, working as a performer, sound mixer, electronics technician, A&R specialist, and record producer. He has served as chief engineer and technical director of a recording studio and has taught several courses in modern recording techniques. He is a member of the Audio Engineering Society.

Trademarks

All terms mentioned in this book that are known to be trademarks or service marks have been appropriately capitalized. Sams Publishing cannot attest to the accuracy of this information. Use of a term in this book should not be regarded as affecting the validity of any trademark or service mark.

ADAT and QuadraVerb are registered trademarks of Alesis Studio Electronics.

Aphex Compellor is a trademark of Aphex Systems Ltd.

Apple; Macintosh Plus, SE, and II; Hypercard; Videoworks; and MacRecorder are registered trademarks of Apple Computer, Inc.

Blue Window is a trademark of Arboretum Systems.

Cakewalk is a trademark of Twelve Tone Systems.

dbx is a registered trademark of dbx, Newton, MA, USA, Division of Carillon Technology.

Digital Domain is a registered trademark of Digital Domain, Inc.

Dolby, Dolby SR, Dolby A, Dolby B, Dolby C, Dolby Surround Sound and Dolby Tone are registered trademarks of Dolby Laboratories Licensing Corporation.

The Flying Fader is a trademark of Neve Electronics International, Ltd.

Harmonizer and Ultra-Harmonizer are registered trademarks of Eventide, Inc.

Multiband Dynamics Tool is a trademark of Jupiter Systems.

PZM is a registered trademark of Crown International, Inc.

The Sonic System, CD Printer, and NoNoise are trademarks of Sonic Solutions.

Sony is a registered trademark of Sony Corporation of America.

Sound Designer II, Sound Tools, Pro Tools, NuBus and DINR are registered trademarks of Digidesign.

Tannoy is a registered trademark of Tannoy LTD. (North America Inc.)

Tube Trap is a trademark of Acoustic Sciences Corp.

Yamaha is a registered trademark of Yamaha Corporation of America.

CHAPTER 1

Introduction

The world of modern music and sound production is multifaceted. It's a world of creative individuals: musicians, engineers, producers, manufacturers, and businesspeople who are experts in such fields as music, acoustics, electronics, production, visual media, multimedia, marketing, law, and the general day-to-day workings of the business of music. The combined efforts of this talent pool work together to create a single end-product: a master recording. Once the recording process is complete, the master can be manufactured into a final, saleable form, be it compact disc, cassette, or a motion picture soundtrack.

For those new to the world of modern multitrack recording, MIDI (musical instrument digital interface), digital audio, and their production environments, years of dedicated practice is often required to develop the skills to successfully master the art and the application of these technologies. A person new to the recording or project studio environment (see Figure 1.1) may be awestruck by the amount and variety of equipment involved. When you begin to familiarize yourself with this environment, however, you soon notice a definite order to the studio's makeup, with each piece of equipment serving a role in the overall scheme of music and audio production.

Figure 1.1. Paisley Park's new *Studio A, Chanhassen, MN. (Courtesy of Paisley Park Studios)*

The goal of this book is to serve as a guide and a reference tool to help you become familiar with the recording and production process. When used in conjunction with mentors, lots of hands-on experience, further reading, and simple common sense, this book can help you understand the toys, tools, and day-to-day practices of music recording and production.

Although it has taken the modern music studio about 70 years to evolve to its present level of technological sophistication, we have just witnessed an important, evolutionary stage in the business of music and music production: the dawning of the digital age. In a time when digital audio and its related technologies are often taken for granted, it's easy to forget that our present position in production history is one that presents us with cost-effective and powerful tools for fully realizing our creative and human potential. Almost always, patience and a nose-to-the-grindstone attitude are needed in order to learn this technology and its language. However, this knowledge can help free you for the really important stuff—making music and sound. Indeed, these are the good ol' days. All the best ....

The Recording Studio

The commercial music studio (see Figures 1.2 and 1.3) is made up of one or more acoustic environments specially designed and tuned for the purpose of getting the best sound possible onto tape while using a microphone pickup. The commercial studio is structurally isolated in order to keep outside sounds from entering the room and getting on tape, as well as keeping internal sounds from leaking out and disturbing the surrounding neighbors.

Figure 1.2. Criteria Recording
Studio's (Miami, FL) premier
mixing suite, Studio B, includes a
96-channel SL4000 G+ console.
Pictured is Criteria's president
and owner, Joel Levy. (Courtesy of
Solid State Logic)

Figure 1.3. Producer, engineer,
songwriter and arranger Gary
Henry using Roland's DM-80
hard-disk recording systems at
Hide-out studios in Hillside, NJ.
(Courtesy of Roland Corporation,
Pro Audio Division)

Studios vary in size, shape, and acoustic design, as you can see in the studios illustrated in Figures
1.4 and 1.5, and are determined by the personal tastes of the owners. They can be designed to best
accommodate certain styles of music or production needs, as shown by the following examples:

◆ A studio that records a wide variety of music (ranging from classical to rock) may have a
large main room with smaller, isolated rooms off to the side for loud or soft instruments,
vocals, and so on.

◆ A studio designed for orchestral film scoring may be larger than studios used for other kinds of recording. This type of studio often has high ceilings to accommodate the sound from a large number of studio musicians.

◆ A studio generally used for audio-for-video, film dialog, vocals, and the like may have a single small room off the control room.

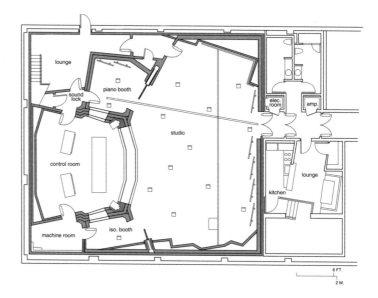

Figure 1.4. *Layout of Bad Animals' Studio X (Seattle, WA).(Courtesy of Bad Animals-Seattle and studio bea:ton)*

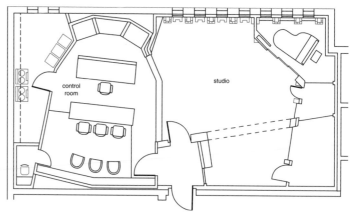

Figure 1.5. *Layout of Sony/Tree's Music Studio (Nashville, TN). (Courtesy of Russ Berger Design Group, Inc.)*

There is no secret formula for determining the perfect studio. Each studio has its own sonic character, layout, "feel," and decor based on the personal tastes of its owners, the designer (if any), and the investment return based on studio rates that can be supported by industry market conditions.

During the 1970s, studios were generally small. Due to the advent of, and reliance on, artificial effects devices, such as spring/reverb plates and digital and acoustic delays, these rooms tended to be acoustically absorptive. The basic concept was to eliminate as much of the original acoustic environment as possible and replace it with artificial ambience.

Fortunately, since the mid-1980s, many commercial studios have begun to move back to the studio design concepts of the 1930s and 1940s when studios were much larger in size. This increase in size (along with the addition of one or more smaller, iso-rooms to accommodate instruments that have to be acoustically isolated) has repopularized the art of recording the room's original acoustic ambience along with the actual sound pickup. In fact, through improved studio design techniques, we have learned how to have the best of both eras: by building a room that disperses sound in a controlled manner (thereby reducing the amount of unwanted sound leaked from other instruments in the room to a single microphone pickup), while at the same time having well-developed sonic and reverberant characteristics. The effect of combining direct and natural reverberant acoustics is often used as a tool for "livening up" an instrument—a popular technique with live rock drums, string sections, electric guitars, choirs, and so on.

The Control Room

A recording studio's control room, illustrated in Figures 1.6 and 1.7, serves several purposes. Ideally, the control room is acoustically isolated from the sounds produced in the studio and the surrounding vicinities. It is optimized to act as a critical listening environment using carefully balanced and placed monitor speakers. It also houses the majority of the studio's recording, control, and effects-related equipment. At the heart of the control room is the recording console.

Figure 1.6. *Floor plan of Paisley Park's new Studio A. (Courtesy of Paisley Park Studios)*

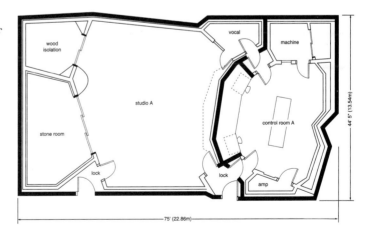

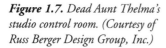
Figure 1.7. *Dead Aunt Thelma's studio control room. (Courtesy of Russ Berger Design Group, Inc.)*

The *recording console* (also referred to as the *board* or *desk*) can be thought of as the equivalent of an artist's palette for a recording engineer and producer. The console enables the engineer to mix together and control the input and output signals of most (if not all) of the devices found in the control room. The console's basic function is to allow for any variable combination of mixing (control over relative amplitude and signal blending between channels), spatial positioning (left/right, as well as possible control over front and back), routing (the capability to send any input from a source to a signal destination), and switching for the multitude of audio input/output signals commonly encountered in an audio production facility.

Tape machines generally are located at the rear or to the side of a control room. Due to the added noise and heat generated by recorders, amplifiers, and other devices, it has become increasingly common for these devices to be housed in a separate, isolated *machine room* that has an adjoining window and door for easy access and visibility. In either case, remote control and *autolocator devices* (used for locating tape and media position cue points) are often situated in the control room, near the engineer so that he or she has easy access to all tape and electronics functions. *Effects devices* (used to electronically affect the character of a sound) and other signal processors are also often placed nearby for easy accessibility.

As with recording studio designs, every control room has its own unique sonic characteristics, "feel," comfort factor, and associated studio rates. Commercial control rooms can vary in design and amenities—from a room that is basic in its form and function to one that is lavishly outfitted with the best toys and design layouts in the business. Again, there is no right or wrong type of room. As you will see throughout this book, however, numerous guidelines are available that can help you "get it right." The important thing to remember is that it is the people—the staff, musicians, and you—that make it all happen, not the equipment.

Recording Studio Marketing Techniques

In recent years, the role of the recording studio has begun to change as a result of upsurges in project studios, audio-for-video/film, and an added emphasis on mixdown. These market forces have made it necessary for certain facilities to rethink their operational business strategies, and often these changes have been met with great advantage and success, as is the case in the following two examples:

◆ Personal production and home project studios have deeply reduced the need for an artist or producer to have constant and costly access to a professional facility. Professional facilities, however, still are often required when a project calls for a larger space or better-equipped recording facility than can be provided by a particular project studio. After a project has been completed in a project studio, a professional facility may be required to mix the tapes down to their final form. Many studios realize these market demands and are only too happy to capitalize on them.

◆ Upsurges in audio-for-video and audio-for-film postproduction have created new markets that enable the professional recording studio to provide services to the local, national, and international broadcast and production community. Studios with both the technical equipment and the creative staff to enter into and be sustained by audio-for-visual and broadcast production can thrive in the tough business of music, when music production alone might not provide enough income for the studio to stay afloat.

These and other market niches (many of which may be unique to your particular area) have been adapted by commercial music and recording facilities to meet the changing markets as we enter into the new millennium. No longer is there only one game in town. Changes in market demands and the often ingenious ways in which they are met has come to count for a great deal in this unique and interesting business.

Another way studios have seized on new opportunities is by joining forces. For example, in 1992, Chris Stone (former founder/owner of the Record Plant) founded the World Studio Group (WSG), which serves as a booking affiliation for premier recording studios (see Figures 1.8 through 1.12). Membership to this group is extended by invitation only to the most highly regarded audio facilities in major geographical markets worldwide.

Figure 1.8. *A & M Studios, Los Angeles, California. (Courtesy of the World Studio Group)*

Figure 1.9. *Studio 301, Sydney, Australia. (Courtesy of the World Studio Group)*

Figure 1.10. Sound Inn, Tokyo, Japan. (Courtesy of the World Studio Group)

Figure 1.11. Arco Studios, Munich, Germany. (Courtesy of the World Studio Group)

Figure 1.12. *Owner Buddy Bruno and his cat Conway (not to be confused with the author's cat Conway) at Conway Recording Studios, Los Angeles, California. (Courtesy of the World Studio Group)*

The Recording Process

The recording process can occur in either of two fundamental forms: multitrack recording or real-time performance recording.

Multitrack Recording

The role of multitrack recording technology is to provide an added degree of flexibility to the recording process by allowing multiple sound sources to be recorded to and played back from isolated tracks in a non-real-time production environment. Because the recorded tracks (with recording systems often being available in groups of eight—that is, 8, 16, 24, 32 and 48 tracks) are independently isolated from one another, any number of sound sources can be recorded and rerecorded without affecting other tracks. In addition, recorded tracks can be altered, added, or erased at any point in time to augment or "sweeten" the original soundtrack.

The common phases of the multitrack recording process—recording, overdubbing, and mixdown—are introduced in the following sections and discussed further in Chapter 15, "Studio Session Procedures."

Recording

The first phase in multitrack production is the recording process. In this phase, one or more sound sources are picked up by a microphone or recorded directly (as often occurs when recording electric or electronic instruments) to one or more of the isolated tracks of a multitrack tape machine.

Isolation is a key concept in multitrack production. After they are recorded, these isolated sound sources (see Figure 1.13) can be independently varied in level, spatial positioning (such as L/R panning), and routing without affecting adjacent tracks. This isolation also makes it possible for tracks to be replaced or manipulated at a later time without affecting adjacent tracks, thus giving the multitrack process its increased flexibility.

Figure 1.13. *Basic representation of how isolated sound sources can be recorded to a multitrack recorder.*

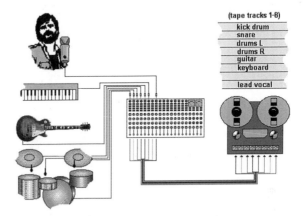

Overdubbing

One of the most important aspects of the multitrack production process is that of overdubbing. Due to the isolated nature of the recorded tracks, it's possible for one or more of the previously recorded tracks to be monitored (generally from the record head in a playback process known as the *sync mode*) while simultaneously recording one or more signals onto other available tracks (see Figure 1.14). This overdub process can be repeated until the song or soundtrack has been built up. If a mistake is made, it generally is a simple matter to re-cue the tape to the desired starting point and repeat the process until you have captured the best take on tape.

Figure 1.14. *Basic representation of the overdubbing process.*

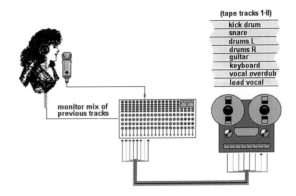

Mixdown

After the phases of recording and overdubbing have been completed, the mixdown process can begin (see Figure 1.15). During mixdown, the separate audio tracks of a multitrack tape machine can be mixed, combined, and routed through the recording console. At this point, volume, tone, special effects, and spatial positioning can be artistically balanced by the engineer to create a stereo or surround-sound mix that is then recorded to a master recording device, such as a digital audio tape (DAT) recorder. After all the songs in a music project have been mixed down, they can be transferred to a hard-disk recording system and organized into a final, sequenced order. The resulting sequenced program is again recorded to a medium (such as DAT), which then serves as the final master recording for manufacturing the project into a commercially saleable product.

Figure 1.15. Basic representation of the mixdown process.

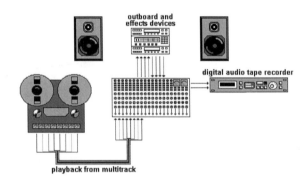

Real-Time Performance Recording: A Different Animal

Real-time performance recording (often referred to as *live* recording) involves the mixing of microphone or direct signal pickups in a live environment during a performance or concert event. This event may take place in a live venue (such as a theater, church, gymnasium, or outdoor arena), as well as in a soundstage or music studio environment.

Unlike the traditional multitrack recording environment in which large amounts of overdubbing are often used to build up a song over time, live recording is created on the spot—often during a single, onstage performance, with little or no studio postproduction other than mixing.

A live recording may be very simple—possibly recorded using two microphones sent directly to a two-track recorder. Alternatively, a live recording may call for a more elaborate multitrack setting, requiring the use of a temporary control room or fully equipped mobile recording van or truck (see Figures 1.16a and 1.16b). This latter setting obviously allows for a greater degree of control over the individual instruments during the mixdown phase.

Figure 1.16. *Le Mobile remote recording studio. (Courtesy of Le Mobile remote recording studio - North Hollywood, California)*

a. Control room.

b. Truck.

a

b

Although the equipment and the setup may be quite familiar to any studio engineer, live recording differs from its more controlled studio counterpart in that it exists in a world in which the motto is "you only get one chance." When you're recording an event in which the artist is spilling his or her guts to hundreds or even thousands of fans, it's critical for everything to run smoothly. Live recording is a facet of the recording industry that requires its own set of expertise, choice of equipment and system setup, degree of preparedness, and, above all, experience.

The Project Studio

In recent decades, the business of manufacturing cost-effective professional and semiprofessional recording and production-related equipment has virtually exploded. As a result, it's now possible, if not downright common, for musicians, engineers, or producers to have a high-quality MIDI and/or recording facility in their homes, apartments, or personal places of business for the purpose of recording their own compositions. This proliferation has grown to such a point that these facilities, known as *project studios*, have become a driving force in the music and communications industry (see Figure 1.17).

Figure 1.17. Example of a project studio. (courtesy of Walt Wagner Productions)

Project studios have become important for the following reasons:

◆ *Cost-effective power.* The obvious reason for the proliferation of these facilities is that with the advent of the VLSI (very large scale integrated-circuit), the price of mass-duplicating highly sophisticated electronic systems has dropped significantly. Studio equipment that two decades ago cost hundreds of thousands of dollars, for example, can now routinely be purchased at one-tenth the price.

Studio costs can add up quickly. So having your own facility and learning its technology and techniques pays off almost immediately in financial savings. Knowing when to make full use of your own facility and when to employ the services of an outside professional studio can result in a product that costs significantly less to produce.

◆ *Setting your own schedule and saving money while you're at it!* An equally obvious advantage is the capability to create your own music on your own schedule. The expense incurred in using a professional studio requires that you be practiced and ready to roll on a specific date or range of days. A project studio can free you up to record when the mood hits, without having to worry about punching the studio's time clock.

◆ *Creative and functional advantages.* With the advent of MIDI as a medium for creating music using electronic instruments, hard-disk recorders, modular digital multitrack recorders, and so on, the project studio offers creative and functional advantages over the commercial studio for creating and producing your own personal style of music. It must be stressed, however, that some projects require the guidance and experience of a professional. Being aware of the production needs of the overall project is an important aspect in ensuring its overall success.

Audio-for-Visual Production and Postproduction

Within the fields of video, film, and broadcast production, audio has become increasingly recognized as an important and integral part of the overall process of creating a more exciting and higher-quality end-product. In past decades, broadcast audio was almost an afterthought. With the advent of MTS (multichannel television sound, which gives many TV sets stereo capabilities), the music video, and the MTV generation, audio has grown from its relatively obscure position to the highly respected field of audio-for-video production (see Figure 1.18).

With the resurgence of quadraphonic and spatialized sound in the form of the Dolby Surround Sound and other processing systems, audio-for-video and audio-for-film has experienced heightened expectations from its public to produce quality soundtracks that complement the overall visual component. Today, the fields of audio-for-video, audio-for-film, and broadcast audio include MIDI, hard-disk recording, time code, automated mixdown, and advanced recording systems in their day-to-day production habits in order to meet these demands.

Figure 1.18. SSL Scenaria installed in the new audio and video postproduction facility of NOB at Hilversum in the Netherlands. (Courtesy of Solid State Logic).

Multimedia

With the integration of text, graphics, MIDI, and digital audio sound into an ever-increasing number of personal computer systems, the field of multimedia has come to embrace high-quality digital audio and quality music production as an integral part of its newly emerging media. In fact, the professional audio community almost instantly embraced the production and distribution of educational, entertainment, and data storage for PC users as an area with enormous growth potential. This area represents an important and lucrative source of income for both creative individuals and production facilities alike.

The People Who Make It All Happen

The recording industry is a service industry. As such, it's the people in the recording industry who make the business of music happen. Recording studios, as well as the other businesses in the industry, aren't known only for the kind of equipment they have, but often are judged by the quality, knowledge, vision, and personalities of their staffs.

The following areas are some of the ways in which a person can be involved in this industry:

◆ Studio management
◆ Music law
◆ Graphic arts and layout
◆ Artist management
◆ A&R (artist and repertoire)
◆ Manufacturing
◆ Music and print publishing

Of course, this is just a partial listing. The list is far too lengthy to fully explore within the confines of this chapter. The following sections, however, describe some of the key positions found in a commercial recording studio.

The Artist

The strength of a recorded performance begins and ends with the artist. All the technology in the world is of little use without the existence of the central ingredients of human creativity, emotion, and technique.

Just as the overall sonic quality of a recording is no better than its weakest link, it is the performer's job to see that the foundation of all music—its inner soul—is laid out for all to experience and hear. After this has been done, a well produced, high-quality recording can act as a framework for the music's original drive, intention, and emotion.

Studio Musicians

A project often requires additional musicians to add to the artist's recorded performance. This can take a number of forms. A project may require musical ensembles (such as a choir, string section, or background vocals), for example, to add a necessary part or to give a piece a "fuller" sound. If a large ensemble is required, it may be necessary to call in a professional music contractor to coordinate all the musicians and make all the financial arrangements, as well as a music arranger to notate and possibly conduct the various musical parts.

It's also possible that a specific member of a group may not have the musical ability to perform a particular part or may not be up to the overall musical standards required by the project. In such a situation, it's not uncommon for a professional studio musician to be called in to fill the part. An entire group of studio musicians may even be called on to provide the best musical support possible for a high-profile artist or vocalist.

The Producer

Beyond the scheduling and budgetary factors that go into coordinating a recording project, it is the job of a producer to help create—according to his/her vision—the best possible recorded performance and final product.

In reality, a producer can be hired onto a project to fulfill a number of possible duties. The producer may be contracted to have complete control over a project's artistic, financial, and programmatic content. More likely, however, the producer acts collaboratively with an artist or group to guide the artist or group through the process of recording. The producer assists in the selection and focus of musical arrangements to best fit the targeted audience, brings out the best performance possible, and then translates that performance (through the recording medium) into a final, saleable product.

A producer is often chosen for his/her ability to understand the process of creating a final recorded project from several perspectives: business, musical performance, creative insight, and mastery of the recording process. As artists, engineers, and others in the industry have become more knowledgeable about the numerous aspects that go into producing a project, this important role may be handled by the artist him-/herself or collaboratively with the engineer. Conversely, as producers become increasingly more knowledgeable about recording technology, it's becoming more and more common to see them sitting behind the console at the controls.

The Engineer

The engineer's job can best be described as "an interpreter in a techno-artistic field." It is the engineer's job to express the artist's music and the producer's concepts through the medium of recording. This job actually is an art form because both music and recording are subjective in nature and rely on the tastes and experience of those involved.

During a recording session, the engineer generally places the musicians in the desired studio positions, chooses and places the microphones, sets levels and balances on the recording console, and records the performance onto tape. In an electronic music setting, the engineer may also set up the numerous MIDI sequencers, musical instruments, hard-disk recorders, and so on. During an overdubbing or mixdown session, the engineer uses his or her talent and knowledge of the art and technology of the recording media to get the best sound possible.

Assistant Engineer

Larger studios often train future staff engineers by allowing them to work as assistants to engineers. The assistant engineer often does microphone and headphone setups, runs tape machines, does session breakdowns, and, in certain cases, performs rough mixes and balance settings for the engineer on the console.

With the proliferation of *freelance* engineers (those not employed by the studio but retained by the artist or record company to work on a particular project), the role of the assistant engineer has become even more important. It is the assistant engineer's role to help the freelance engineer with the technical aspects and quirks that are peculiar to that studio.

Maintenance Engineer

The maintenance engineer's job is to see that the equipment in the studio is maintained in top condition, regularly aligned, and repaired when necessary. Larger organizations—those with more than one studio—often employ a full-time staff maintenance engineer. Many of the smaller studios, however, are serviced by freelance maintenance engineers on an on-call basis.

Studio Management

Running a music or production studio is a serious business that requires the special talents of businesspeople who are knowledgeable about the inner workings of the music studio, the music business, and—above all—about people. It requires the constant attention of a studio manager (who may or may not be the owner), booking department (who keeps track of most of the details relating to studio usage, billing, and possibly marketing), and last but not least, a competent secretarial staff. Although any or all of these functions may vary from studio to studio, these and other equally important roles are required in order to successfully operate a commercial production facility on a day-to-day basis.

Women in the Industry

Ever since its inception, the recording industry has been dominated by males. I remember many a session in which the only women on the scene were female artists or studio groupies. Fortunately in recent years, women have begun to play a more prominent role "behind the glass." It has become increasingly more common for women to reach levels of prominence, both as engineers and in production. No matter who you are, where you're from, or what your race or gender, remember this universal truth: If your heart is in it and you're willing to work hard enough, you'll make it.

The Transducer

One concept that bears looking into before you jump into the heart of this book is a concept that is central to all music, sound, and electronics: the transducer. If any single conceptual tool can help you to understand the technological underpinnings of the recording process, this is it!

A *transducer* is any device that changes one form of energy into another, corresponding form of energy. For example, a guitar is a transducer. It takes the vibrations of picked or strummed strings (the media), amplifies them through a body of wood, and converts these vibrations into corresponding sound-pressure waves. You then perceive these waves as sound (see Figure 1.19).

A microphone is another example of a transducer. Sound-pressure waves (the media) act upon the diaphragm of the microphone and are converted into corresponding electrical voltages. The electrical signal from the microphone can then be amplified (not a process of transduction because the media stays in its electrical form) and fed to a recording device. The recording device then converts them into analogous magnetic flux signals on magnetic tape or into representative digital data that can be encoded onto tape, computer disk, or compact disc.

Figure 1.19. *The guitar and microphone as transducers.*

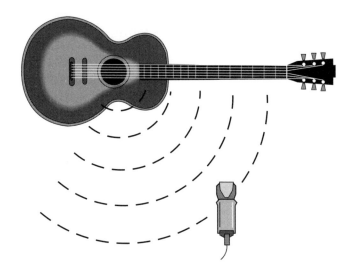

On playback, the stored magnetic signal or digital data can be reconverted to its original electrical form, amplified, and then fed to a speaker system. The speaker system converts the electrical signal into a mechanical motion (by way of magnetic induction), which, in turn, re-creates the original air-pressure variations that were sensed by the microphone.

As you can see in Table 1.1, transducers can be found practically everywhere in the audio environment. In general, transducers and the media they use are often weak links in the chain of an audio system. As I stated earlier, a transducer changes the energy in one medium into a corresponding form of energy in another medium. Given our present technology, this process can't be accomplished perfectly. Noise, distortion, and (often) coloration of the sound are introduced to some degree. These effects can only be minimized, not eliminated. Differences in design are another factor that can affect sound quality. Even a slight design change between two microphones, speaker systems, digital audio converters, guitar pickups, or other transducers can cause them to sound quite different. This factor, combined with the complexity of music and acoustics, makes the field of recording a very subjective and personal one.

Table 1.1. Media used by transducers in the studio to transfer energy.

Transducer	From	To
Ear	Sound waves in air	Nerve impulses in the brain
Microphone	Sound waves in air	Electrical signals in wires
Record head	Electrical signals in wires	Magnetic flux on tape
Playback head	Magnetic flux on tape	Electrical signals in wires
Phono cartridge	Grooves cut in disc surface	Electrical signals in wires
Speaker	Electrical signals in wires	Sound waves in air

Digital recordings often are vastly reduced in noise and distortion because fewer transducers are introduced into the playback chain. In a totally digital system (see Figure 1.20), the acoustic waveforms are picked up by a microphone and converted into electrical signals. These signals are then converted into digital form by an analog-to-digital (A/D) converter. The A/D converter changes continuous electrical waveforms into corresponding discrete numeric values that represent the waveform's instantaneous, analogous voltage levels.

Arguably, digital information has the distinct advantage over analog in that the data can be transferred between electrical, magnetic, and optical media with virtually no degradation in quality because the information isn't transduced; it continues to be stored in its original, discrete binary form. In other words, only the medium changes; the data containing the actual information doesn't. Therefore, if you are listening to a digital recording on a home compact disc (CD) player over decent speakers, the lack of distortion (that would have been introduced by a vinyl record and playback stylus) yields a sonic clarity of near or equal quality to that of the original master recording.

Figure 1.20. *The all-digital recording chain.*

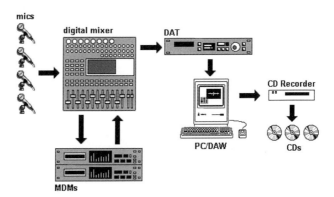

CHAPTER

Sound and Hearing

When you record, what you are really interested in doing is capturing and storing sound so that an original sound event can be re-created at a later date, or you are using the medium to create a totally new sound event. Starting from the concept that the word *sound* is only a name for the brain's interpretation of a certain type of physical stimulus arriving at the ears, the examination of sound can be divided into three areas: the nature of the stimulus, the characteristics of the ear as a transducer, and the psychoacoustics of hearing. This last area, psychoacoustics, deals with how and why the brain interprets a particular stimulus from the ears in a certain way. By understanding the physical nature of sound and how the ears change sound from a physical phenomenon to a sensory one, you can discover what is needed to make your recordings convey a particular effect.

Sound-Pressure Waves

Sound arrives at the ear in the form of a periodic variation in atmospheric pressure—the same atmospheric pressure measured by the weather service with a barometer. The pressure variations corresponding to sound, however, are too small in magnitude and vary too rapidly to be observed on a barometer. These variations in pressure are called *sound-pressure waves*. You can visualize these waves by imagining the waves you see in a pool of water when you drop in a stone. The motion of the waves on the water's surface as they move away from the spot where a stone has hit (see Figures 2.1a and 2.1b) can be used to represent the motion of sound-pressure waves in the air as they move away from a sound source. The only difference between these analogies is that sound-pressure waves radiate outwardly in a three-dimensional, spherical pattern.

Figure 2.1. *A representation of wave movement on the surface of water as it moves away from its source of origin.*
a. Top view.
b. Side view of the water's surface.

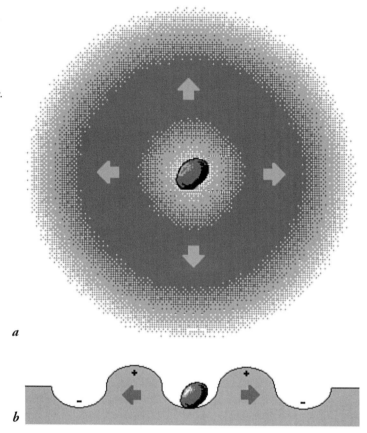

Sound-pressure waves are generated by a vibrating body that is in contact with the air. This may be a loudspeaker, a person's vocal cords, or a guitar string that vibrates the instrument's body, which, in turn, vibrates the surrounding atmosphere, and so on. The atmospheric pressure is proportional to the number of air molecules in the area being measured. A vibrating mass squeezes additional air molecules into a space as it moves toward the space, which creates an area having a greater than normal atmospheric pressure called a *compression* (see Figure 2.2a). As the compressed area continues to move away from the sound source, a vacuous area with lower than normal atmospheric pressure, called a *rarefaction,* is created (see Figure 2.2b). It's interesting to note that the involved molecules themselves don't move through air at the velocity of sound, but the sound wave moves through the atmosphere in the form of compressed waves that struggle to travel outward from areas of higher pressure to areas of lower pressure. This process is known as the *propagation* of a wave.

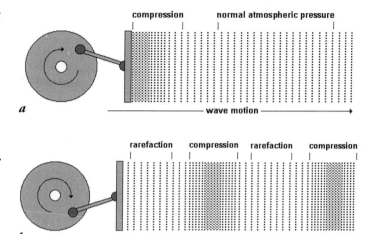

Figure 2.2. *Effects of a vibrating mass on air molecules and their propagation.*

a. Compression—Air molecules are forced together to form a compression.

b. Rarefaction—As the compression moves away from the originating source, a rarefacted area of lower atmospheric pressure is created.

Waveform Characteristics

A *waveform* is a graphic representation of a signal's sound-pressure level or voltage level over time. In short, a waveform enables you to visually see, study, and explain the actual phenomena of wave propagation in your physical environment. A waveform has the following fundamental characteristics:

◆ Amplitude
◆ Frequency
◆ Velocity
◆ Wavelength
◆ Phase
◆ Harmonic content
◆ Envelope

These characteristics allow one waveform to be distinguished from another. The most fundamental of these characteristics are amplitude and frequency. The following sections describe each characteristic.

Amplitude

The distance above or below the centerline of a waveform, such as the pure sine wave shown in Figure 2.3, represents the amplitude (level) of that signal. The greater the distance or displacement from the centerline, the more intense the pressure variation, electrical signal, or physical displacement within a medium.

Figure 2.3. Graph of a waveform showing various amplitude values.

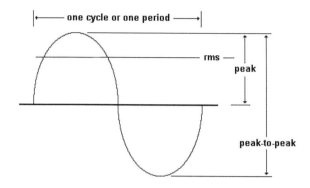

Waveform amplitudes can be measured using various standards. The measurement between the maximum positive and negative signal levels of a wave is called its *peak amplitude value*, and the difference between the positive and negative peak signal levels is called the *peak-to-peak value*. The *root-mean-square* (rms) value was developed to arrive at a meaningful average of these values and to more closely approximate the signal level perceived by your ears. For a sine wave, the rms value is arrived at by squaring the amplitude of the wave at each point along the waveform and then taking the mathematical average and square root of the combined results. These calculations are equal to 0.707 times the instantaneous peak amplitude level. Because the square of a positive or negative value is always positive, the rms value will always be positive. The following mathematical formulas detail the relationships that exist between the peak and rms values of a waveform:

$$\text{peak} = \sqrt{2} \times \text{rms} = 1.414 \times \text{rms}$$
$$\text{rms} = \text{peak} \div \sqrt{2} = 0.707 \times \text{peak}$$

Frequency

The rate at which an acoustic generator, electrical signal, or vibrating mass repeats a cycle of positive- and negative-going amplitude is known as the *frequency* of that signal. One completed excursion of a wave, which is plotted over the 360° axis of a circle, is known as a *cycle* (see Figure 2.4). The number of cycles that occurs over the period of one second (frequency) is measured in hertz (Hz).

In the diagram shown in Figure 2.5, the value of the waveform starts at zero. At time t=0, the value increases to a maximum in the positive direction and then decreases through zero. The process then begins again. A cycle can begin at any angular degree point on the waveform; to be complete, however, it must pass through the zero line and end at a point moving in the same direction (positive or negative) that has the same value as the starting point. Thus, the waveform from t=0 to t=2 constitutes a cycle, and the waveform from t=1 to t=3 is also a cycle.

Figure 2.4. *The cycle is divided into the 360° of a circle.*

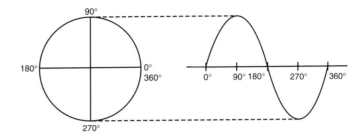

Figure 2.5. *Graph of waveform amplitude over time.*

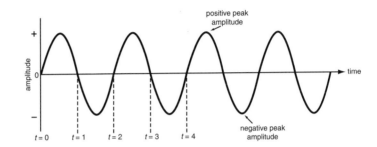

Velocity

The velocity of a wave is the speed at which it travels through a medium and is given by the equation:

$$V = \frac{d}{t_2 - t_1}$$

(This assumes that t_2 is later than t_1.)

where

> V is the wave velocity of propagation in the medium,
> d is the distance from the source,
> t is the time in seconds.

For sound waves, the medium is air molecules; for electricity, the medium is electrons. The wave velocity determines how fast a particular cycle of a waveform can travel a designated distance. At 70°F, the speed of sound waves in air is approximately 1130 feet per second (ft/sec) or 344 meters per second (m/sec). This speed is temperature-dependent and increases at a rate of 1.1 ft/sec for each degree Fahrenheit increase in temperature.

Wavelength

The wavelength (λ) of a wave is the actual distance in the medium between the beginning and the end of a cycle, or between corresponding points on adjacent cycles, and is equal to

$$\lambda = \frac{V}{f}$$

where

λ is the wavelength in the medium,
V is the velocity in the medium,
f is the frequency in hertz.

The time it takes to complete 1 cycle is called the *period* of the wave. To illustrate, a 30-Hz sound wave completes 30 cycles each second or 1 cycle every one-thirtieth of a second (approximately every 0.0333 second). The period of the wave is expressed using the symbol T:

$$T = \frac{1}{f}$$

where

T is the number of seconds per cycle.

Assuming that sound propagates at the rate of 1130 ft/sec, a 30-Hz waveform will have traveled over its entire 360° cycle in 0.0333 of a second, which works out to a distance of 1130 x 0.0333, or 37.6 feet ('). Therefore, you can say that the wavelength of a 30-Hz sound wave in air is 37.6' long, as shown in Figure 2.6. As the frequency of the waveform increases, each cycle is completed in a shorter amount of time (the period of the wave gets smaller), and the beginning of the waveform can't travel as far before the end of that cycle is reached. The wavelength, therefore, decreases as the frequency increases.

Figure 2.6. *Two cyclic wavelengths.*

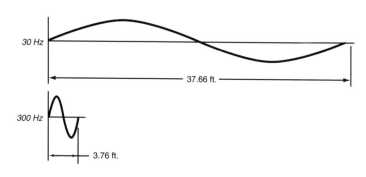

For example, ten times as many cycles occur per second in a 300-Hz wave as in a 30-Hz wave, so each cycle of the 300-Hz wave occurs in one-tenth the time of a 30-Hz cycle, permitting the beginning of a cycle to travel only one-tenth the distance, or 3.76', before the cycle ends. Because cycles can be measured between any corresponding points on adjacent waveforms, the distance between peaks of adjacent waveforms having the same relative degree of angular rotation is also one wavelength. The distance between positive and negative peaks of the same cycle would be one-half cycle. The concept of wavelength shows you that the perception of a wave depends on the distance from the source as well as time.

Reflection of Sound

Much like light waves, sound reflects off a surface boundary at an angle equal to (and in an opposite direction to) its initial angle of incidence. This basic property is one of the cornerstones of the complex study of acoustics. Figure 2.7a, for example, shows how a sound wave reflects off a single planed, solid, smooth surface in a simple and straightforward manner. Figure 2.7b shows how a convex surface splays the sound outward from its surface, radiating the sound in a wide dispersion pattern. In Figure 2.7c, the concave surface can be used to focus a sound at a single focal point, while a 90° corner, as shown in Figure 2.7d, serves to reflect patterns back in their original directions. This holds true for both the corners of a wall, as well as for the 90° intersections where the wall and floor angles meet. This equal angle of reflection can also provide insight into the frequency build ups that often occur at the corners—and particularly at wall-to-floor intersections. Of course, corners or wall intersections with angles less than or greater than 90° will reflect signals according to their designated angle.

Figure 2.7. *Incident sound waves striking surfaces with varying shapes.*

a. A single-planed, solid, smooth surface.

b. A convex surface.

c. A concave surface.

d. A 90° corner reflector.

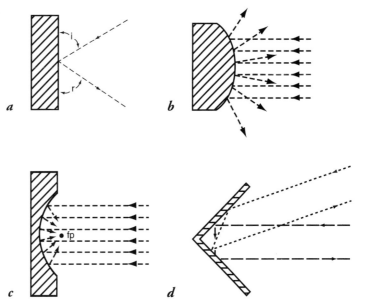

Diffraction of Sound

Sound has the inherent capability to diffract around or through a physical acoustic barrier. In other words, sound can bend around an object in a manner that reconstructs the original waveform in both frequency and amplitude (relative to the size of the obstructing obstacle).

In Figure 2.8a, for example, you can see how an obstacle that is small relative to the initial source of a large wavelength scarcely impedes the signal. Instead, it bends around the obstruction in a way that fully reconstructs the waveform. Figure 2.8b shows how a larger obstacle can obstruct a larger portion of a radiated signal. Past this obstruction, however, the signal begins to bend around the area in the barrier's wake and begins to reconstruct the waveform. In Figure 2.8c, you can see how a signal can radiate through an opening in a large barrier. Although the signal is greatly impeded, it can, nevertheless, begin to reconstruct itself in wavelength and amplitude (relative to the size of the opening) and can radiate outward from the opening as though it were the point of origin. And finally, Figure 2.8d shows how a large opening in a barrier lets much of the waveshape pass through unimpeded. On passing through, the waveform begins to bend outwards in order to reconstruct itself back into its original shape.

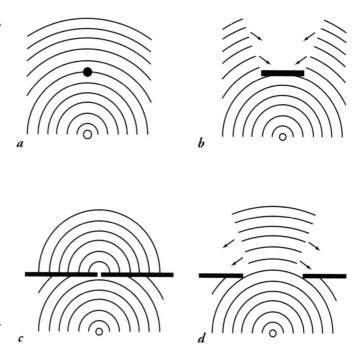

Figure 2.8. *The effects of obstacles on sound radiation and diffraction.*

a. A sound having a large wavelength compared to a small obstacle size will scarcely be impeded by the obstacle.

b. An obstacle that is large compared to the signal's wavelength will impede the signal.

c. An opening that is small in source (large boundary) will impede the signal and will thereafter act as a new radiation point-of-source.

d. An opening that is large (smaller boundary) compared to the wavelength will allow sound to pass through and readily diffract to its original waveshape.

Frequency Response

Now take a look at the *frequency-response curve* of a device such as a microphone or an equalizer (see Figures 2.9a and 2.9b). In this case, the y-axis represents the average amplitude of the signal at the output of the device being measured; the x-axis represents the frequency (or pitch) of the signal. If the input of the device is fed a constant amplitude signal, which rises from the low end to the high end of the scale on the x-axis, the graph shows how the amplitude at the output of the device varies as the frequency of the signal at its input changes. If the output amplitude is the same at all frequencies, the curve will be a flat, straight line from left to right. This is where the term *flat frequency response* comes from. It indicates that the device passes all frequencies equally; no frequency is emphasized. If the curve were to dip at certain frequencies, you would know that those frequencies have lower amplitudes than the other frequencies and vice versa. A frequency-response curve graphically indicates the effect that a device has on the tone of an instrument.

Figure 2.9. *Frequency response curves.*

a. A curve showing a bass boost.

b. A rising high-frequency-response curve.

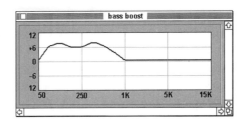

a

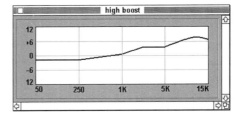

b

Phase

A cycle can begin at any point on a waveform, so it's possible to have two wave generators producing sine waves of the same frequency and peak amplitude that have different amplitudes at any one point in time. These waves are said to be *out-of-phase* with respect to each other. *Phase* is measured in degrees, and a cycle can be divided into 360°. The sine wave (so named because its amplitude follows the trigonometric sine function) is usually considered to begin at 0° with 0 amplitude, increase to a positive maximum at 90°, decrease to zero at 180°, increase again to a maximum (but in a negative direction) at 270°, and again return to zero at 360°.

You can add waveforms by summing their signal amplitudes at each instance in time. When two waveforms that are completely in-phase (0° phase difference) and of the same frequency, shape, and peak amplitude are added, the resulting waveform is of the same frequency, phase, and shape but has twice the original peak amplitude (see Figure 2.10a). If two waves are the same as the ones just described, except that they are completely out-of-phase (having a phase difference of 180°), they cancel each other when added. This results in a straight line of zero amplitude (see Figure 2.10b). If the second wave is only partially out-of-phase (not exactly 180° or $(2n - 1) \times 180°$ out-of-phase, where $n = 1, 2, 3 ...$), it *interferes constructively* at points where the amplitudes of the two waves have the same sign (that is, both are positive or both are negative). This results in a greater amplitude in the combined wave than in the first wave at that point in time, and *interferes destructively* at points where the signs of the two wave amplitudes are opposing (see Figure 2.10c). The waves can be said to be in-phase, or correlated, at points where the combined signs are the same and out-of- phase, or uncorrelated, where the signs are opposing.

Figure 2.10. *Adding sine waves.*

a. *The amplitude of in-phase waves add when they are mixed together.*

b. *Waves of equal amplitude cancel completely when mixed 180° out-of-phase.*

c. *When mixed together, the amplitudes of partially out-of-phase waves add in certain places and subtract in others.*

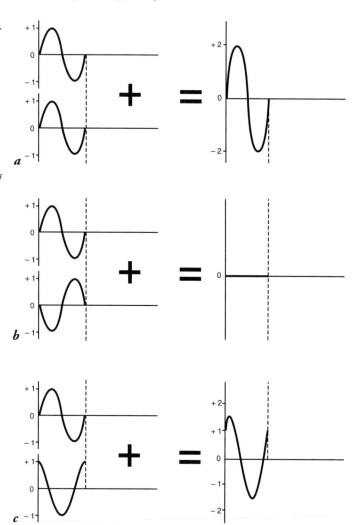

Phase shift is a term that describes the amount of lead or lag in one wave with respect to another. It results from a time delay in the transmission of one of the waves. For example, a 500-Hz wave completes one cycle every 0.002 second. If you start with two in-phase 500-Hz waves and delay one of them by 0.001 second (half the period of the wave), the delayed wave will lag the other by one-half a cycle, or 180°. The number of degrees of phase shift introduced by a time delay can be computed by the following formula:

$$\varnothing = \Delta t \times f \times 360°$$

where

Ø is the phase shift in degrees,
Δt is the time delay in seconds,
f is the frequency in Hz.

From the formula, you can see that the amount of phase shift resulting from a fixed time delay varies in direct proportion to the frequency involved. Plugging in the values of a few different frequencies shows that for a 1-millisecond (0.001 second) time delay, the following phase shifts at different frequencies will occur: 250 Hz, 90°; 500 Hz, 180°; 1000 Hz, 360°; 1500 Hz, 540° − 360° = 180°; 2000 Hz, 720° − (2 × 360°) = 0°; 2500 Hz, 900° − (2 × 360°) = 180°; and so forth. Every thousand hertz increase is a whole-number multiple of 360° and appears in-phase with the original frequency. Every increase of 500 hertz is 180°, or a whole-number multiple of 360° plus 180°, and therefore appears out-of-phase with the original.

If you combine a signal at equal amplitude with the same signal that was delayed by 1 millisecond (ms), the amplitude increases by a factor of two at the frequencies with 0° phase shift and cancels completely at those frequencies with 180° phase shift. Frequencies that are phase shifted exactly 90° combine with the nondelayed signal with equal amounts of constructive and destructive interference, resulting in the same amplitude of the combined waves as for either wave alone. Those frequencies that are shifted between 90° and 180° have more destructive interference and produce a sum smaller than either signal alone. Those frequencies shifted between 0° and 90° have more constructive interference and produce a sum greater than either signal alone. Except for the 0° and 180° cases, the combined signal will be shifted in phase somewhere between the two original signals.

Any time a signal follows different paths to the same point, such that the energy from one path is delayed in time with respect to the energy from another path, a frequency-dependent phase difference exists between the two signals. If the energy from the different paths is added together, peaks and dips in the frequency response are created. These peaks and dips occur because some frequencies are boosted by constructive interference while others are lowered in level by destructive interference.

Distance is the most common source of this type of time delay. For example, if the same source is picked up by two microphones at different distances from the source, a time delay corresponding to the path-length difference will exist. A second source of time delay occurs when the distance traveled by a reflected sound is picked up by the same microphone that picks up the direct sound. The signals will be in-phase at frequencies where the path-length difference is equal to the wavelength of the signals, and out-of-phase at those frequencies where the path-length difference is one-half the wavelength of the signals. The result of these combined time delays is an altered frequency response.

With continuous tones, the interference created by the phase shifts at different frequencies occurs for long delays as well as for short ones. For the majority of sounds in the studio, however, there is a certain time delay beyond which the interference is no longer noticeable due to changes in the signal. At this point, delayed signals that are longer than about 3 to 5 ms (depending on the character and frequency of the sound), begin to sound like a second source playing in unison with the original. To keep the interference above 20 kHz and thus out of the audio range, the path-length difference must be less than 0.34 inch, which corresponds to a time delay of 0.03 ms. Because this is such a small amount of delay, you can see that virtually any reflection or time-delayed signal of sufficient level will cause extreme frequency-response degradation. To avoid this form of distortion, you must either eliminate the reflections or reduce their level to the point at which they cannot produce audible cancellations. This is one reason you should try to avoid leakage between instruments whenever possible while recording.

Harmonic Content

Up to this point, the discussion has centered around the sine wave, which is composed of a single frequency and produces a pure sound at a certain pitch. Musical instruments rarely produce pure sine waves, however, and it's good that they don't. If they did, all instruments playing the same musical note would sound exactly the same, and music would be pretty boring. The factor that enables us to differentiate between instruments is the presence of several different frequencies in the sound wave, in addition to the one that corresponds to the note being played (called the *fundamental*). The frequencies present in a sound, other than the fundamental, are called *partials*. Partials that are higher than the fundamental frequency are called *upper partials* or *overtones*. For most musical instruments, the frequencies of the overtones are whole-number multiples of the fundamental frequency and are called *harmonics*. For example, the frequency corresponding to concert A is 440 Hz (see Figure 2.11a). An 880-Hz wave is a harmonic of the 440-Hz wave because it is two times the 440-Hz frequency (see Figure 2.11b). In this case, the 440-Hz wave is called the *fundamental* or *first harmonic* because it is one time the fundamental frequency, and the 880-Hz wave is called the *second harmonic* because it is two times the fundamental. The *third harmonic* would be three times 440 Hz, or 1320 Hz (see Figure 2.11c). Some instruments, such as bells, xylophones, and other percussion instruments, have partials that are not harmonically related to the fundamental.

Figure 2.11. *An illustration of harmonics.*

a. Fundamental waveform (first harmonic).

b. Second harmonic.

c. Third harmonic.

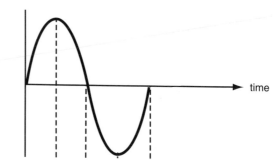

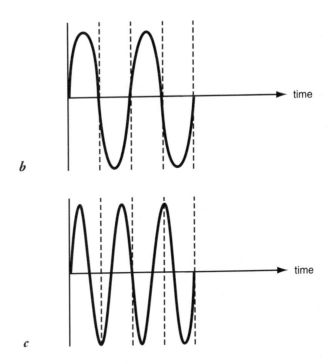

b

c

The ear perceives those sounds with a frequency ratio of 2:1 to be specially related, and this relationship is the basis of the musical octave. For example, because concert A is 440 Hz, the ear hears 880 Hz as having a special relationship to concert A; namely, that it is the first tone higher than concert A that sounds most like concert A. The next note above 880 Hz that sounds most like 440 Hz is 1760 Hz. Therefore, 880 Hz is said to be one octave above 440 Hz, and 1760 Hz is said to be two octaves above 440 Hz. Two notes that have the same fundamental frequency and are played at the same time are said to be in unison, even if they have different harmonics. The human ear doesn't respond to all waveform frequencies. Its range is limited to the 10 1/2 octaves from about 15 Hz to 20 kHz. Some young people can hear as high as 23 kHz, but the ear's high-frequency response drops off with increased age. For example, few people over age 60 can hear above 8 kHz.

Because the sound waves produced by musical instruments contain harmonics in various amplitude and phase relationships, the waveforms bear little resemblance to the shape of the single-frequency sine wave. Musical waveforms can be divided into two categories: simple and complex. Square waves, triangle waves, and sawtooth waves are examples of simple waves containing harmonics (see Figures 2.12a through 2.12c). These waveforms are called *simple* because they are continuous and repetitive. One cycle of a square wave looks exactly like the next, and they are all symmetrical about the zero line. The seven wave characteristics mentioned earlier apply to simple waves containing harmonics as well as to sine waves.

Figure 2.12. *Simple waveforms.*
a. *Square waves.*
b. *Triangle waves.*
c. *Sawtooth waves.*

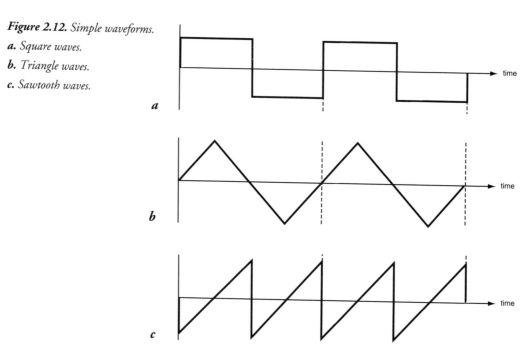

a

b

c

Complex waves, on the other hand, are waves that do not necessarily repeat and are not necessarily symmetrical about the zero line. An example of a complex waveform (see Figure 2.13) is one created by music. Complex waves don't always repeat, so it may be difficult to divide them into cycles or categorize them as to frequency simply by looking at the waveshape.

Figure 2.13. *A complex waveform.*

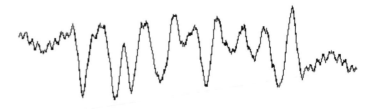

Regardless of the shape or complexity of the waveform reaching the eardrum, the inner ear separates the sound into its component sine waves before transmitting the stimulus to the brain. For this reason, it isn't the shape of a waveform that should interest you as much as the components that cause it to have that shape, for these components determine the character of sound perceived by the brain. This can be illustrated by passing a square wave through a bandpass filter that is set to pass only a narrow band of frequencies at any one time. This would show that the square wave is composed of a fundamental frequency plus all the harmonics whose frequencies are odd-number multiples of the fundamental, with the amplitude of the harmonics decreasing as their frequency increases. In Figures 2.14a through 2.14d, you can see how individual sine-wave harmonics combine to form a square wave, subtracting from the fundamental where they are uncorrelated and adding to the fundamental where they are correlated.

Figure 2.14. *Obtaining a square wave by adding odd harmonics.*

a. *A square wave with frequency f.*

b. *A sine wave with frequency f.*

c. *Sum of a sine wave with frequency f and a lower amplitude sine wave of frequency 3f.*

d. *Sum of a sine wave of frequency f and lower amplitude sine waves of 3f and 5f; beginning to resemble a square wave.*

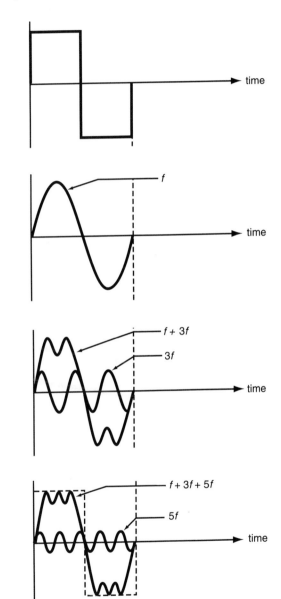

If you were to analyze the harmonic content of the waves produced by a violin and compare them to the content of the waves produced by a viola—when both are playing concert A (440 Hz)—you would obtain the results shown in Figures 2.15a and 2.15b. Notice that the violin has a set of harmonics differing in both degree and intensity from that of the viola. The harmonics present, and their relative intensities, determine the characteristic sound of each instrument and are called the *timbre* of the instrument. If you changed the balance of the harmonics, you would change the sound character of the instrument. For example, if the level of violin harmonics 4 through 10 were reduced and the harmonics above the tenth were eliminated, the violin would sound just like the viola.

Figure 2.15. *Harmonic structure of concert A-440.*
a. Played on a violin.
b. Played on a viola.

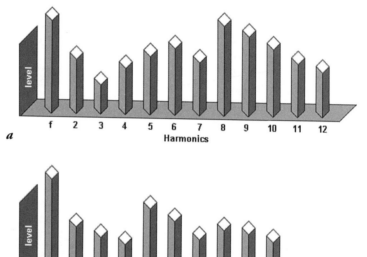

In addition to variations in harmonic balance that can exist between instruments and instrument families, it's common for both the fundamental and harmonic frequencies to changes in direction as they emanate from an instrument over its performance range. Figure 2.16, for example, shows the principal radiation patterns for the cello as seen from both the side and top views.

The significance of harmonics to our perception of tone quality was best summarized as follows by Russel Hamm in the *Journal of the Audio Engineering Society*, May 1973:

> The primary color characteristics of an instrument are determined by the strength of the first few harmonics. Each of the lower harmonics produces its own characteristic effect when it is dominant, or it can modify the effect of another dominant harmonic if it is prominent.

In the simplest classification, the lower harmonics are divided into two tonal groups. The odd harmonics (third and fifth) produce a *stopped* or *covered* sound. The even harmonics (second, fourth, and sixth) produce *choral* or *singing* sounds…. Musically, the second is an octave above the fundamental and is almost inaudible; yet it adds body to the sound, making it fuller. The third is termed a quint or musical twelfth. It produces a sound many musicians refer to as *blanketed*. Instead of making the tone fuller, a strong third actually makes the tone softer. Adding a fifth to a strong third gives the sound a metallic quality that gets annoying in character as its amplitude increases. A strong second with a strong third tends to open the *covered* effect. Adding the fourth and fifth to this changes the sound to an *open horn*-like character…. The higher harmonics, above the seventh, give the tone *edge* or *bite*. Provided the edge is balanced to the basic musical tone, it tends to reinforce the fundamental, giving the sound a sharp attack quality. Many of the edge harmonics are musically unrelated pitches, such as the seventh, ninth, or eleventh. Therefore, too much edge can produce a raspy dissonant quality. Since the ear seems very sensitive to the edge harmonics, controlling their amplitude is of paramount importance. (The study of a trumpet tone) shows that the edge effect is directly related to the loudness of the tone. Playing the same trumpet note loud or soft makes little difference in the amplitude of the fundamental and the lower harmonics. However, the sixth increases and decreases in amplitude in almost direct proportion to the loudness. This edge balance is a critically important loudness signal for the human ear.

Because the relative balance of an instrument's harmonics is so important to its sound, the frequency response of microphones, amplifiers, speakers, and all other elements in the signal path can have an effect on the timber or harmonic balance of a sound. If the frequency response isn't flat, the timbre of the sound will be changed. For example, if the high frequencies are amplified less than the low and middle frequencies, the sound will be duller than it should be. Equalizers can be used to vary the timbre of instruments, thus changing their subjective effect on the listener.

Figure 2.16. *The radiation patterns of a cello as viewed from the side (left) and top (right).*

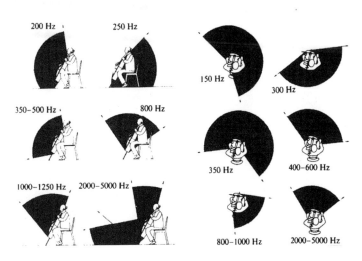

Acoustic Envelope

Timbre is not the only characteristic that enables us to differentiate between instruments. Every instrument produces its own characteristic *envelope* which works in combination with its timbre to determine the subjective sound of the instrument. The envelope of a waveform describes the way its intensity varies and can be viewed on a graph by connecting the peak points of the same polarity over a series of cycles.

The envelope of an acoustic signal is composed of three sections: attack, internal dynamics, and decay. *Attack* is the manner in which the sound begins and increases in intensity; *internal dynamics* describes volume increases, decreases, and sustentations after the attack period; and *decay* is the manner in which the sound ceases. Each of these sections has three variables: time duration, amplitude, and amplitude variation with time. Figure 2.17a illustrates the envelope of a clarinet note. The attack and decay times are long, and the internal dynamics consist of sustain, producing a smooth, flowing sound. Figure 2.17b illustrates the envelope of a snare drum beat. Notice that the initial attack has a much greater amplitude than the internal dynamics, and the attack, initial decay, and final decay are fast, resulting in a sharp crack at the start and a short percussive sound. A cymbal crash, as shown in Figure 2.17c, has similar high-amplitude fast attack with a fast initial decay. It sustains longer and decays slower, however, combining the sharp impulse sound with a smooth lingering shimmer.

Figure 2.17. *Various musical waveforms.*

a. *Envelope of a clarinet.*

b. *Envelope of a snare drum.*

c. *Envelope of a cymbal crash.*

Legend:

    A = Initial attack

    B = Initial decay

    C = Internal dynamics

    D = Final decay

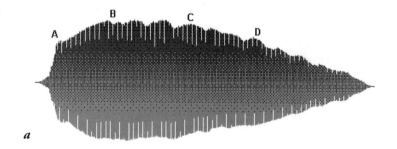

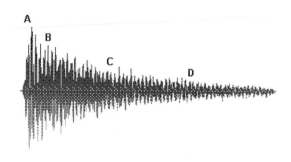

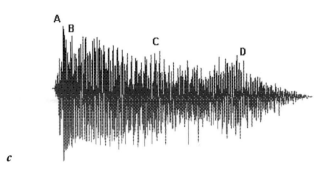

An organ tone has very rapid attack and decay times and a constant internal amplitude, unless the volume pedal is used to vary the envelope. It can produce sounds varying from a click, if sustain is very short, to smooth sounds, if the attack and decay times are made long by the volume pedal. Envelopes with short attack times, followed by fast initial decays, are characterized as sounding percussive or *punchy*, while slow attacks and decays have gentler, smoother sounds.

It's important to note that the concept of envelopes uses peak waveform values, while human perception of loudness is proportional to the average wave intensity over a period of time. Therefore, high-amplitude portions of the envelope will not make an instrument sound loud unless the high amplitude is maintained for a long enough period. Short high-amplitude sections contribute to sound character rather than to loudness. Through the use of amplitude controllers, such as compressors, limiters, and expanders, the sound character of an instrument can be modified by changing its envelope without changing the timbre of the sound.

The waveform envelope of an electronic musical instrument is similar in most respects to its acoustic counterpart and is measured with respect to its initial attack time, decay time (from the initial attack), sustain time, and final release time. It is most commonly referred to in its abbreviated form, ADSR.

Loudness Levels: The dB

The ear operates over an energy range of approximately 10^{13}:1 (10,000,000,000,000:1)—an extremely wide range. Because such a wide range is difficult for humans to deal with easily, a logarithmic scale has been adopted to compress the measurements into more workable figures. The system used for measuring sound-pressure level (SPL), signal level, and changes in signal level is the *decibel* (dB).

To develop an understanding of the decibel, you first should examine logarithms and the logarithmic scale (see Figures 2.18a and 2.18b). The *logarithm* (*log*) is a mathematical function that reduces large numeric values into smaller, more manageable numbers. Because it increases exponentially, it also expresses our sense of perception more precisely than can a linear curve.

Figure 2.18. *Linear and logarithmic scales.*

a. *Linear.*

b. *Logarithmic.*

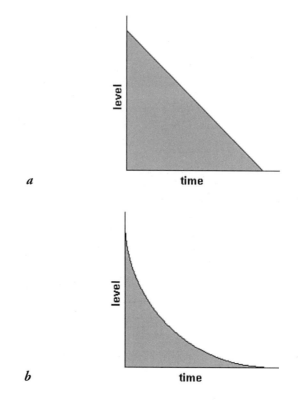

a

b

The log of a value is the number which, when applied to 10 as an exponent, produces that value as a result. The log of the value 2 must be memorized, looked up in a set of log tables, or punched up on your calculator; but when the value is an integral power of 10, the logarithmic value can be found easily. Simply write the number in exponential form, and the exponent is the log. For example, the number 10,000 is 10^4 in exponential form, so its log is 4.

log 2 = .3
log 1 = 0
log 10 = 1
log 100 = 2
log 1000 = 3

Log values are negative for powers of 10 that are less than 1. For example:

log 0.1 = −1
log 0.01 = −2
log 0.001 = −3

The following two sections about sound-pressure level and signal level are taken from Appendix A of *Practical Recording Techniques* by Bruce and Jenny Bartlett (Howard W. Sams & Co., 1992).

Sound-Pressure Level

Sound-pressure level is the pressure of sound vibration measured at a point. It's usually measured with a sound-level meter in dB SPL (decibels of sound-pressure level). The higher the sound-pressure level, the louder the sound, as shown in Figure 2.19. The quietest sound we can hear, the *threshold* of hearing, is 0-dB SPL. Average conversation at one foot is 70-dB SPL. The average home stereo listening level is around 85-dB SPL. The threshold of pain (so loud that the ears hurt) is 125- to 130-dB SPL.

Figure 2.19. *Chart of sound-pressure levels.*

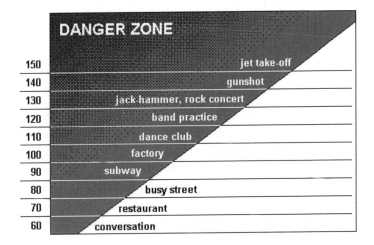

Sound-pressure level in decibels is 20 times the logarithm of the ratio of two sound pressures:

$$dB\ SPL = 20 \log \frac{P}{P_{ref}}$$

where

P is the measured sound pressure in dynes/cm²,
P_{ref} is a reference sound pressure—0.0002 dyne/cm² (the threshold of hearing).

Signal Level

Signal level also is measured in dB. The level in decibels is 10 times the logarithm of the ratio of two power levels:

$$dB = 10 \log \frac{P}{P_{ref}}$$

where

P is the measured power in watts,
P_{ref} is a reference power in watts.

Recently, it has become common to use the decibel to refer to voltage ratios as well:

$$dB = 20 \log \frac{V}{V_{ref}}$$

where

V is the measured voltage,
V_{ref} is a reference voltage.

This expression is mathematically equivalent to the previous one because power equals voltage squared divided by the circuit resistance:

$$dB = 10 \log \frac{P_1}{P_2}$$

$$= 10 \log \frac{(V_1^2/R)}{(V_2^2/R)}$$

$$= 10 \log \frac{V_1^2}{V_2^2}$$

$$= 20 \log \frac{V_1}{V_2}$$

Signal level in decibels can be expressed in various ways—dBm, dBu, dBv, and dBV.

dBm: Decibels referenced to 1 milliwatt.
dBu or dBv: Decibels referenced to 0.775 volt. (dBu is preferred.)
dBV: Decibels referenced to 1 volt.

The following explains each one of these.

If you're measuring signal power, the decibel unit to use is dBm.

$$dBm = 10 \log \frac{P}{P_{ref}}$$

where

P is the measured power,
P_{ref} is the reference power, 1 milliwatt.

For example, convert 0.01 watt to dBm:

$$dBm = 10 \log \frac{P}{P_{ref}} = 10 \log \frac{0.01}{0.001} = 10$$

So 0.01 watt is 10 dBm (10 decibels above 1 milliwatt).

Now convert 0.001 watt into dBm:

$$dBm = 10 \log \frac{P}{P_{ref}}$$

$$= 10 \log \frac{0.001}{0.001}$$

$$= 0$$

So 0 dBm = 1 milliwatt.

Any voltage across any resistance that results in 1 milliwatt is 0 dBm.

$$0\ dBm = \frac{V^2}{R}$$

$$= 1 \text{ milliwatt}$$

where

V is the voltage in volts,
R is the circuit resistance in ohms.

For example, 0.775 volt across 600 ohms is 0 dBm. One volt across 1000 ohms is 0 dBm.

Some voltmeters are calibrated in dBm. The meter reading in dBm is accurate only when you're measuring across 600 ohms. For an accurate dBm measurement, measure the voltage and circuit resistance, and then calculate:

$$dBm = 10 \log \frac{(V^2/R)}{0.001}$$

Another unit of measurement is called dBu or dBv. This means decibels referenced to 0.775 volt. The 0.775 volt figure comes from 0 dBm. 0 dBm = 0.775 volt across 600 ohms, where 600 ohms used to be a standard impedance for audio connections. Thus,

$$dBu = 20 \log \frac{V}{V_{ref}}$$

where V_{ref} = 0.775 volt.

Signal level also is measured in dBV, or decibels referenced to 1 volt. The equation for this is

$$dBV = 20 \log \frac{V}{V_{ref}}$$

where V_{ref} = 1 volt.

For example, let's convert 1 millivolt to dBV:

$$dBV = 20 \log \frac{V}{V_{ref}}$$

$$= 20 \log \frac{0.001}{1}$$

$$= -60$$

So 1 millivolt = –60 dBV (60 decibels below 1 volt). Now let's convert 1 volt to dBV, as follows:

$$dBV = 20 \log \frac{1}{1}$$

$$= 0$$

So 1 volt = 0 dBV. To convert dBV to voltage, use the formula

$$Volts = 10^{(dBV / 20)}$$

The Ear

A sound source produces sound waves by alternately compressing and rarefying the air between it and the listener. These compressions cause fluctuations of pressure above and below the normal atmospheric pressure. The ear is a sensitive transducer that responds to these pressure variations by way of a series of related processes occurring within the auditory organs that make up the ear.

When they arrive at the listener, sound-pressure waves are collected into the aural canal by way of the outer ear's pinna and are then directed to the eardrum, a stretched drum-like membrane (see Figure 2.20). The sound waves are then changed into mechanical vibrations and transferred to the inner ear by way of three bones called the *hammer, anvil,* and *stirrup.* These bones act as both an amplifier (by significantly amplifying the vibrations given them by the eardrum) and as a limiting protection device (reducing the level of loud transient sounds, such as thunder or fireworks explosions). The vibrations are then applied to the inner ear (cochlea)—a tubular, snail-like organ that contains two fluid-filled chambers. Within these chambers are tiny hair receptors lined in a row all along the cochlea. Vibrations transmitted to the hairs, which respond to certain frequencies depending on their placement along the organ, result in neural stimulation that gives us the sensation of hearing. Hearing loss generally occurs when these hairs are damaged or deteriorate with age.

Figure 2.20. *Diagram showing the outer, middle, and inner ear.*

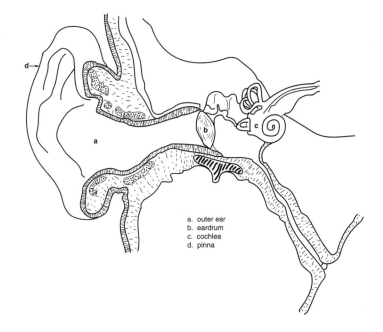

a. outer ear
b. eardrum
c. cochlea
d. pinna

Threshold of Hearing

In the case of SPL, a convenient pressure-level reference is that of the *threshold of hearing*, which is the minimum sound pressure that produces the phenomenon of hearing in most people. It is equal to 0.0002 microbar. One microbar is equal to one-millionth of normal atmospheric pressure, so it is apparent that the ear is extremely sensitive. In fact, if the ear were any more sensitive, the thermal motion of the molecules in the air would be audible. When referencing sound-pressure levels to 0.0002 microbar, this threshold level usually is denoted as 0-dB SPL.

The threshold of hearing is defined as the SPL for a specific frequency that the average person can hear only 50 percent of the time.

Threshold of Feeling

The SPL that will cause discomfort in a listener 50 percent of the time is called the *threshold of feeling*. It occurs at a level of about 118-dB SPL between the frequencies of 200 Hz and 10 kHz.

Threshold of Pain

The SPL that causes pain in a listener 50 percent of the time is called the *threshold of pain* and corresponds to an SPL of 140 dB in the frequency range between 200 Hz and 10 kHz. Figure 2.21 shows typical SPLs for different sounds. The levels are weighted to take into account the reduced sensitivity of human hearing to low frequencies.

Figure 2.21. Typical SPLs for common sounds. The reference level of 20N/m² is equivalent to 0.0002 microbar or 0.002 dynes/cm². (Courtesy of General Radio Company).

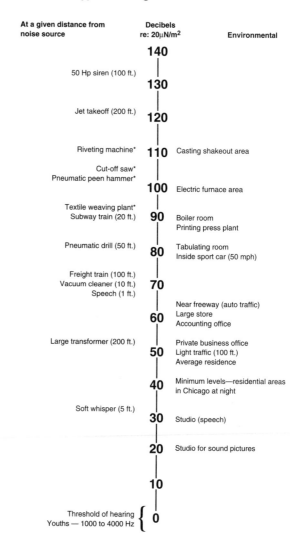

Typical A-Weighted Sound Levels

At a given distance from noise source	Decibels re: 20μN/m²	Environmental
	140	
50 Hp siren (100 ft.)	**130**	
Jet takeoff (200 ft.)	**120**	
Riveting machine*	**110**	Casting shakeout area
Cut-off saw* Pneumatic peen hammer*	**100**	Electric furnace area
Textile weaving plant* Subway train (20 ft.)	**90**	Boiler room Printing press plant
Pneumatic drill (50 ft.)	**80**	Tabulating room Inside sport car (50 mph)
Freight train (100 ft.) Vacuum cleaner (10 ft.) Speech (1 ft.)	**70**	
	60	Near freeway (auto traffic) Large store Accounting office
Large transformer (200 ft.)	**50**	Private business office Light traffic (100 ft.) Average residence
	40	Minimum levels—residential areas in Chicago at night
Soft whisper (5 ft.)	**30**	Studio (speech)
	20	Studio for sound pictures
	10	
Threshold of hearing Youths — 1000 to 4000 Hz {	**0**	

*Operator's position

Auditory Perception

The ear is a *nonlinear* device; as a result, it produces harmonic distortion when it is subjected to sound waves above a certain loudness. *Harmonic distortion* is the production of waveform harmonics that do not exist in the original signal. Thus, the ear can cause a loud 1-kHz sine wave to be heard as a combination of waves of 1 kHz, 2 kHz, 3 kHz, and so on. Although the ear may receive the overtone structure of a violin, if the listening level is loud enough, the ear produces additional harmonics, thus changing the perceived timbre of the instrument. This means that sound monitored at very loud levels may sound quite different when played back at low levels.

The terms *linear* and *nonlinear* are used to describe the output-amplitude versus input-amplitude characteristics of transducers and signal-processing equipment. A linear device or medium is one whose input and output amplitudes have the same ratio at all signal levels. For example, if an amplifier is linear, doubling the input signal amplitude will double the output signal amplitude regardless of the original input signal level. If at certain signal levels, however, doubling the input signal amplitude increases the output signal amplitude by more or less than a factor of two, the amplifier would be called nonlinear at those amplitudes. The use of the term, linear, comes from a graph that is often called the *transfer characteristic* of the device in question. Linear means that the graph is a straight line, while nonlinear means that it either has curves or one or more angles in it. A device can have both linear and nonlinear operating regions. Restricting operation to the linear section, however, avoids distortion.

In addition to being nonlinear with respect to amplitude, the ear's frequency response (that is, its perception of timbre) changes with the loudness of the perceived signal. The loudness compensation switch found on many hi-fi preamplifiers is an attempt to compensate for the decrease in the ear's sensitivity to low-frequency sounds at low levels.

The curves in Figure 2.22 are the Fletcher-Munson equal-loudness contours. They indicate the ear's average sensitivity to different frequencies at various levels. The horizontal curves indicate the sound-pressure levels that are required to produce the same perceived loudness at different frequencies. Thus, to equal the loudness of a 1.5-kHz tone at a level of 110-dB SPL (which is the level typically created by a trumpet-type car horn at a distance of 3'), a 40-Hz tone has to be 2 dB greater in sound-pressure level, while a 10-kHz tone must be 8 dB greater than the 1.5-kHz tone to be perceived as being equally as loud. At 50-dB SPL (the noise level present in the average private business office), the level of a 30-Hz tone must be 30 dB greater and a 10-kHz tone must be 14 dB greater than a 1.5-kHz tone to be perceived as being at the same volume. Thus, if a piece of music is monitored so that the signals produce a sound-pressure level of 110 dB, and it sounds well balanced, it will sound both bass and treble deficient when played at a level of 50-dB SPL.

From the standpoint that changes in apparent frequency balance are less apparent above and below 85-dB SPL than for any other level, this figure appears to be the best average sound monitoring level. For example, if the monitoring level is 120-dB SPL and you are satisfied with the musical balance, you will notice that as you begin to decrease the playback level, the ear's low-frequency sensitivity decreases and the bass frequencies appear less pronounced. As the playback level is lowered further, the ear's low-frequency sensitivity decreases even further and the bass begins to disappear. Ear sensitivity in the upper frequencies will likewise fall. Thus, a mix made at a monitoring level of 120-dB SPL will sound bass deficient, distant, and lifeless at lower levels. If you monitor the mix at 100-dB SPL and are satisfied with the balance, decreasing the monitoring level will again cause a loss of perceived clarity, although to a lesser extent.

Figure 2.22. The Fletcher-Munson equal loudness contour for pure tones as perceived by humans having average hearing acuity. These perceived loudness levels are charted relative to sound pressure levels at 1000 Hz.

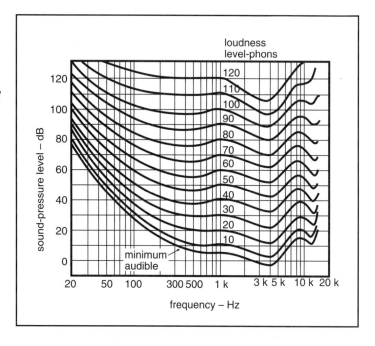

If you monitor at 85-dB SPL and are happy with the balance, you can play the mix back at any level between 90- and 60-dB SPL and hear very little change in balance except at the extreme high and low ends of the frequency spectrum where the changes are less than 5 dB. Conveniently, home listening levels generally fall in the 75- to 85-dB SPL range, so 85-dB SPL can be considered the optimum monitoring level for mixdowns.

The loudness of a tone can also affect the pitch that the ear perceives. For example, if the intensity of a 100-Hz tone is increased from 40- to 100-dB SPL, the ear perceives a pitch decrease of about 10 percent. At 500 Hz, the pitch changes about 2 percent for the same increase in sound-pressure level. This is one reason why musicians find it difficult to tune their instruments while listening through headphones. The headphones are often producing higher SPLs than might be expected.

As a result of the nonlinearity of the ear, tones can interact with each other rather than being perceived separately. Three types of interaction effects occur: beats, combination tones, and masking. These three effects are described in the following sections.

Beats

Two tones that differ only slightly in frequency and have approximately the same amplitude will produce *beats* (repetitive volume surges) at the ear equal to the difference between the two frequencies. The phenomenon of beats can be used as an aid in tuning instruments because the beats slow down and stop as the two notes approach and reach the same pitch. In a properly tuned piano, not all notes are in perfect tune, and the piano tuner will slightly off-tune the instrument by listening to the beat relationships. These beats are the result of the ear's inability to separate closely pitched notes. The resulting synthesis of a third wave represents the addition of the two waves when they are in-phase and the subtraction of their intensities when they are out-of-phase.

Combination Tones

Combination tones result when two loud tones differ by more than 50 Hz. The ear produces an additional set of tones that is equal to both the sum and the difference of the two original tones and is also equal to the sum and difference of their harmonics. The formulae for computing the tones are

difference tone frequencies = $f_1 - f_2$

sum tone frequencies = $f_1 + f_2$

where

f_1 and f_2 are positive integers

The difference tones can be easily heard when they are below the frequency of both the original tones. For example, 2000 and 2500 Hz produce a difference tone of 500 Hz.

Masking

Masking is the phenomenon by which loud signals prevent the ear from hearing softer sounds. The greatest masking effect occurs when the frequency of the sound and the frequency of the masking noise are close to each other. For example, a 4-kHz tone will mask a softer 3.5-kHz tone but has little effect on the audibility of a quiet 1000-Hz tone. Masking can also be caused by harmonics of the masking tone, so a 1-kHz tone with a strong 2-kHz harmonic could mask a 1900-Hz tone. The masking phenomenon is one of the main reasons that stereo placement and equalization are so important in the mixdown process. An instrument that sounds fine by itself can be completely hidden or changed in character by louder instruments with a similar timbre. Equalization may be required to make the instruments sound different enough to overcome the masking.

Perception of Direction

Although one ear can't discern the direction from which a sound originates, two ears can. This capability of two ears to localize a sound source within an acoustic space is called *spatial* or *binaural localization*. This effect results from using the following three cues received by the ears:

- ◆ Interaural intensity differences
- ◆ Interaural arrival-time differences
- ◆ The effects of the pinnae (outer ears)

Middle- to higher-frequency sounds originating from the right side reach the right ear at a higher intensity level than the left ear, causing an *interaural intensity difference*. This difference occurs because the head casts an acoustic block, or shadow, allowing only reflected sound from surrounding surfaces to reach the left ear (see Figure 2.23). Because the reflected sound travels farther and loses energy at each reflection, the intensity of sound perceived by the left ear is reduced. The resulting signal is perceived as originating from the right.

Figure 2.23. *Acoustic shadow thrown by the head causes interaural intensity differences at middle to upper frequencies.*

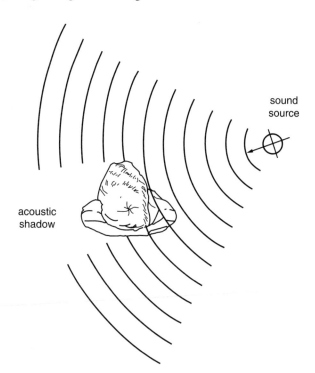

sound source

acoustic shadow

This effect is relatively insignificant at lower frequencies where wavelengths are large compared to the diameter of the head and easily bend around its acoustic shadow. A different method of localization, known as *interaural arrive-time differences,* is employed at lower frequencies. In both Figure 2.23 that you just saw and Figure 2.24 (shown next), small time differences occur because the acoustic path length to the left ear is slightly longer than the path to the right ear. The sound pressure is therefore sensed by the left ear at a later time than by the right ear. This method of localization, in combination with interaural intensity differences, gives us lateral localization cues over the entire frequency spectrum.

Figure 2.24. *Interaural arrive-time differences occurring at lower frequencies.*

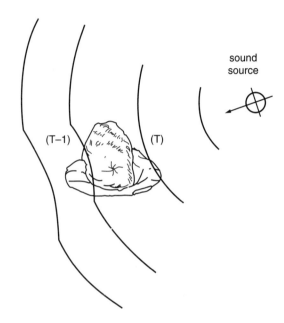

The intensity and delay cues enable us to perceive the angle from which a sound originates, but not whether the sound originates from front, behind, or below. The pinna (shown in Figure 2.25), however, makes use of two ridges that reflect the incident sound into the ear. These ridges introduce time delays between the direct sound (which reaches the entrance of the ear canal) and the sound reflected from the ridges (which varies according to source location).

It's interesting to note that from beyond about 130° from the front axis there can be no ridge reflections because they are blocked by the pinna. Unreflected sounds that are delayed between 0 and 80 microseconds (μsec) will still be perceived as originating from the rear. Ridge number two will produce delays of between 100 and 330 μsec, corresponding to a source located within the vertical plane. The delayed reflections from both ridges combine with the direct sound to produce the characteristic frequency response colorations that are due to constructive and destructive interference at differing frequencies. The brain is able to compare these colorations at each ear and use this information to determine source location. Small movements of the head provide additional position information, due to the changing source perspective. This latter cue is minor, however, compared with that of the other localization cues.

Figure 2.25. The pinna and its reflective ridges for determining vertical location information.

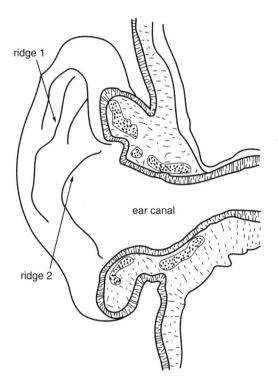

If there are no differences between what the left and right ears hear, the brain assumes that the source is the same distance from each ear. It is this phenomenon that enables the audio engineer to position sound not only in the left and right loudspeakers, but also monophonically between the loudspeakers. By feeding the same signal to both loudspeakers, the brain perceives the sound identically in both ears and deduces that the source must originate from directly in front of the listener. By changing the proportional level to each loudspeaker, the engineer changes the interaural intensity differences and thus creates the illusion that the sound source is positioned at any desirable point between these two loudspeakers. The source positioning may even be caused to move from point to point between these loudspeakers. This placement technique is known as *panning* (see Figure 2.26). Although it is the most widely used method, it isn't the most effective positioning technique because only those listeners who are equidistant from left and right loudspeakers will perceive the desired effect. A listener located close to the left loudspeaker will tend to locate the source as coming from that source even though the signal may be panned towards the right. The engineer can use other, more effective tools of localization, such as a digital delay line (DDL), phase shifter, filter, or stereophonic microphone techniques to assign a location point between two loudspeakers.

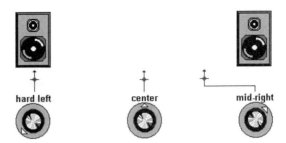

Figure 2.26. *Pan pot settings versus spatial positioning.*

Perception of Space

In addition to perceiving the direction of sound, the ear and brain combine to perceive distance and a physical sense of the acoustic space in which a sound occurs. When a sound is generated, it simultaneously propagates away from the source in all directions. The angles of propagation are determined by the nature of the source. A percentage of the sound reaches the listener directly, without encountering any obstacles. A much greater portion, however, is propagated to the many surfaces of an acoustic enclosure. If these surfaces are reflective, they bounce the sound waves back into the room, where some of these reflections will reach the listener. If the surfaces are absorptive, very little energy is reflected back to the listener.

In air, sound travels at a constant speed of about 1130 feet per second, so the wave that travels a straight line from the source to the listener follows the shortest path and arrives at the listener's ear first. This is called *direct sound.* Those waves that bounce off surrounding surfaces must travel further to reach the listener and therefore arrive after the direct sound. These waves form what is called the *reflected sound,* which, in addition to being delayed, can also arrive from different directions than the direct sound. As a result of these additional longer path lengths, the ear hears the sound even after the source stops emitting it. Highly reflective surfaces absorb less of the wave energy at each reflection and enable the sound to persist longer after the source stops than highly absorptive surfaces that dissipate the wave energy. The sound heard in a room can be divided into three successively occurring categories: direct sound, the early reflections, and reverberation (see Figure 2.27). Direct sound determines our perception of the sound source location and size and conveys the true timbre of the source. The amount of absorption that occurs when sound is reflected from a surface is not equal at all frequencies. As a result, the timbre of the reflected sound is altered by the characteristics of the surfaces it encountered.

Early reflections reach the ear within 50 ms after the direct sound. These reflections are the result of waves that encountered only a few boundaries before reaching the listener, and they may arrive from a different direction than the direct sound. The time elapsed between hearing the direct sound and the beginning of the early reflections provides information about the size of the performance room. The further the surfaces are from the listener, the longer it takes the sound to reach them and be reflected back to the listener. The psychoacoustic phenomenon known as the *precedence* or *Haas effect* suppresses our perception of these reflections by as much as 8 to 12 dB, depending on how late after the direct sound they arrive at the ear. The precedence effect applies equally, whether you consider a sound source and its reflection or two spatially separate sound sources—such as two loudspeakers—producing the same sound. The sound will appear to originate at the earlier source even if the delayed source is 8 to 12 dB louder.

Figure 2.27. *The soundfield generated at a listening position within an enclosed space.*

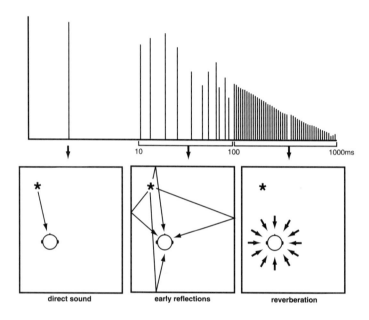

Another aspect of the precedence effect is called *temporal fusion*. Early reflections arriving at the listener within 30 ms of the direct sound are not only suppressed in audibility, but they are also fused with the direct sound. The ear can't separately distinguish the closely occurring sounds and considers the reflections to be part of the direct sound. The 30-ms time limit for temporal fusion is not an absolute; rather it depends on the envelope of the sound. Fusion breaks down at 4 ms for transient clicks, whereas it can extend beyond 80 ms for slowly changing sounds, such as legato violin passages. Despite the fact that the early reflections are suppressed and fused with the direct sound, they do modify our perception of the sound, making it both louder and fuller.

Sounds reaching the listener more than about 50 ms after the direct sound have been reflected from so many different surfaces that they begin to reach the listener in a virtually continuous stream and from all directions. These densely spaced reflections are called *reverberation*. Reverberation is characterized by its gradual decrease in amplitude and the warmth and body it gives the sound as well as the loudness it adds to the sound. Because it has undergone multiple reflections, the timbre of the reverberation is quite different from that of the direct sound, with the most notable difference being a rolloff of high frequencies and a resultant emphasis of bass. The time it takes for the persisting sound to decrease to 60 dB below its original level is called its *decay time* or *reverberation time* and is abbreviated RT_{60}. The absorption characteristics of a room's surfaces determine its reverberation time. The brain perceives the reverberation time and the timbre of the reverberation and then uses this information to form an opinion on the hardness or softness of the surrounding surfaces. The loudness of the direct sound perceived by the ear increases rapidly as the listener moves closer to the source. Meanwhile, the loudness of the reverberation remains the same because it is so well diffused throughout the room. The perception of the ratio of the loudness of the direct sound to the loudness of the reflected sound enables the listener to judge his distance from the sound source.

To summarize, the direct sound provides information about source location, size, and timbre. The time between perception of the direct sound and the early reflections determines our impression of the size of the performance room. The reverberation decay time provides information about the performance room surfaces, and the proportion of reverberation to direct sound determines our perception of distance from the sound source.

Through the use of artificial reverberation and delay units, the engineer can generate the cues necessary to convince the brain that a sound was recorded in a huge stone wall cathedral when it was, in fact, recorded in a small, dead sounding room. To do this, the engineer feeds the dry or unreverberated signal into a console fader where it is split into two paths (see Figure 2.28). One path leads directly to the console's output signal path to provide the direct sound, while the second is routed to a signal processor unit to provide the necessary reverberation and delays. The processor's stereo outputs are then fed back into the output signal path via an available pair of faders on the console, or by specially designated effects inputs called *effects returns*. Adjusting the number and amount of delays on the effects processor gives the engineer control of the time that has elapsed between hearing the dry, or direct, sound and the early reflections, thus determining the listener's perception of the size of the room. Adjusting the reverberation unit's decay time determines the listener's perception of the room surfaces. A long decay time would indicate a hard-surfaced room, while a short decay time would indicate a soft-surfaced room.

By increasing or decreasing the proportion of direct sound to the delayed and reverberated signal, the listener can be fooled into believing the sound source is placed either at the front or rear of the artificially created performance room.

Time-delay units can be used independent of reverberation to stimulate the dry single-echo effect heard at outdoor live performances or to stimulate additional instruments performing in unison. This is possible because our perception that more than one instrument is playing depends on the lack of exact synchronization between the sounds. No matter how good the musicians are, the fact that each instrument is a different distance away from the listener, no matter how slight, ensures a lack of synchronization due to the different amounts of time required for the waves to reach the listener's ear. By repeating a signal after a short delay of 4 ms or so, the apparent number of instruments being played is doubled. This process is called *doubling* or *automatic double tracking (ADT)*. Often, acoustic doubling and tripling is done by the artist alone through multiple overdubbing so that one violin can sound like an ensemble or two vocalists can sound like a chorus. Both automatic and acoustic doubling can be used to strengthen weak vocal performances because the slight irregularities in one performance are hidden by the other. If the delay is long enough (more than about 35 ms) that the repeat is heard discreetly, the repeat is often called *slap echo*, or *slap back*, and it has the effect of causing the rhythm to bounce or double.

Figure 2.28. Example of a signal routing path for an electronically reverberated signal.

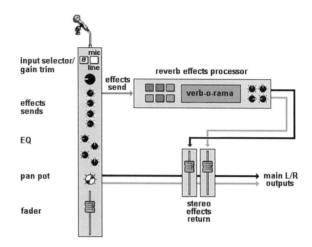

In addition to this bag of placement and psychoacoustic tricks, systems are now available that are able to digitally manipulate phase and time-delay parameters in ways that allow the ears to be fooled into thinking that a signal is coming from a source that is actually wider than the speaker's actual position, or that it is emanating from a position that is behind the listener. These processing systems, generically known as *spatial signal processors* (see Figure 2.29) have become increasingly popular with the growth of the home entertainment and video industries, as have surround matrix systems that can encode and decode semi-discrete 4- or 6-channel information within two encoded audio tracks.

Figure 2.29. The Spatializer spatial audio manipulator. (Courtesy of Desper Products, Inc.)

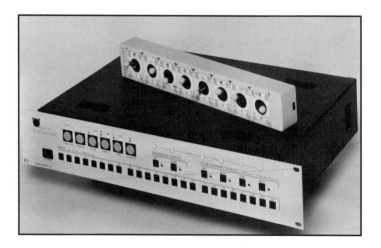

CHAPTER 3

Studio Acoustics and Design

The Audio Cyclopedia defines the term *acoustics* as "…a science dealing with the production, effects, and transmission of sound waves; the transmission of sound waves through various mediums, including reflection, refraction, defraction, absorption, and interference; the characteristics of auditoriums, theaters, and studios, as well as their design."

We see from this description that the proper acoustic design of a music recording, project, audio-for-visual, or broadcast studio is no simple matter. A wide range of complex variables and interrelationships often come into play in the making of a successful studio design. The following basic requirements must be considered:

◆ *Acoustic isolation.* Prevents external noises from entering the studio environment, as well as preventing the neighborhood feuds that often occur when excessive volume levels leak out into the surrounding environment.

◆ *Frequency balance.* The frequency components of recorded sound should maintain their relative level intensities. Simply stated, the acoustic environment shouldn't adversely affect or alter the sound quality of the originally recorded performance.

◆ *Acoustic separation.* The acoustical environment should offer a high degree of intelligibility, as well as acoustic separation (often a requirement for ensuring that acoustic sounds emanating from an instrument aren't unduly picked up by another instrument's mic). These characteristics are often best controlled through the absorption of sound and its reflections.

◆ *Reverberation.* The control of sonic reflections within a space is also an important factor in maximizing the intelligibility of music and speech. In addition, reverberation adds an important psychoacoustic sense of "space" in the sense that it can give our brain subconscious cues as to a room's size, the number of reflective boundaries, the distance between the source and listener, and so on.

◆ *Cost factors.* Not the least of all design and construction factors is cost. Multimillion-dollar facilities often employ studio designers and construction teams to create a plush decor that has been acoustically tuned to fit the needs of both the owners and the clients. Owners of project studios and budget-minded production facilities, however, can also take full advantage of basic acoustic principles and construction techniques and apply them in cost-effective ways.

This chapter discusses many of the basic acoustic principles and construction techniques that must be considered in the design of a music production facility. I want to emphasize that any or all of these acoustical topics can be applied to any type of audio production facility; they aren't limited only to professional recording studio designs. For example, even owners of modest project studios should know the importance of making the control room symmetrical. If one speaker is in a corner and the other is on a wall, the perceived center image will be off balance.

As with many techno-artistic endeavors, studio acoustics and design are a mixture of fundamental physics (in this case, mostly dimensional mathematics) and good ol' common sense.

The Project Studio

The design and construction considerations for creating a privately owned recording or MIDI project studio (see Figure 3.1) often differ from the design considerations for a professional music facility in two fundamental ways: building constraints and cost.

Generally, a project studio's room or series of rooms are built into an artist's home or a rented space in which the construction and dimension details already are defined. This fact, combined with the inherent cost considerations, often lead the owner/artist to employ cost-effective techniques for sonically treating a room. Even if the room has little or no sonic treatment, or if treatment isn't deemed necessary, keep in mind that a basic knowledge of acoustical physics and room design can be a handy and cost-effective tool as your experience, production needs, and (hopefully) your business grow.

Figure 3.1. *Rick Rottenberg's home studio where the music for* The Simpsons *TV show is created. (Courtesy of Rick Rottenberg and Mackie Designs)*

The Music Studio

The professional recording studio (see Figures 3.2a, 3.2b, 3.3a, and 3.3b) is first and foremost a commercial business, so its design, decor and acoustical construction requirements often are much more demanding than its privately owned counterpart, the project studio. In some cases, a professional acoustical designer and construction team are in charge of the overall building phase of a professional facility. In most cases, however, the studio's budget precludes the hiring of such professionals, and the studio owners and staff are charged with both designing and constructing the facility.

Figure 3.2. *Bad Animals' Studio X. (Courtesy of Bad Animals, Seattle, Washington and Studio bea:ton)*

a. *Control room.*

a

Figure 3.2. *(continued)*
b. *Recording studio.*

b

Figure 3.3. *Sony Music Studios (New York, NY). (Courtesy of Russ Berger Design Group, Inc.)*
a. *Control Room C.*
b. *Tracking Room B.*

a

b

If you have the luxury of building a new facility from the ground up or within an existing shell, you probably would benefit from a professional studio designer's experience and skills—even if you are building or renovating the facility yourself. Such expert advice often proves to be cost-effective in the long run. Errors in design judgment can lead to cost overruns, lost business due to unexpected delays, or the unfortunate fact of having to live with a mistake that could have been avoided.

The Audio-for-Visual Production Environment

An audio-for-visual production studio is an environment used for video or film audio postproduction (see Figures 3.4a and 3.4b) that includes such facets as Foley (the replacement and creation of on- and off-screen sound effects), automatic dialog replacement (ADR), music recording (scoring), and mixdown. As with music studios, audio-for-visual production facilities can range from a simple budget-minded room equipped with project-studio-class equipment to a high-end, specially designed facility that can accommodate the postproduction needs of network video or feature film productions. As with the music studio, construction design and techniques often span a wide range to fit the budget requirements at hand.

Figure 3.4. *Audio-for-visual postproduction systems.*

a. SSL Scenaria installed at Sunders & Gordon, London. (Courtesy of Solid State Logic)

b. Studer-Editech Post Trio Audio-for-visual and digital audio workstation. (Courtesy of Studer-Editech Corporation)

a

b

Many audio-for-visual production facilities are designed to perform a specific function. For example, a Foley studio is used to rerecord certain on- and off-screen sound effects (such as shoes and doors), whereas an ADR facility is designed expressly for rerecording on- and off-screen dialog. Such facilities pale in size compared to the dubbing and mixing stages often used to mix the various facets of a first-run feature film.

Primary Factors Governing Studio and Control Room Acoustics

In this section, you take a close look at the most important and relevant aspects of acoustics as they pertain to the *sound studio* (acoustic space used for recording music or speech) as well as to the *control room* (acoustic space used for the production and mixdown of music or speech). Important factors in this discussion include acoustic isolation, frequency balance, absorption, reflection, and reverberation.

Acoustic Isolation

Under certain conditions, it may be necessary for modern day recording studios and audio-for-visual sound stages to utilize effective isolation techniques that reduce external noise—whether that noise is transmitted through the medium of air (as might occur with nearby auto, train, or jet traffic) or through solids (such as noise transmission from air conditioner rumble or underground subways through building foundations or structures). In order to reduce such extraneous noises, special construction techniques may be required.

If you have the luxury of building a studio facility from the ground up, you should put a great deal of thought into determining the studio's location. If a location has considerable neighborhood noise, you may have to resort to extensive—and expensive—construction techniques to isolate the sound. If you have no choice in a studio's location, however, and the studio happens to be located next to a factory, just under the airport's main landing path, or over the subway's uptown line, you simply have to give in to destiny and build acoustical barriers to these outside interferences.

The reduction in the sound pressure level (SPL) of a sound source as it passes through an acoustic barrier of physical mass (see Figure 3.5) is termed the transmission loss (TL) of a signal. This attenuation can be expressed (in dB) using

$$TL = 14.5 \log M + 23$$

where

TL is the transmission loss in decibels,
M is the surface density (or combined surface densities) of a barrier in pounds per square foot (lbs/ft^2).

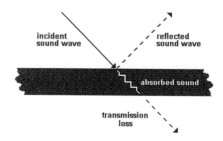

Figure 3.5. *Transmission loss refers to the reduction of a sound signal (in dB) as it passes through an acoustic barrier.*

And because transmission loss is frequency dependent, you can use this second equation to calculate with relative accuracy transmission loss at various frequencies:

$$TL = 14.5 \log Mf - 16$$

where

f is the frequency in Hertz.

Both common sense and the preceding two equations tell us that heavier acoustic barriers yield a higher transmission loss. For example, from Table 3.1, you see that a 12-inch-thick wall of dense concrete (yielding a surface density of 150 lb/ft^2) offers a much greater resistance to the transmission of sound than a 4-inch cavity filled with sand (yielding a surface density of 32.3 lb/ft^2). From the second equation ($TL = 14.5 \log Mf - 16$), you can also draw the conclusion that for a given acoustic barrier, higher transmission losses will be encountered as the frequency rises. You can prove this fact easily by closing the door of a car that has its sound system turned up or by shutting a single door to a music studio's control room. The high frequencies will be reduced in level, while the bass frequencies will be impeded to a much lesser extent.

Table 3.1. The surface densities (lbs/ft^2) of common building materials.

Material	Thickness (inches)	Surface Density (lbs/ft^2)
Brick	4	40.0
	8	80.0
Concrete (light weight)	4	33.0
	12	100.0

Material	Thickness (inches)	Surface Density (lbs/ft²)
Concrete (dense)	4	50.0
	12	150.0
Glass	1/4	3.8
	1/2	7.5
	3/4	11.3
Gypsum wallboard	1/2	2.1
	5/8	2.6
Lead	1/16	3.6
Particle board	3/4	1.7
Plywood	3/4	2.3
Sand	1	8.1
	4	32.3
Steel	1/4	10.0
Wood	1	2.4

From this, the goal would seem to be to build a studio wall, floor, ceiling, window, or door out of the thickest and densest material that you can get. Practicality, however, as much as expense, often plays a role in determining just how much of a barrier is required to achieve the desired isolation. In most situations, you can strike a balance to get the job done while using both space- and cost-effective building materials.

Generally, an equal amount of isolation is required between the studio and the control room as is required between the studio's interior and exterior environments. This is important so that an accurate tonal balance can be heard over the control-room monitors without leakage from the studio coloring the audible signal.

Walls

When you build a studio wall or reinforce an existing structure, your primary goal should be to reduce transmission (increase the transmission loss) through a wall at most or all frequencies. Generally, this is accomplished by building a wall structure that is as massive as possible (in terms of both cubic and square density) and that is highly *damped* (meaning well supported by its reinforcement structures and relatively free of resonances) as is shown in Figures 3.6a and 3.6b.

Figure 3.6. *Two typical gypsum wallboard constructions.*

a. *Single stud design.*

b. *Double, staggered stud construction with a higher transmission loss value.*

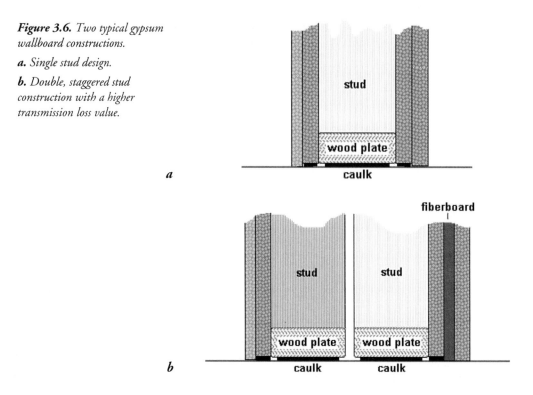

a

b

The following guidelines can be beneficial in the construction of framed walls with a high degree of transmission loss:

♦ If at all possible, the inner and outer wallboards should not be directly attached to the same stud. The best way to avoid this is to attach each wall facing to its respective staggered studs (refer again to Figure 3.6b).

♦ Each wall facing should have a different density to reduce the likelihood of increased transmission due to resonant frequencies that are sympathetic to both sides. For example, one wall might be constructed of two, 5/8" thick gypsum wallboards, while the other wall might be composed of soft fiberboard surfaced with two, 1/2" thick gypsum wallboards.

♦ If you are going to attach a number of gypsum wallboards to a single wall face, you can increase transmission loss by mounting the additional layers (not the first layer) with adhesive caulking rather than screws or nails.

♦ Spacing the studs 24" on center instead of using the traditional 16" spacing yields a slight increase in TL.

♦ To reduce leakage that might make it through the cracks, apply a bead of non-hardening caulk sealant to the inner gypsum wallboard layer at the wall-to-floor, wall-to-ceiling, and corner junctions.

Floors

For many recording facilities, the isolation of floor-borne noises is an important consideration. For example, a building that is on a busy street and whose concrete floor rests directly on ground foundation may experience severe low-frequency rumble. Alternatively, a second-floor facility might have undue leakage from a noisy downstairs neighbor or, even more likely, may interfere with a quieter neighbor's business. In each of these situations, some form of isolation from floor-borne sound is essential. One of the most common ways to isolate floor-related noise is to construct a "floating" floor, which is a floor that is structurally decoupled (to a certain degree) from its sub-floor foundation.

Two of the most common construction methods for floating a floor make use of either neoprene "hockey puck" isolation mounts (see Figure 3.7) or a continuous underlayment, such as rubberized floormat coverings (see Figure 3.8). In either of these approaches, a flexible underlayment is laid over the existing floor foundation. The underlayment is covered with 1/2" plywood, followed by plastic sheeting (for insulation and to prevent seepage), a layer of reinforcing wire mesh, and is finished off with a 4" top layer of concrete. This floating substructure is then ready for carpeting, wood finishing, painting, or any other desired surface.

Figure 3.7. Construction details *of a floating floor using neoprene mounts.*

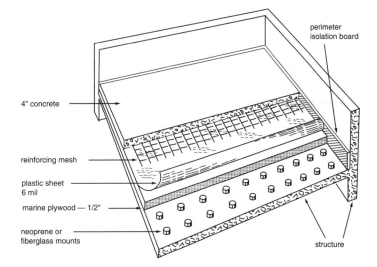

Figure 3.8. *Construction details of a floating floor using a roll-out mat system.*

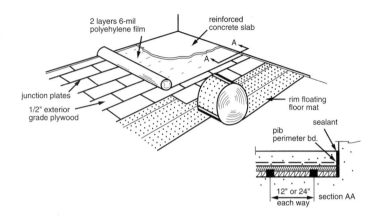

Be sure to isolate the superstructure from both the underflooring and the outer or perimeter wall. Failing to isolate these can transmit sound both through the walls to the floor and from the subfloor to the upper layer—defeating the whole purpose of floating the floor. This perimeter isolation can be made from such pliable decoupling materials as widths of soft mineral fiber boards or neoprene (see Figure 3.9).

Figure 3.9. *Decoupling between the floated floor and the surrounding wall perimeters should be accomplished using a pliable boundary, such as soft fiber boards or neoprene.*

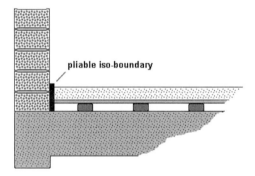

A less expensive method for decoupling a floor is to layer the initial layer with a carpet foam pad. A 1/2" or 5/8" layer of tongue-and-groove plywood boards can be laid on top of the pad (these should not be nailed to the sub-floor; instead, they can be stabilized by locking the pieces together with thin, metal braces). Another foam pad can be laid over this structure and then topped with carpeting or any other desired finishing (see Figure 3.10).

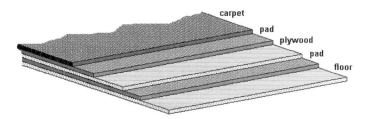

Figure 3.10. *A cost-effective, alternative method for floating an existing floor.*

Risers

As you saw earlier in the equation TL = 14.5 log Mf – 16, low-frequency sound travels through barriers much more easily than do high-frequency sounds. It stands to reason, then, that strong, low-frequency energy transmits best—between studios, from studio to the control room, and to outside locations. And what is the most likely offender for traveling through barriers? Live drums, of course. By decoupling much of the drum set's low-frequency energy from a studio floor, you can greatly reduce this leakage problem. In most cases, you can fix the problem by constructing a drum riser, as shown in Figure 3.11.

Figure 3.11. *Construction detail for a typical drum riser.*

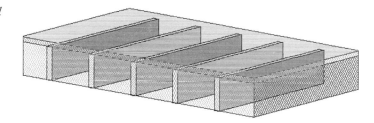

In order to reduce unwanted resonances, drum risers should be constructed using 2" x 6" or 2" x 8" beams for both the frame and the supporting joists (spaced at 16" or 12" on-center intervals). Sturdy 1/2" or 5/8" tongue-and-groove plywood panels should be glued to the supporting frames with carpenter's glue (or a similar wood glue) and then nailed down. To the bottom of the frame, attach neoprene strips, neoprene coasters, or (at the least) strips of carpeting. The riser is ready for action.

Ceilings

Foot traffic or other noises from above a sound studio or production room is another common source of external leakage. Ceiling noise can be isolated in a number of ways. If foot traffic is your problem and you own the floors above you, footstep noise can be reduced by simply carpeting the overhead hallway or floating the upper floor. If you don't have this luxury, one approach to deadening ceiling-borne sound is to float a false structure from the existing ceiling or from the overhead ceiling joists (as is often done when a new room is being constructed). This technique can be fairly cost-effective if you use "Z" suspension channels, as shown in Figure 3.12. The "Z" channels can be screwed to the ceiling joists to provide a flexible, yet strong support to which a newly floated wallboard ceiling can be attached. If necessary, you can even place fiberglass or other sound-deadening materials in the cavities between the overhead structures. Other more expensive methods are also available that use spring support systems to hang false ceilings from existing structures.

Figure 3.12. *"Z" channels can be used to hang a floating ceiling from an existing overhead structure.*

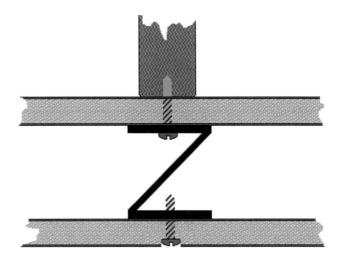

Windows and Doors

Access to and from a studio or production room area (in the form of windows and doors) can also be potential sources of sound leakage. For this reason, you need to give strict attention to their design and construction.

Visibility within a studio is extremely important in a music production environment. Visibility promotes good communication between the producer or engineer and the studio musician, as well as between the musicians themselves when multiple rooms are involved. For this reason, windows have been an important factor in studio design for decades. The design and construction details for windows varies with studio needs and budget requirements. Windows can range from deep double-plate cavities built into double-wall constructions (see Figure 3.13a), to less deep and more modest constructions built into a single wall (see Figure 3.13b). Other, more modern designs range from floor-to-ceiling windows that create a virtual "glass wall," to designs in which the windows have been built into studio/control room walls that employ 3' of solid, poured concrete.

Figure 3.13. *Details for practical window construction between the control room and studio.*

a. *A window suitable for a high transmission loss wall.*

b. *A window for a more modestly framed wall.*

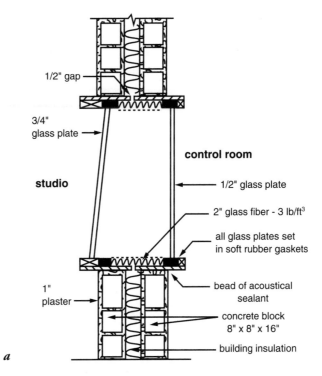

1/2" gap

3/4" glass plate

control room

studio

1/2" glass plate

2" glass fiber - 3 lb/ft^3

all glass plates set in soft rubber gaskets

1" plaster

bead of acoustical sealant

concrete block 8" x 8" x 16"

building insulation

a

3/8" glass plate

studio

1/2" glass plate

control room

2" glass fiber - 3 lb/ft^3

all glass plates set in soft rubber gaskets

double 5/8" gypsum board

bead of acoustical sealant

5/8" gypsum board outer layer on resilent strips

building insulation

b

The glass panels used in window construction typically are 3/8" to 3/4" thick and are seated into the window frame with a rubber or similar type of elastic damping seal to prevent structure-borne oscillations. It is important that at least one of these panels be tilted at a 5° angle (minimum), with respect to the other, in order to eliminate standing waves in the sandwiched air space. The existence of standing waves (which are discussed later in this chapter in the "Frequency Balance" section) often serves to break down the transmission loss at specific frequency intervals. Similar to wall construction techniques, using glass panels of varying thickness reduces the possibility of sympathetic vibration. Other windows—such as observation and tape machine room windows— should be designed with similar isolation considerations in mind.

Access doors to and from the studio, control room, and exterior areas should be constructed using a double-door design to form a *sound lock* (see Figure 3.14). This construction technique is used because of the high TL value air has whenever it is sandwiched between two isolated barriers. Solid doors generally offer a higher TL value than their cheaper, hollow counterparts. No matter which door type is used, the appropriate seals, weather stripping, and door jams should be used throughout so as to reduce leakage around the edges.

Figure 3.14. *Sound lock design examples.*

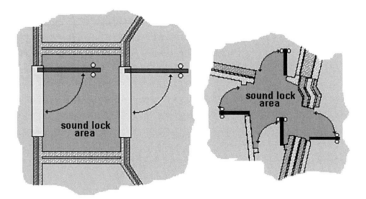

Iso-Rooms and Iso-Booths

Isolation rooms (iso-rooms) and the smaller iso-booths are acoustically sealed areas built into and easily accessible from the studio. These areas improve separation between loud and soft instruments, vocals, and so on. The construction of these areas depends on the studio's size, design, and sonic requirements (see Figure 3.15). They can vary in acoustics from the main studio area, offering an environment that is more "live" or more "dead" and can be used to better fit the acoustical needs of the instrument. Iso-booths and iso-rooms may be designed as totally separate areas that can be accessed from the main studio, or they can be tied directly to the main studio by way of sliding walls or glass sliding doors.

Figure 3.15. *Example of an iso-booth design.*

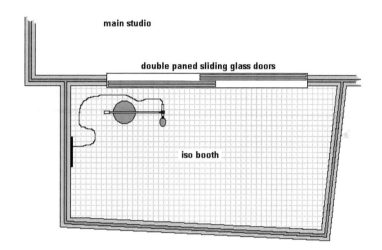

Movable partitions (known as *flats* or *gobos*) are commonly used in studios to provide on-the-spot sound barriers (see Figure 3.16). These partitions generally provide the least amount of isolation, but they are the most flexible and permit the isolation changes that many situations require. Often, an adequate degree of isolation can be attained by partitioning off a musician and/or instrument on one or more sides and by placing the mic inside. It's important, however, to be aware of the musician's need to see other musicians, the conductor, or the producer. Musicality almost always takes precedence over technical protocol.

Figure 3.16. *Acoustic partition flats.*

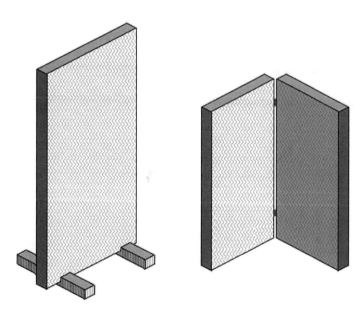

Frequency Balance

The second prerequisite for a properly designed room is that the frequency components of the acoustic signals should maintain their relative levels. In other words, the room should exhibit a relatively flat frequency response over the entire audio range without adding its own particular coloration to the sound. The most common way to control the tonal character of an acoustic environment is to use materials and design techniques that govern the ways by which sound is both reflected and absorbed within the room.

Reflections

One of the most important phenomena of sound is that it is reflected off a boundary's surface at an angle that is equal to (and opposite) its original angle of incidence (see Figure 3.17). Just as light bounces off a mirrored surface or any number of images can be seen in a room with mirrored walls, sound reflects throughout the various surfaces of that room in ways that are often just as complex, but can nonetheless be controlled in ways that add to (or detract from) the room's sonic character.

Figure 3.17. *Sound reflects off a surface at an angle equal to (and opposite) its original angle of incidence.*

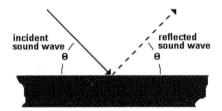

In Chapter 2, "Sound and Hearing," you learned that sonic reflections can be controlled in ways that can disperse the sound outward in a wide-angled pattern (through the use of a convex surface) or that can focus sound at a specific point (through the use of a concave surface). Other designs can reflect sound back at the same angle as its original incident angle. For example, a 90°-corner will reflect sound back in the same direction—a fact that accounts for the additive acoustic buildups at various frequencies at or near a wall corner or corner-to-floor intersection. However, the all-time winner of the "avoid at all possible costs award" goes to constructions that include opposing parallel walls in their designs. Such conditions give rise to a phenomenon known as *standing waves*. Standing waves (also known as *room modes*) occur when sound reflects off parallel surfaces and travels back on its own path, interfering with the amplitude response characteristics of the room. Walking around the room produces the sensation of an increase or decrease in the sound's perceived level. This perceived increase or decrease is due to cancellations and reinforcements of the combined reflected waveforms at the listener's position. The frequencies that produce standing waves are determined by the distance between parallel surfaces and the wavelength of the signal. This makes the room's response frequency discriminate, possibly producing sharp peaks or dips (up to or beyond 19 dB) in the frequency curve at the affected fundamental frequency(ies) and harmonic intervals (see Figure 3.18). This condition exists not only for opposing parallel walls, but for all parallel surfaces, such as between floor and ceiling or between two reflective flats.

Nonparallel walls, however, do not eliminate low-frequency standing waves. A better solution is to use bass traps tuned to room resonance frequencies or proper room dimensions. Nonparallel walls do prevent flutter echoes.

Figure 3.18. *The existence of reflective, parallel walls creates an undue number of standing waves, which occur at various frequency intervals (f1, f2, f3, f4, and so on).*

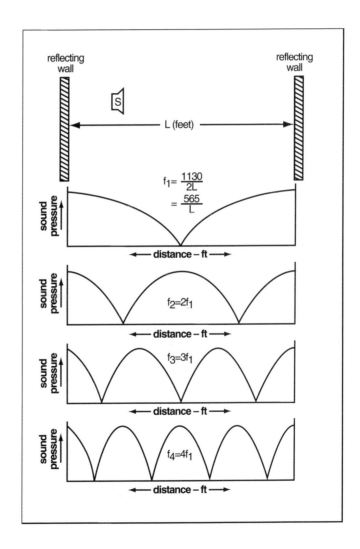

The frequencies at which standing waves occur within a room depend on the dimensions of the room: its height, width, and length. The formula for basic room standing wave resonances is

$$f = \frac{c}{2} \sqrt{ \left(\frac{p}{L}\right)^2 + \left(\frac{q}{W}\right)^2 + \left(\frac{r}{H}\right)^2 }$$

where

     f is the resonance frequency in Hz,
     L is the length of the room in feet (or meters),
     W is the width of the room in feet (or meters),
     H is the height of the room in feet (or meters),
     p, q and r are the room mode integers (1, 2, 3,…).

Or when only a single dimension is to be considered

$$f = \frac{N \times 565}{D}$$

where

     N is the room mode (1, 2, 3,…).
     D is the room dimension (L, W or H) in feet (or meters).

Various room dimensions, of course, have a direct effect on which standing wave frequencies are affected. If two or more room dimensions are identical, the modal or upper harmonic frequencies will be greatly exaggerated. If the dimensions aren't multiples of each other, however, these modes will be de-emphasized, smoothing the response characteristics of a room.

The most effective way to ensure against standing waves is to construct walls, boundaries, and ceilings that are nonparallel. Figure 3.19, for example, shows the various standing wave modes as they occur within two rooms having equal square footage areas. The rectangular room shows the unwanted standing waves as being uniform throughout the entire length or width (and probably the ceiling for that matter) of the room, while the modes within the nonrectangular room have been broken up and appear at irregular spaces and intervals within the room.

If the room is rectangular, or if further dispersion of the sound waves within the space is desired, diffusers can be attached to the wall and/or ceiling boundaries. *Diffusers* (see Figure 3.20) are acoustical boundaries that reflect the sound wave back at various angles that are wider than that of the original incident angle, thereby breaking up the energy-destructive standing-wave condition. In addition, the use of both nonparallel and diffusion wall construction will reduce a condition known as *flutter echo* and will smooth out the reverberation characteristics of the room by adding further and more complex acoustical pathways.

Figure 3.19. *A comparison of two dimensional soundfields in both rectangular and nonrectangular rooms having the same square surface area.*

a. *The 1,0,0 mode of the rectangular room (34.3 Hz) compared to the nonrectangular room (31.6 Hz).*

b. *The 3,1,0 mode of the rectangular room (81.1 Hz) compared to the nonrectangular room (85.5 Hz).*

c. *The 4,0,0 mode in the rectangular room (98 Hz) compared to the nonrectangular room (95.3 Hz).*

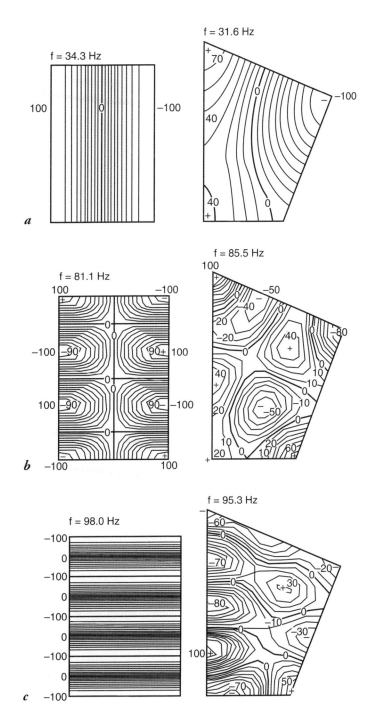

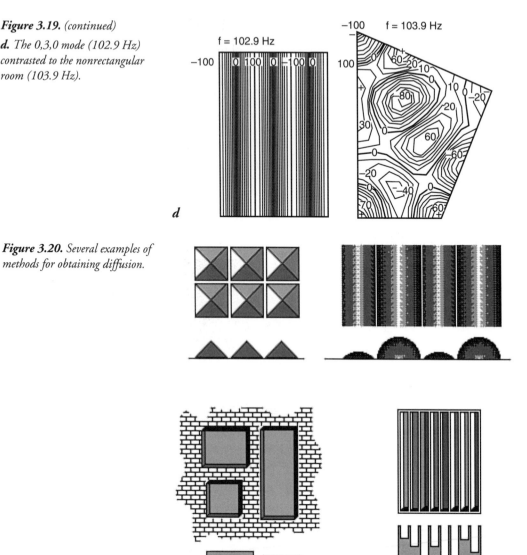

Figure 3.19. (continued)
d. The 0,3,0 mode (102.9 Hz) contrasted to the nonrectangular room (103.9 Hz).

Figure 3.20. Several examples of methods for obtaining diffusion.

Flutter (also called *slap*) *echo* is a condition that occurs when parallel boundaries are spaced far enough apart for the listener to discern many discrete echoes. Flutter echo often gives a smaller room a tube-like hollow sound that affects the sound character as well as the frequency response. A larger room, which may contain delayed echo paths of 50 msec or more, has its echoes spaced far enough apart in time that the discrete reflections actually interfere with the intelligibility of the direct sound, resulting in a jumble of noise.

Symmetry in Studio and Control Room Design

Although some prestigious designers of control rooms and studios have worked hard to create a successful, standardized acoustic environment, most control room and studio designs have as many forms and express as many design philosophies as there are studio owners. Certain ground rules of acoustical physics, however, must be followed in order to create a proper working environment. Of these ground rules, one guideline that is even more important than control over amplitude response is the need for symmetrical reflections in all axes within the room design. Should any primary boundaries of a control room (especially those wall or ceiling boundaries near the mixing position) be asymmetrical, sounds heard by one ear of the mixing engineer receive one combination of direct and reflected sound, while the other ear hears a different balance (see Figure 3.21).

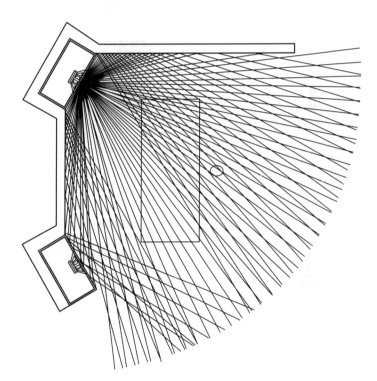

Figure 3.21. *Asymmetrical side reflections cause an acoustic imbalance at the listener's position. (Courtesy of Acoustical Physics Laboratories).*

This condition alters the sound's center image characteristics, so when a sound is actually panned between the two monitor speakers, the sound appears to be off-center. To avoid this problem, you should take care to ensure that both the side and ceiling boundaries are symmetrical with respect to each other. The splayed side wall shown in Figure 3.22 is one example of a symmetrical construction pattern which would reduce the amount of acoustic reflections that potentially could interfere with the direct acoustic energy at the listener's position.

Figure 3.22. *Symmetrical splayed side walls give a proper acoustic balance. (Courtesy of Acoustical Physics Laboratories).*

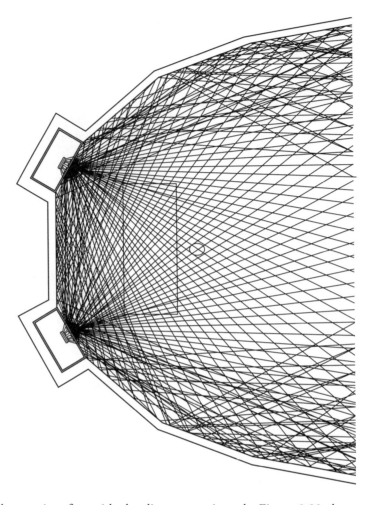

Large ceiling reflections also can interfere with the direct acoustic path. Figure 3.23 shows reflections radiating back to the listener from an average level ceiling, while the splayed ceiling design in Figure 3.24 shows the reduction of such unwanted reflections.

Figure 3.23. Ceiling reflections cause acoustic interference at the listener's position. (Courtesy of Acoustical Physics Laboratories).

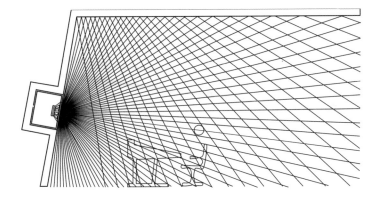

Figure 3.24. Splayed ceiling reduces unwanted reflections. (Courtesy of Acoustical Physics Laboratories).

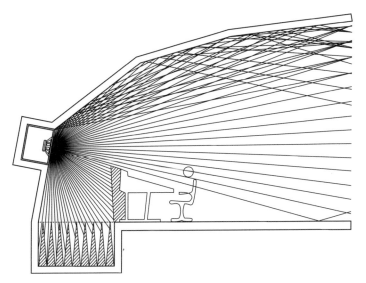

Absorption

Another method you can use to alter the soundfield within an acoustic space involves the incorporation of surface materials and designs that can absorb unwanted sound—either across the entire audible band or at specific frequencies.

The absorption of acoustic energy is, effectively, the inverse of reflection (see Figure 3.25). Whenever sound strikes a material, the amount of acoustic energy that is absorbed (often in the form of physical heat dissipation), relative to the amount that's reflected, can be expressed as a simple ratio known as the *absorption coefficient*. For a given material, the absorption coefficient is

$$a = \frac{I_a}{I_r}$$

where

    I_a is the sound level (in dB) that is absorbed by the surface,
    I_r is the sound level (in dB) that is reflected back from the surface.

Figure 3.25. *Absorption occurs when only a portion of the incident acoustic energy is reflected back from a material's surface.*

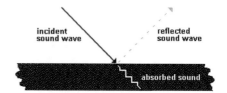

The factor $(1 - a)$ is the reflected sound. This makes the coefficient *a* a decimal percentage value between 0 and 1. A sample listing of these coefficients is provided in Table 3.2.

Table 3.2. Absorption coefficients for various materials.

Material	125 Hz	250 Hz	Coefficients 500 Hz	1000 Hz	2000 Hz	4000 Hz
Brick, unglazed	0.03	0.03	0.03	0.04	0.05	0.07
Carpet (heavy, on concrete)	0.02	0.06	0.14	0.37	0.60	0.65
Carpet (with latex backing, on 40-oz. hairfelt of foam rubber)	0.08	0.27	0.39	0.34	0.48	0.63
Concrete block, coarse	0.36	0.44	0.31	0.29	0.39	0.25
Light velour (10 oz./sq. yd in contact with wall)	0.03	0.04	0.11	0.17	0.24	0.35
Concrete or terrazzo	0.01	0.01	0.015	0.02	0.02	0.02
Wood	0.15	0.11	0.10	0.07	0.06	0.07
Glass, large heavy plate	0.18	0.06	0.04	0.03	0.02	0.02
Glass, ordinary window	0.35	0.25	0.18	0.12	0.07	0.04

Material	125 Hz	250 Hz	Coefficients 500 Hz	1000 Hz	2000 Hz	4000 Hz
Gypsum board, nailed to 2 x 4 studs on 16-inch centers	0.29	0.10	0.05	0.04	0.07	0.09
Plaster, gypsum, or lime (smooth finish on tile or brick)	0.013	0.015	0.02	0.03	0.04	0.05
Plywood, 3/8 inch	0.28	0.22	0.17	0.09	0.10	0.11
Air, Sabins/1000 cu. ft.	—	—	—	—	2.3	7.2
Audience seated in upholstered seats	0.44	0.54	0.60	0.62	0.58	0.50
Wooden pews	0.57	0.61	0.75	0.86	0.91	0.86
Chairs, metal or wooden, seats unoccupied	0.15	0.19	0.22	0.39	0.38	0.30

If you say that a surface material has an absorption coefficient of 0.25, you actually are saying that the material absorbs 25 percent of the incident acoustic energy while reflecting 75 percent of the total sound energy at that frequency.

In order to determine the amount of absorption obtained by the total number of absorbers within a total volume area, it is necessary to calculate the average absorption coefficient for all the surfaces together. The average absorption coefficient (a_{ave}) of a room or total area can be expressed as

$$-a- = \frac{s_1 a_1 + s_2 a_2 + ... s_n a_n}{S}$$

where,

$s_{1,2...n}$ are the individual surface areas,
$a_{1,2...n}$ are the individual absorption coefficients of the individual surface areas,
S is the total surface area in square feet.

(*Note:* These coefficients were obtained by measurements in the laboratories of the Acoustical Materials Association. Coefficients for other materials may be obtained from Bulletin XXII of the Association.)

High-Frequency Absorption

The absorption of high frequencies is accomplished through the use of dense porous materials, such as cloth, fiberglass, and carpeting. These materials are capable of having high absorption values at higher frequencies, thus allowing reflections within the room to be controlled in a frequency-dependent manner. Specially-designed foam treatments also are available that can be attached easily to recording studio, production studio, or control room walls as a means to correct for multiple room reflections or to dampen high-frequency reflections (see Figure 3.26).

Figure 3.26. Project studio incorporating AlphaSorb wall panels in the control room and Alpha Pyramid Acoustical foam in the voice-over room. (Courtesy of Acoustical Solutions, Inc.)

Low-Frequency Absorption

As Table 3.2 showed, materials having a degree of absorption in the high frequencies often provide little resistance to the low-frequency end of the spectrum, and vice versa. This is due to the fact that low frequencies may be damped by materials that are compliant, meaning that the low-frequency energy is absorbed by the material's capability to bend and flex with the incident waveform (see Figure 3.27).

Figure 3.27. Low-frequency absorption.

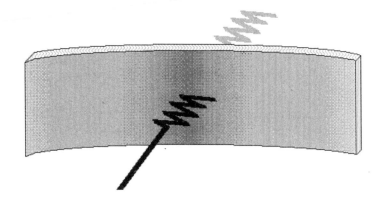

Other design methods can be used to reduce low-frequency buildup at specific frequencies (and their multiples) within a room. These low-frequency attenuation devices, known as *bass traps*, are available in a number of design types. This section discusses three types of bass traps: the quarter-wavelength trap, the pressure zone trap, the functional trap, and the Helmholtz resonator trap.

Quarter-Wavelength Trap

The quarter-wavelength bass trap, shown in Figure 3.28, is an enclosure with a depth one-fourth the wavelength of the offending frequency's fundamental frequency and often is built into a wall (often the rear wall), the ceiling, or the floor structure (covered by a metal grating to allow foot traffic). The physics behind the absorption of a calculated frequency (and many of the harmonics that fall above it) lay in the fact that at the rear boundary of the trap, the pressure component of a sound wave will be at its maximum, while the *velocity component* of the wave (the portion that measures the kinetic energy of atmospheric molecular motion) will be at a minimum. At the mouth of the bass trap (which is at a one-fourth wavelength distance from this rear boundary), the overall acoustic pressure is at its lowest, while its velocity component is at its highest potential. Because the wave's molecular motion is greatest at this point, much of the signal can be absorbed by placing an absorptive material at the bass trap's opening. Low density fiberglass lining can also be placed inside the trap to increase the amount of absorption (especially at harmonic intervals of the calculated fundamental).

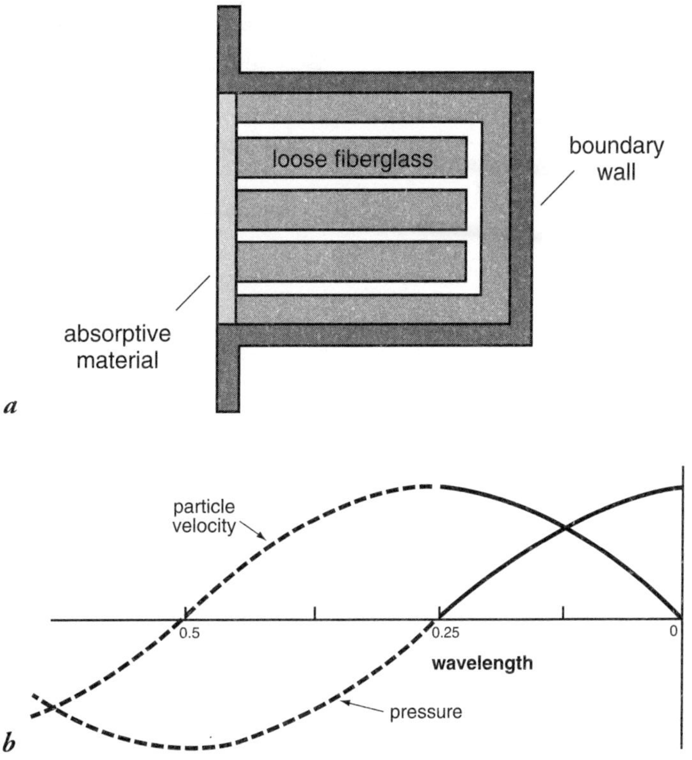

Figure 3.28. A quarter-wavelength bass trap.

a. Physical design.

b. Acoustic cancellation of the wavelength that is one-fourth the dimension of the trap.

Pressure Zone Trap

The pressure zone bass trap absorber (see Figure 3.29) makes use of the physical fact that sound pressure is doubled (+3 dB) at physical boundary points (such as walls and ceilings). By rigidly supporting a number of medium-density fiberglass boards (commonly referred to as 703) directly on a surface boundary, the built-up pressure can be partially absorbed. Wood slats or other non-resonant reflective structures may also be built out from the fiberglass boards, so as to reflect mid and high frequencies back into the room. Take care not to place the slats flush with the absorption material as this would choke off the trap's bass-breathing capability.

Figure 3.29. *Example of a pressure zone bass trap absorber.*

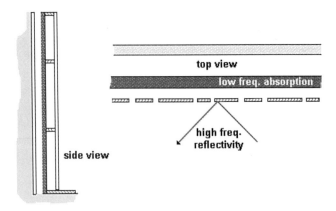

Functional Trap

Originally created in the 1950s by Harry F. Olson (Director of RCA Labs) and currently manufactured by Acoustic Sciences Corporation under the trade name "Tube Trap," the functional bass trap (see Figure 3.30) also uses a medium-density fiberglass board. This board generally is formed into a tube or half-tube structure that is rigidly supported so as to reduce structural vibrations. By placing these boards in the corners and boundaries of a room, a large portion of the undesired bass buildup frequencies can be absorbed. By placing a reflective membrane over the portion of the trap that faces into the room, frequencies above 400 Hz can be dispersed back into the listening environment. Figure 3.31 shows a studio that has incorporated tube traps into its acoustic design.

Figure 3.30. *The functional bass trap.*

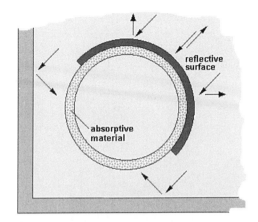

Figure 3.31. *Example of a studio incorporating Tube Traps into its acoustic design. (Courtesy of Acoustic Sciences Corporation and Recording Arts)*

Helmholtz Resonators

Another type of absorber is the slat-type of the Helmholtz resonator. As you have learned, absorption comes from the dissipation of sound energy in the form of heat due to frictional losses.

A slat-type Helmholtz resonator design uses spaced slats that have been supported off a wall or ceiling to create an air cavity of a specific depth and mass. Because this air mass (at the calculated frequencies) reacts with the "springiness" of the air in the cavity, a resonance is set up that effectively serves to absorb the desired frequency according to the following equation:

$$f_o = 2160 \sqrt{\frac{r}{(d \times D) + (w + r)}}$$

where:

    f_o is the frequency or resonance in Hertz,
    r is the slot width in inches,
    w is the slat width in inches,
    D is the air space depth in inches,
    d is the effective depth of the slot in inches (which is approximately 1.2 × the thickness of the slat in inches).

In addition to absorbing a single frequency, a variable-depth Helmholtz resonator can be designed that can effectively absorb a range of frequencies by varying the cavity depth, slat, and slot dimensions (see Figure 3.32).

Figure 3.32. A slat-type Helmholtz resonator with and without varying depths to broaden its absorption curve.

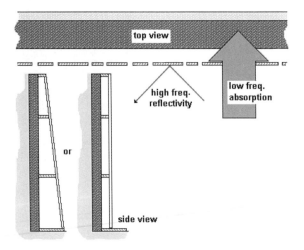

Reverberation

Another criterion for studio design is the need for a desirable room ambience and intelligibility, combined with the need for good acoustic separation when recording various instruments onto the tracks of a multitrack tape recorder. Each of these factors is governed by the careful control and tuning of the reverberation constants throughout the frequency spectrum within the studio environment.

Reverberation (*reverb*) is the persistence of a signal, in the form of reflected waves within an acoustic space, once the original sound has ceased. The effect of these closely spaced and random multiple echoes gives us perceptible cues as to the size, density, and nature of a space. Reverb also adds to the perceived warmth and depth of recorded sound and plays an extremely important role in the enhancement of our perception of music.

The reverberated signal itself (see Figures 3.33 and 3.34) can be broken down into three components: direct signal, early reflection, and reverberation. The direct signal is the point at which the original sound wave is received by the listener. Early reflections consist of the first few reflections that are projected to the listener from major boundaries within a determined space. These reflections generally give the listener subconscious cues as to the size of a space. The last set of signal reflections are those that make up the actual reverberation characteristic. These signals can be broken down into the many random reflections that travel from boundary to boundary in a room. These reflections are so closely spaced that the brain can't discern the individual reflections; taken together, they are perceived as a single decaying signal.

Figure 3.33. *The three components of reverberation.*

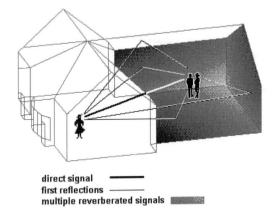

direct signal   ━━━━━
first reflections   ────────
multiple reverberated signals   ▨▨▨▨

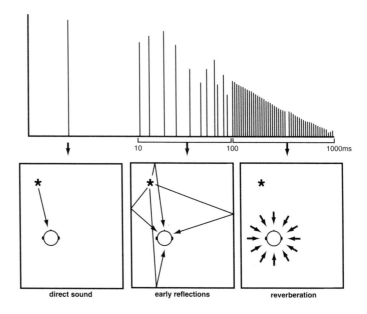

Figure 3.34. *Simplified behavior of the reverberation components within an acoustic space.*

Technically, reverberation is calculated as the time required for a sound to die away to a millionth of its original intensity, resulting in a decrease of 60 dB, as shown by the following formula:

$$RT_{60} = \frac{V \times 0.049}{AS}$$

where

> RT is the reverberation time in seconds,
> V is the volume of the enclosure in cubic feet,
> A is the average absorption coefficient of the enclosure,
> S is the total surface area in square feet.

As you can see in this equation, reverberation time is directly proportional to two major factors: the volume of the room and the absorption coefficients of the studio surfaces. A large environment with a relatively low absorption coefficient, such as a concert hall, yields a relatively long RT_{60} decay time; whereas a small studio, incorporating a heavy amount of absorption, gives a very short RT_{60}.

The style of music and the room application determines the optimum RT_{60} for an acoustical environment. Figure 3.35 shows a basic guide to reverb times for different applications and musical styles. Reverberation times can range from 0.25 second in a smaller absorptive recording studio environment to 1.6 seconds in a larger music or scoring studio. In certain designs, the RT_{60} of a room can be made to fit the desired application by using movable panels or louvers (see Figure 3.36) or by placing carpets in a room. Other designs may have sections of the studio environment that exhibit different reverb constants. One side of the studio (or separate iso-room) may be relatively nonreflective or "dead," while another section or room may be much more acoustically "live." The more reflective, live section is often used to bring certain instruments to life, such as strings, which rely heavily on room reflections and reverberation. The recording of any number of instruments, including drums and percussion, can also greatly benefit from an acoustically "live" environment.

Figure 3.35. Reverberation times for various types of production studios at 512 Hz.

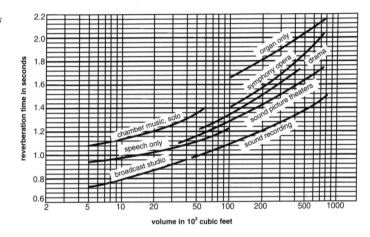

Figure 3.36. Example of a hinged panel design that can be moved so as to vary the acoustics of a room.

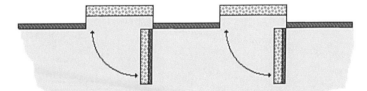

Isolation between different instruments and microphone channels is extremely important in the studio environment. If leakage isn't controlled, the room's effectiveness becomes severely limited for a multitude of applications. The studio designs of the 1960s and 1970s brought about the rise of the "*sound sucker*" era in studio design, whereby the absorption coefficient of many rooms was raised to an anechoic condition. With the advent of the music styles of the 1980s and a return to the respectability of live studio acoustics, the modern recording studio and control room have again begun to increase in "liveness" and size, with a corresponding increase in the studio's RT_{60}. This has reintroduced the buying public to the thick, live-sounding music production of earlier decades when studios were larger, more live, acoustic structures.

Acoustic Echo Chambers

Before this chapter concludes, I want to discuss one other studio design that has been used extensively in the past (before the invention of artificial effects devices) for re-creating room reverberation: the acoustic echo chamber.

The *echo chamber* (see Figure 3.37) is an isolated room with highly reflective surfaces in which a speaker and microphone are placed. The speaker is fed the signal to be reverberated and the mic picks up a combination of sound from the speaker and reflections off the walls, ceiling, and floor. By using one or more directional mics that have been pointed away from the room speakers, the pickup of direct sound can be minimized. Movable partitions also can be used to vary the room's decay time. When properly designed, acoustic echo chambers have a very natural sound quality to them. The disadvantage is that they take up space (typically 18' x 15' x 12') and require isolation from external sounds. Smaller rooms can be used, but the bass response generally suffers. Size and cost often make it unfeasible to build a new echo chamber, especially ones that can match the caliber and quality of the electronic reverb devices available today. However, this shouldn't discourage you from experimenting with the placement of mics—either in the studio, iso-room, or other room in your facility or home—to, in effect, build a temporary echo chamber that can add an interesting degree of acoustic "spice" to your next project (see Figure 3.38).

Figure 3.37. *Basic layout of Les Paul's design for the highly-prized acoustic echo chambers at Capitol Records in LA.*

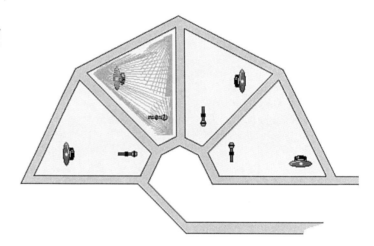

Figure 3.38. *Example of how a room or studio space can be used as a temporary echo chamber.*

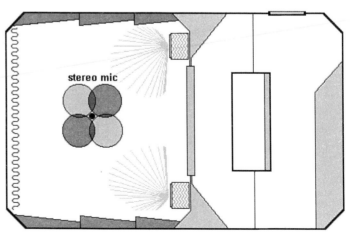

CHAPTER 4

Microphones: Design and Application

The Microphone Pickup: An Introduction

A microphone is often the first device in the recording chain. The microphone (mic) is a transducer that changes one form of energy (sound waves) into another corresponding form of energy (electrical signals). The microphone's pickup quality depends on many external variables, such as placement and acoustic environment, and internal variables, such as microphone design. These interrelated elements work together to affect the mic's overall sound quality. In order to satisfy the requirements of a wide range of applications and personal taste, an equally wide range of microphones are available to the professional user. Because the particular sound characteristics of each mic often best suit a specific range of applications, a user can derive the best possible sound from an acoustic source by carefully matching mic choice to the pickup application at hand.

In choosing the best microphone placement, you need to consider the following two rules:

> Rule 1. *There are no rules, only guidelines.*
> Although guidelines can help you achieve a good pickup, don't hesitate to experiment in order to get a sound that best suits your personal tastes.
>
> Rule 2. *The overall sound of an audio signal is no better than the weakest link in its signal path.*

Microphone Design

A microphone can convert acoustic energy into corresponding electrical voltages in a number of ways; however, the two transducer types used most often today in recording are the dynamic and condenser mics.

The Dynamic Microphone

In principle, the *dynamic* pickup system operates by electromagnetic induction in order to generate an output signal. When an electrically conductive metal is made to cut across the flux lines of a magnetic field, a current of specific magnitude and direction is generated within that metal.

Dynamic microphones are of two types: moving-coil and ribbon. These two types are described in more detail in the following sections.

The Moving-Coil Microphone

The moving-coil microphone, shown in Figure 4.1, generally consists of a Mylar diaphragm of roughly 0.35-mil thickness. To this is attached a finely wrapped core of wire called a *voice coil* that is precisely suspended within a high-level magnetic field. When an acoustic pressure wave hits the face of this diaphragm (A), the attached voice coil (B) is displaced in proportion to the amplitude and frequency of the wave, causing the coil to cut across the lines of magnetic flux supplied by a permanent magnet (C). In doing so, an analogous electrical signal (of a specific magnitude and direction) is generated across the voice coil leads.

Figure 4.1. *Moving-coil microphone.*

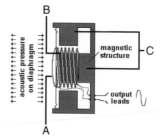

The Ribbon Microphone

Like the moving-coil microphone, the ribbon microphone, shown in Figure 4.2, operates on the principle of electromagnetic induction. This system, however, uses a diaphragm of extremely thin (2 micrometers) aluminum ribbon. Often, this diaphragm is corrugated transversely (along its length) and is suspended within a strong field of magnetic flux. As sound pressure variations displace the metal diaphragm in accordance with air-particle velocity, the ribbon is made to cut across the magnetic lines of flux. This induces a current in the ribbon of proportional amplitude and frequency to the acoustic waveform.

Figure 4.2. *Cutaway detail of a ribbon microphone.*

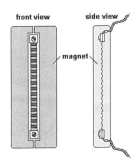

Due to the short length of the ribbon diaphragm (compared to the moving-coil), its electrical resistance is on the order of 0.2 ohm ( ). This impedance rating is too low to directly drive a microphone input stage, so a step-up transformer must be employed to bring the output impedance to an acceptable 150- to 600- range.

Recent Developments in Ribbon Technology

Over the last three decades, certain microphone manufacturers have made major strides toward miniaturizing and improving the operating characteristics of ribbon mics. Beyerdynamic, for instance, designed the Beyerdynamic M260 and M160 systems. In the case of the M260, Beyer uses a rare-earth magnet to produce a magnetic structure small enough to fit into a 2" grill ball—much smaller than traditional ribbon mics, such as the RCA 44 or 77. The ribbon, which is corrugated lengthwise to give it added strength and at each end to give it flexibility, is 3 microns thick, about 0.08" wide, 0.85" long, and weighs only 0.000011 ounce. A plastic throat is fitted above the ribbon, which houses a pop-blast filter. Two additional filters and the grill ball greatly reduce the ribbon's potential for blast and wind damage, making it suitable for outdoor and handheld use.

Another recent advance in ribbon technology has been the development of the printed ribbon mic. In principle, the printed ribbon operates in precisely the same manner as the conventional ribbon pickup. The diaphragm is made of a polyester film upon which is printed a spiral aluminum ribbon. The magnetic structure is produced by two ring magnets at the diaphragm's front and two in back. This creates a *wash* of magnetic flux, such that displacement causes the ribbon to cut across magnetic lines of force.

The Condenser Microphone

Condenser microphones, such as the ones shown in Figures 4.3 and 4.4, operate on an electrostatic principle rather than the electromagnetic principle used by the dynamic and ribbon mics. The *head*, or *capsule*, of the condenser mic consists of two very thin plates—one movable and one fixed. These two plates form a *capacitor* (formerly called a *condenser*, hence, the name *condenser microphone*). A capacitor is an electrical device that is capable of storing an electrical charge. The amount of charge a capacitor can store is determined by its value of capacitance and the applied voltage, according to the formula

$$Q = CV$$

where

Q is the charge, in coulombs,
C is the capacitance, in farads,
V is the voltage, in volts.

Figure 4.3. The Neumann U-67 condenser microphone shell. (Courtesy of Neumann USA)

Figure 4.4. Inner detail of the Neumann U-67 condenser microphone. (Courtesy of Neumann USA)

The capacitance of the capsule is determined by the composition and surface area of the plates (which are fixed values), the dielectric or substance between the plates (which is air and fixed), and the distance between the plates (which varies with sound pressure). Therefore, the plates of a condenser mic capsule form a sound-pressure-sensitive capacitor (see Figure 4.5).

In the design used by most manufacturers, the plates are connected to opposite sides of a DC power supply, which provides a polarizing voltage for the capacitor (see Figure 4.6). Electrons are drawn from the plate connected to the positive side of the power supply and forced through a high-value resistor onto the plate connected to the negative side of the supply. This process continues until the charge on the capsule (that is, the difference between the number of electrons on the positive and negative plates) is equal to the capacitance of the capsule times the polarizing voltage. When this equilibrium is reached, no further appreciable current flows through the resistor. If the mic is fed a sound-pressure wave, the capacitance of the head changes. When the distance between the plates decreases, the capacitance increases; when the distance increases, the capacitance decreases. According to the previous equation, Q, C, and V are interrelated; so if the charge (Q) is constant and sound pressure moves the diaphragm-changing capacitance (C), the voltage (V) must change in direct proportion.

Figure 4.5. Output and potential relationships as a result of changing capacitance.

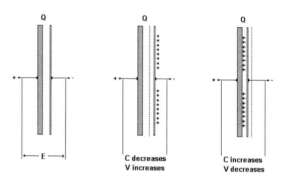

Figure 4.6. As a sound wave decreases a condenser element's spacing by d, the capacitance increases by C and the voltage across the plates falls conversely by V.

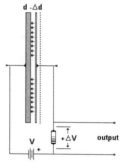

In conjunction with the capacitance of the plates, a high-value resistor produces a circuit time constant that is longer than a cycle of an audio frequency. The time constant of a circuit is a measure of the time needed for a capacitor to charge or discharge. Because the resistor prevents the capacitor charge from varying with the rapid changes in capacitance caused by the applied sound pressure, the voltage across the capacitor changes according to V = Q/ C. The resistor and capacitor are in series with the power supply, so the sum of the voltage dropped across them equals the supply voltage. When the voltage across the capacitor changes, the voltage across the resistor changes equally—but in the opposite direction. The voltage across the resistor then becomes the output signal. Because the signal off the diaphragm has an extremely high impedance, it is fed through an impedance conversion amplifier, which is placed into the circuit at a very short distance (often 2" or less) from the diaphragm. This amplifier is placed within the microphone's body in order to prevent hum, noise pickup, and the signal-level losses that otherwise would occur. It is also another reason why a condenser microphone requires a power supply voltage in order to operate.

Although most modern day microphones employ a FET (field effect transistor) to reduce the capsule impedance, certain highly prized older designs (and newer remakes of these older models) use a vacuum tube housed within the mic housing itself (see Figure 4.7).

Figure 4.7. The AKG C12VR
vacuum tube microphone.
(Courtesy of AKG Acoustics, Inc.)

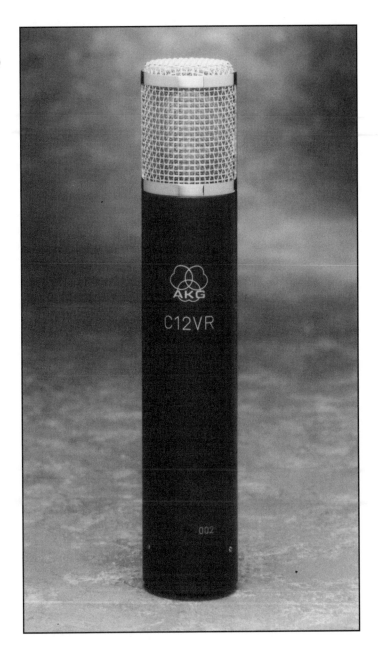

These mics generally are prized by studios and collectors alike for their "tube-like sound." They often exhibit a sonically favorable coloration that results from the physical design (they often have a large case and grill mesh enclosure), as well as from the even-harmonic distortion and other characteristics that occur when tubes are used.

The Electret-Condenser Microphone

Electret-condenser microphones operate like externally polarized condensers except the polarizing charge is stored permanently within the diaphragm or on the backplate. Because of this electrostatic charge, no external powering is required to charge the diaphragm. The high-output impedance of the capsule still requires an impedance-changing amplifier, however, so an internal supply or phantom power supply tap is needed.

Microphone Characteristics

In order to satisfy the wide range of applications encountered in studio and on-location recording, microphones often differ in their overall physical and electrical characteristics. The following information highlights many of these characteristics to help you choose the best mic for a particular application.

Directional Response

The *directional response* of a microphone refers to its variation in sensitivity (output level) at various angles of incidence with respect to the on-axis (front) side of the microphone (see Figure 4.8). This chart, known as the *polar response* or *polar pattern* of a microphone, is used to graphically plot a microphone's sensitivity with respect to direction and frequency over 360°.

Figure 4.8. *Directional axis of a microphone.*

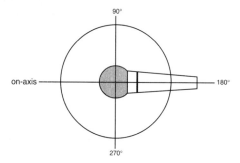

Microphone directionality can be classified into two broad types:

◆ The omnidirectional polar response
◆ The directional polar response

The omnidirectional microphone (see Figure 4.9) is a pressure-operated device that is responsive to nondirectional acoustic sound pressure. The diaphragm reacts equally to all sound pressure fluctuations at its surface, regardless of the source's location. A pickup that displays a directional property is a *pressure-gradient* device. This means that the system is responsive to differences in pressure between the two faces of a diaphragm. A pure pressure-gradient microphone exhibits a *bidirectional polar pattern* (cosine or figure-of-eight), as shown in Figure 4.10.

Figure 4.9. Typical pickup pattern of an omnidirectional microphone.

Figure 4.10. Typical pickup pattern of a bidirectional microphone.

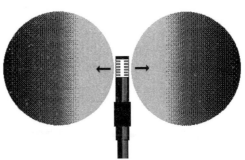

Most ribbon mics have a bidirectional pattern. Because the ribbon's diaphragm is exposed to sound waves from both the front and rear axes, it is equally sensitive to sounds emanating from either direction. Sounds from the rear produce a voltage that is 180° out-of-phase with the equivalent on-axis signal (see Figure 4.11a). Sound waves arriving 90° off-axis produce an equal but opposite pressure at both the front and rear of the ribbon (see Figure 4.11b) and cancel at the diaphragm, resulting in no output signal.

Figure 4.12 graphically illustrates how the outputs of bidirectional (pressure-gradient) and omnidirectional (pressure) pickups can be combined to obtain various other directional patterns. In effect, an infinite number of directional patterns can be obtained from this mixture. The most widely known patterns resulting from these combinations are the *cardioid, supercardioid,* and *hypercardioid* polar patterns (see Figure 4.13).

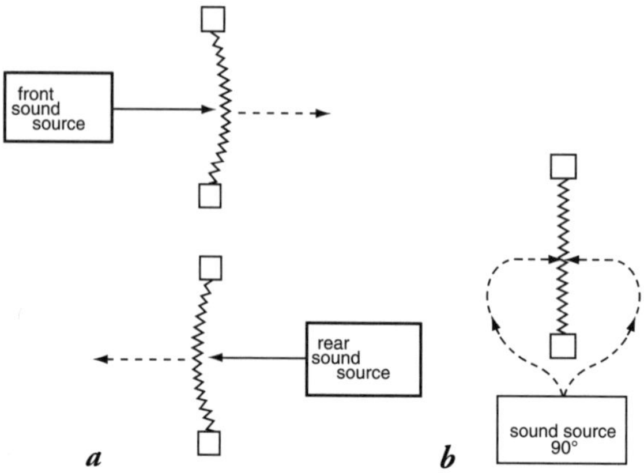

Figure 4.11. *Sound sources on-axis and 90° off-axis at the diaphragm of a ribbon microphone.*
a. *The ribbon is sensitive to sounds at front and rear.*
b. *Sound waves from 90° off-axis.*

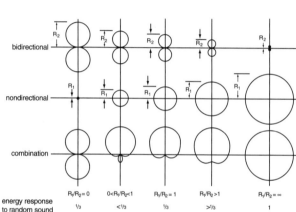

Figure 4.12. *Directional diagrams of various bidirectional and nondirectional pickup patterns.*

The moving-coil microphone achieves a cardioid response by incorporating a rear port into its design. This rear port serves as an acoustic labyrinth that creates an acoustic delay or resistance. A light felt or nylon screen is often used to dampen the diaphragm resonances over the entire frequency range.

In Figure 4.14a, a dynamic mic having a cardioid polar response is shown receiving a sound signal on-axis (at 0°). The diaphragm receives two signals: the incident front signal and the acoustically delayed rear signal. In this instance, the on-axis signal exerts a positive pressure on the diaphragm and also travels 90° to the side port where it is again delayed by 90° (totaling 180° at the rear of the diaphragm). During this delay period, the on-axis signal begins to exert a negative pressure, which is now in-phase with the delayed rear signal. Because the pressures between the two opposing sides of the diaphragm are in-phase, an output signal is produced. Figure 4.14b shows a sound originating 180° off-axis to the mic. Here, the signal travels 90° to the side of the mic where it enters the delay labyrinth and is delayed another 90° (totaling 180°). The sound hitting the front of the diaphragm has also been delayed by 180° (due to the time required to travel around the pickup) and is therefore acoustically out-of-phase with the diaphragm's back side. This results in an acoustic cancellation, so there is little or no output signal. The attenuation of such an off-axis signal, with respect to an equal on-axis signal, is known as the *front-to-back discrimination* of a microphone and is rated in decibels.

Figure 4.13. *Various polar patterns, with output sensitivity plotted versus angle of incidence.*

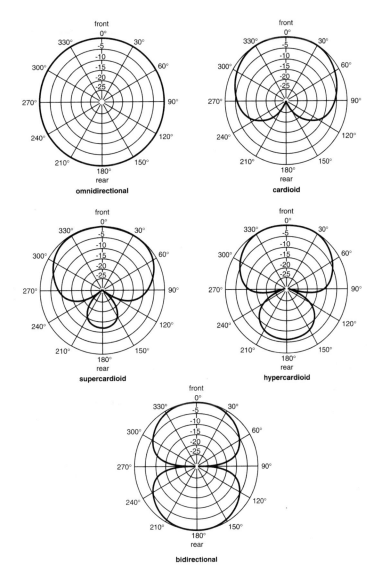

Figure 4.14. *The directional properties of a cardioid mic.* **a.** *Signals arrive at the diaphragm in-phase, producing a full-level output.* **b.** *Signals arrive at the diaphragm 180° out-of-phase, canceling each other out and resulting in a reduced output.*

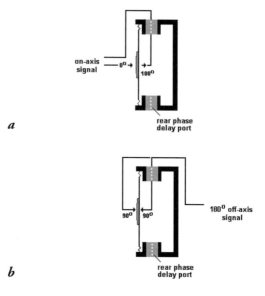

a

b

Certain condenser microphones can be switched electrically from one pattern to another by using a dual-membrane capsule system mounted around a central backplate (see Figure 4.15). Configuring these capsules electrically in-phase results in the creation of an omnidirectional pattern; configuring them out-of-phase results in a bidirectional pattern. Varying electrically (in either continuous or stepped degrees) between these two polar states yields a number of patterns, such as hypercardioid and cardioid.

Figure 4.15. *Variety of polar patterns that can result from combining two cardioid capsules back to back.*

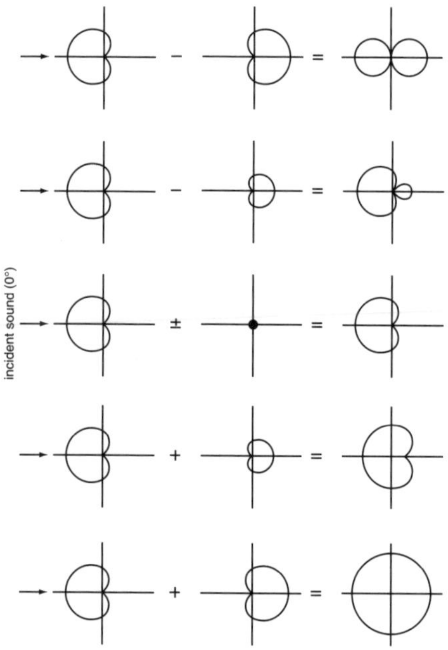

Frequency Response

The on-axis *frequency-response curve* of a microphone is the measurement of its output over the audible frequency range given a constant on-axis input signal at the microphone. This response curve, which is plotted in output level (dB) versus frequency, often yields valuable information and can give clues as to how a microphone will react at specific frequencies.

A mic can be designed to respond equally to all frequencies. Such a device is said to exhibit a *flat frequency response* (see Figure 4.16a). Other microphones may be designed to emphasize or de-emphasize the high-, middle-, or low-end response of the audio spectrum (see Figure 4.16b).

Figure 4.16. Frequency-response curves. *(Courtesy of AKG Acoustics, Inc.)*
a. Response curve of the AKG C-460B/CK61 ULS.
b. Response curve of the AKG D321.

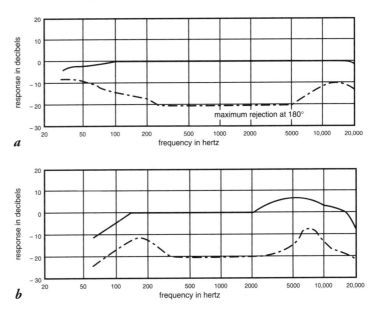

The frequency-response curves shown in Figure 4.17 were measured on-axis and exhibit totally acceptable responses. Certain designs, however, have a "peaky" or erratic curve when measured off-axis. This signal coloration may become evident when the mic is operating in an area where off-axis sounds (in the form of leakage) arrive at the pickup, often resulting in a change in tone quality. The *off-axis frequency response* of a microphone (which indicates the sound character of this leakage) may be charted along with the on-axis curve (as can be seen in the lower, dotted response curves in Figures 4.16a and 4.16b and the off-axis curves in Figure 4.17).

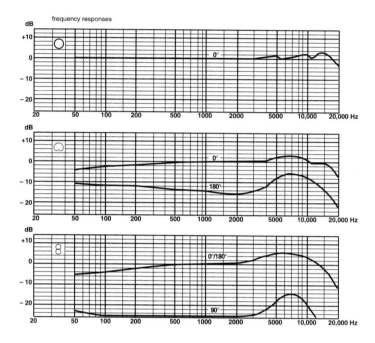

Figure 4.17. *Frequency response of the Microtech UM70S at various angles of sound incidence. (Courtesy of Microtech Gefell GMBH)*

Low-Frequency Response Characteristics

At low frequencies, rumble (high-level vibrations in the 3- to 25-Hz region) may be transmitted in a studio or hall, or along the surface of a large unsupported floor space. You can eliminate this adverse effect in any of the following three ways:

- ◆ Use a shock mount to isolate the mic from the vibrating surface and floor stand.
- ◆ Choose a mic that displays a restricted low-frequency response.
- ◆ Restrict the response of a wide-range mic through the use of a low-frequency rolloff filter.

Another low-frequency phenomenon inherent in most directional mics is known as *proximity effect*. This effect causes an increase in bass response as the signal source is brought closer to the microphone; it is most noticeable when the source is within 1' of the microphone. This bass-boost effect increases proportionately as the distance decreases, and it is somewhat greater for a bidirectional microphone than for one displaying a cardioid polar pattern. To compensate for this effect, a low-frequency rolloff filter is often provided. This filter restores the bass response to a flat and natural sounding balance; it usually is activated by a switch located on the microphone body or optionally within the equalization section of a mixer or console.

Another way to eliminate proximity effect and the associated "popping" of the letters *p* and *b* is to replace the directional microphone with an omnidirectional (pressure) microphone in close-miking applications.

On a more positive note, this increase in bass response has long been appreciated by vocalists for giving a full, "larger-than-life" quality to a thin voice. In many cases, the directional mic has become an important component of the artist's sound.

Transient Response

A significant piece of data, which presently has no accepted standard of measure, is the *transient response* of a microphone (see Figures 4.18a through 4.18c). Transient response is the measure of how quickly the diaphragm of a mic reacts to a waveform. This figure varies wildly among microphones and is a major reason for the difference in sound quality among the three transducer types. The diaphragm of a dynamic microphone can be quite large (up to 2 1/2"). With the addition of the coil of wire and its core, the combination can make for a large mass when compared to the power of the sound wave that drives it.

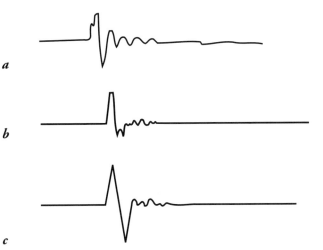

Figure 4.18. Typical transient responses of various microphone types.
a. Dynamic microphone.
b. Double ribbon.
c. Studio-quality condenser.

A dynamic mic can be very slow in reacting to a waveform, giving a rugged, gutsy sound. By comparison, the diaphragm of a ribbon mic is much lighter, so its diaphragm can react more quickly to a sound waveform, resulting in a clearer sound. The diaphragms of a ribbon mic are quite light in comparison, providing for an improved transient response (provided that the ribbon itself is in good working condition and isn't deformed from abuse). The condenser mic has an extremely light diaphragm, which varies in diameter from 2 1/2" to 1/4", with a thickness of about 0.0015". This means that the diaphragm offers very little mechanical resistance to a sound-pressure wave, enabling it to track the wave accurately over the entire frequency range.

Output Characteristics

A microphone's output characteristics refer to its measured sensitivity, equivalent noise, overload characteristics, impedance, and other output responses.

Sensitivity Rating

The *sensitivity rating* is the output level (in volts) that a microphone will produce given a specific and standardized input signal (rated in dB SPL). This specification implies the degree of amplification required to raise the microphone signal to line level (–10 dBV or +4 dBm). This value also allows the recordist to easily judge the output level differences between two mics. A microphone with a higher sensitivity rating produces a stronger output-signal voltage than a microphone with a lower sensitivity, given that both are driven by the same sound-pressure level.

Equivalent Noise Rating

The *equivalent noise rating* of a microphone can be viewed as the device's electrical self-noise. It expresses the equivalent dB SPL that would produce a voltage equal to the mic's self-noise voltage.

As a general rule, the microphone itself doesn't contribute much noise to a system compared to the amplification stages and to the tape used in an analog recording chain. With the advances in digital technology, however, these noise ratings are of increasing interest.

In the dynamic or ribbon pickup, the noise is generated by electrons moving within the coil or ribbon itself. In a condenser, most of the noise is generated by the built-in preamplifier. Certain microphone designs display a greater degree of self-noise than do other designs; thus, care must be taken in the choice of microphones for critical applications.

Overload Characteristics

As the use of a microphone is limited at low levels by its inherent self-noise, it also is limited at high-SPL levels by overload distortion.

In terms of distortion, the dynamic microphone is an extremely rugged pickup, often capable of an overall dynamic range of 140 dB. The typical condenser microphone will not distort—except under the most severe sound-pressure levels. However, the condenser system differs from that of the dynamic in that at high acoustic levels the capsule's output signal may be high enough to overload the microphone's preamplifier. To prevent this condition, many condenser mics contain a switchable attenuation pad that immediately follows the transducer output and serves to reduce the signal level and, therefore, overload distortion. When inserting such an attenuation pad in the microphone circuit, keep in mind that the mic's signal-to-noise ratio is degraded by the amount of attenuation. Therefore, it is wise to remove the inserted pad when you use the microphone under normal SPLs.

Microphone Impedance

Microphones are available with different *output impedances*. Output impedance is a rating used to match the signal-providing capability of one device with the signal-drawing (*input impedance*) requirements of another device. Impedance is measured in ohms and its symbol is Ω or Z.

Commonly used microphone output impedances are 50Ω, 150 to 250Ω (low), and 20 to 50kΩ (high). Each impedance range has its advantage. In the past, high-impedance mics were less expensive to use because the input impedance of tube-type amplifiers was high. To be used with low-impedance mics, tube-type amplifiers required expensive input transformers. All dynamic mics, however, are low-impedance devices, and those with high-impedance outputs achieve them through the use of a built-in impedance step-up transformer. A disadvantage of high-impedance mics is the susceptibility of their high-impedance cables to the pickup of electrostatic noise, such as that caused by motors and fluorescent lights. This makes the use of shielded cable necessary. In addition, the use of a conductor surrounded by a shield creates a capacitor which is, in effect, connected across the output of the microphone. As the length of the cable increases, the capacitance increases until, at about 20' to 25', the cable capacitance begins to short out much of the high-frequency information picked up by the mic. For these reasons, high-impedance microphone pickups are rarely used in the professional recording process.

Very-low-impedance (50Ω) mics have the advantage that their lines are fairly insensitive to electrostatic pickup. They are sensitive, however, to induced hum pickup from electromagnetic fields, such as those generated by AC power lines. This extraneous pickup can be eliminated by using twisted-pair cable, whereby the currents magnetically induced into this cable flow in opposite directions and cancel each other out at the console or mixer's balanced microphone input stage.

150- to 250-Ω mic lines have low signal losses and can be used with cable lengths up to several thousand feet. They are less susceptible to electromagnetic pickup than 50-Ω lines but are more susceptible to electrostatic pickup. As a result, a shielded twisted-pair cable is used and the lowest noise is attained through the use of a balanced signal line. Within such a line, two wires carry the signal voltage, while a third lead and/or shield is used as a neutral ground wire. Neither of the two signal conductors is directly connected to the signal ground. High-impedance mic and instrument lines, on the other hand, use unbalanced circuits (see Figure 4.19) whereby one signal lead carries a positive current potential to a device, while a second ground shield is used to complete the signal's returning circuit path.

Figure 4.19. Unbalanced microphone circuit.

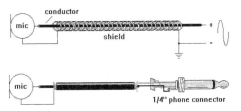

Balanced lines operate on the principle that the alternating current of an audio signal will be presented in opposite polarity between the two conductors, whereas any electrostatic or electro-magnetic pickup will simultaneously be induced into both leads in polarity with respect to each other (see Figure 4.20). The input transformer or balancing amplifier responds only to the difference in voltage between the two leads, resulting in the cancellation of the unwanted signal while the audio signal is unaffected. The mic lines used in most recording studios are 200-Ω balanced lines with the shield grounded only at the preamp end and the mic handle.

Figure 4.20. Balanced micro-phone circuit.

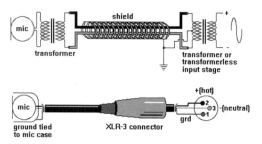

The standard that has been adopted for proper polarity of two-conductor, balanced XLR connector cables specifies pin #2 as the positive (+) or "hot" lead, whereas pin #3 is the negative (–) or "neutral" pin. The outer shield and cable ground are connected to pin #1.

If the hot and neutral pins of balanced mic cables are haphazardly pinned in a music or production studio, it's possible that any number of mics (or other equipment, for that matter) could be wired in opposite polarity. For example, if a single instrument is picked up by two mics that are improperly phased, the instrument may end up being either totally or partially canceled when mixed to mono.

Phantom Power

Most modern-day professional condenser microphones don't require internal batteries, external battery packs, or individual AC-operated power supplies to operate. They are designed to be powered directly from the console through the use of a phantom power supply.

Phantom powering operates by supplying a positive DC supply voltage of +48V to both conductors (pins 2 and 3) of a balanced mic line. This voltage is equally distributed through identical value resistors (4.7kΩ, ±1% values work nicely) so that no differential exists between the two leads. The positive side of the DC voltage is therefore not electrically "visible" to the input stage of a balanced mic preamp. Instead, only the alternating audio signal that's being simultaneously carried on the two audio leads is detected (see Figure 4.21). The DC circuit is completed by supplying the negative side of the supply to the cable's third grounding wire or shield.

The resistors used in distributing power to the signal leads can also provide a degree of power isolation between other mic inputs on a console. If a signal lead were accidentally shorted to ground (as could happen if defective cables or unbalanced XLR cables are used), the power supply should still be able to deliver power to other mics in the system. If two or more inputs are shorted, however, the phantom voltage could drop to levels that are too low to be usable.

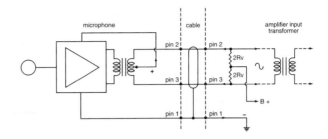

Figure 4.21. *Phantom powering system.*

Microphone Techniques

Each microphone has a distinctive sound character based on its specific type and design. A large number of types and models can be used for a variety of applications, and it's up to the engineer to choose the right one for the job. Choosing the appropriate microphone for the job, however, is only half the story. The placement of the microphone (either within a room or closely miked near an instrument) can play just as important a role in getting the right sound.

I should emphasize right from the start that microphone placement is an art form and one of the engineer's most important tools.

Placement techniques currently considered "bad" may be the accepted standard five years from now. As new musical styles develop, new recording techniques also tend to evolve. This helps to breathe new life into musical sound and production. Recording is an art form; as such, it is totally open to change and experimentation, which keeps both music and the industry alive and fresh.

Pickup Characteristics as a Function of Working Distance

In modern studio and sound-stage recording, four fundamental styles of microphone placement are directly related to the distance of a microphone from its sound source: distant miking, close miking, accent miking, and ambient miking. These styles are described in the following sections.

Distant Microphone Placement

With *distant microphone placement* (see Figure 4.22), one or more microphones are positioned at a distance of 3' or more from a signal source. This technique provides the following two results:

◆ Enables you to pick up a large portion of a musical instrument or ensemble, thus preserving the overall tonal balance of that instrument or ensemble. You can often achieve a natural tone balance by placing the mic at a distance roughly equal to the size of the instrument or sound source.

Figure 4.22. Example of an overall distant pickup.

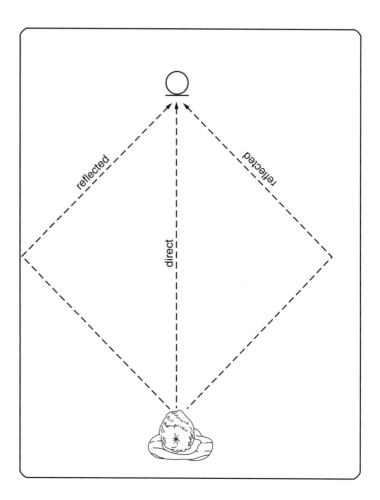

◆ Enables the acoustic environment to be included in the mic's pickup and combined with the direct sound signal.

Distant miking often is used in the pickup of a large instrumental ensemble (such as a symphony orchestra or choral ensemble). In such a situation, the pickup relies on the acoustic environment. The microphone is placed at a distance so as to strike an overall balance between ensemble and environmental acoustics. This balance is determined by a number of factors, including the size of the sound source and the reverberant characteristic of the room.

Distant miking techniques tend to add a live, open feeling to a recorded program. However, this technique could put you at a disadvantage if the acoustics of a hall, church, or studio are not particularly good: improper or bad room reflections often create a muddy or poorly defined recording. To avoid this problem, the engineer can take one of the following actions:

◆ Temporarily correct for bad or excessive room reflections by using absorptive or offset reflective panels

◆ Place the microphone closer to its source and add artificial ambience

If distant miking of an instrument is used to pick up a portion of the room sound, the placement of the distant mic at a random height may result in a hollow sound due to phase cancellation of the direct sound with the sound reflected from the floor (see Figure 4.23). The sound reflected from the floor travels farther than the sound that reaches the mic directly. Frequencies for which this extra path length is one-half a wavelength—or an odd integer multiple of one-half a wavelength—arrive 180° out-of-phase with the direct sound, producing dips in the frequency response of the signal at the mic output. Because the reflected sound is at a lower level than the direct sound (due to its traveling farther and the loss of energy when it hits the floor), the cancellation is often only partially complete.

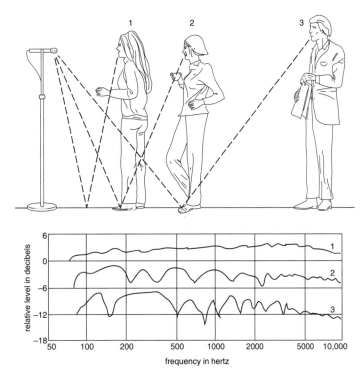

Figure 4.23. Resulting frequency response from a mic receiving direct and delayed sound from one source.

Moving the mic closer to the floor reduces the path length and raises the frequency of cancellation. In practice, a height of 1/8" to 1/16" keeps the lowest cancellation above 10 kHz. One such microphone design type, known as a *boundary microphone* (and shown in Figures 4.24 and 4.25), places an electret-condenser diaphragm well within these height restrictions, which makes it a good choice for use as an overall distant pickup when placed on a floor, wall, or large boundary.

Figure 4.24. *The boundary microphone system.*

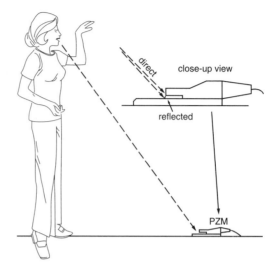

Figure 4.25. *The PZM-30D boundary microphone. (Courtesy of Crown International, Inc.)*

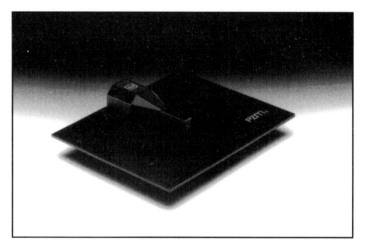

Close Microphone Placement

With *close microphone placement,* the mic is positioned about 1" to 3' from a sound source. This commonly used technique of miking a sound source at such close distances provides two major results:

◆ Creates a tight, present sound quality

◆ Effectively excludes the acoustic environment

Because sound diminishes with the square of its distance from the sound source, a sound originating 6' from a mic is insignificant in level compared to that of the same sound originating 3" from the microphone (see Figure 4.26). As a result, only the desired on-axis sound is recorded on tape; extraneous sounds, for all practical purposes, are not picked up.

Figure 4.26. Close miking reduces the effects of the acoustic environment.

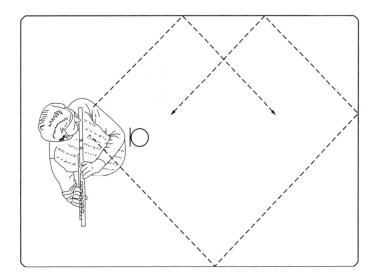

When sounds from one instrument are picked up by a nearby microphone being used to pick up a different instrument, a condition known as *leakage* occurs. Because the microphones contain both direct sound and leakage (see Figure 4.27), control over individual tracks in mixdown may be difficult to maintain without affecting the level and sound character of other tracks. Whenever possible, try to avoid this condition in the studio.

Figure 4.27. *Leakage due to indirect signal pickup.*

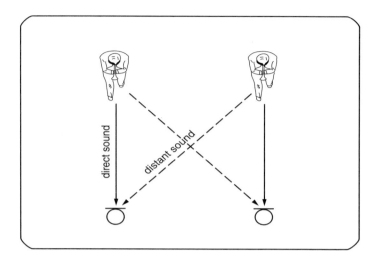

To avoid the problems of leakage, you can use any or all of the following four methods:

◆ Bring the microphones closer to their respective instruments, as shown in Figure 4.28a.

◆ Place an acoustic barrier (known as a flat, gobo, or divider) between the two instruments, as shown in Figure 4.28b.

◆ Use directional mics.

◆ Spread the instruments farther apart.

Whenever individual instruments are being close (or semi-close) miked, it's generally wise to follow the 3:1 distance rule. This principle states that in order to maintain phase integrity for every unit of distance between the mic and its source, the distance between the two mics should be at least three times that distance (see Figure 4.29).

Although the close miking of a sound source offers several advantages, a mic should be placed only as close to the source as necessary, not as close as possible. Miking too close may color the recorded tone quality of a source. Because such techniques commonly involve distances of 1" to 6", the tonal balance (timbre) of the entire sound source can't be picked up. Rather, the mic may be so close to the source that only a small portion of the surface actually is "heard," giving an area-specific tonal balance. At these close distances, moving a mic by only a few inches may change the overall tonal balance. Try using one of the following three remedies:

◆ Move the microphone along the surface of the sound source until the desired balance is achieved.

◆ Place the mic at a greater distance from the sound source to allow for a wider pickup angle, thus improving the overall blend.

◆ Equalize the signal until the desired balance is achieved.

Figure 4.28. Two methods to reduce leakage.
a. Mics placed closer to their sources.
b. An acoustic barrier used to reduce leakage.

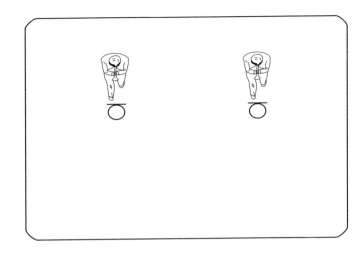

a

b

Figure 4.29. Example of the 3:1 microphone distance rule.

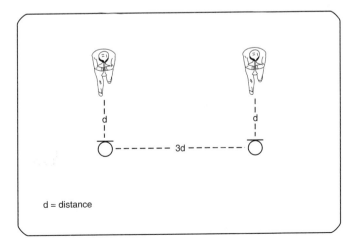

Accent Microphone Placement

The overall pickup and tonal qualities of distant and close miking often sound very different. Under certain circumstances, it's difficult to obtain a naturally recorded balance when mixing these two techniques. For example, if a solo instrument in an orchestra needs an extra mic for more volume and presence, placing the mic too close provides a pickup that sounds overly present, unnatural, and out of context with the distant orchestral pickup. To avoid this pitfall, you have to strike a compromise in distance (and thus pickup balance). A microphone placed within this compromise range to pick up an instrument or section within a larger ensemble is known as an *accent microphone* (see Figure 4.30).

Figure 4.30. *Accent microphone placed at proper, compromised distance.*

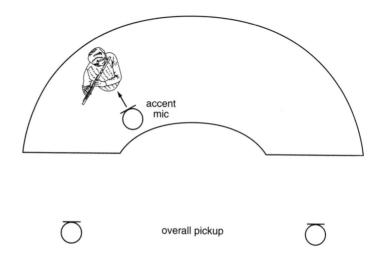

When you use an accent mic, you must exercise care in placement and pickup type. The amount of accent signal introduced into the mix should not change the balance of soloist to the surrounding sounds. A good accent pickup only adds presence to a solo passage and isn't perceptible as separate pickup.

Ambient Microphone Placement

An *ambient* mic is one placed at such a distance that the reverberant or room sound is more prominent than the direct signal. The ambient pickup is often a stereo cardioid pair, and usually is mixed in with closely placed microphones.

To enhance recording, you can use ambient mic pickups in the following ways:

- ◆ In a live concert recording, ambient microphones can be placed in the hall to restore the natural reverberation that may be lost with the use of close miking techniques.

- ◆ In a live concert recording, ambient microphones can be placed over the audience to pick up audience reaction and applause.

◆ In a studio recording, ambient microphones can be used to add the studio's natural acoustics to a recorded sound.

Stereo Miking Techniques

For the purpose of this discussion, stereo miking techniques refer to the use of two microphones to obtain a stereo image. These techniques can be used in either close or distant overall miking of background vocals, a large or small ensemble, or a single instrument on location or in the studio. The only limitation is your imagination. Three fundamental methods are available for stereo miking using two microphones: the spaced pair, the XY technique, and the M-S method.

Spaced microphones can be placed in front of an instrument or ensemble (in a left/right fashion) to obtain an overall stereo image. This technique places the two mics anywhere from a few inches to over 30 feet apart (depending on the size of the instrument or ensemble) and uses time and amplitude cues in order to create a stereo image.

The primary drawback to this technique is the strong potential for the introduction of phase discrepancies between the two channels due to differences in a sound's arrival time at one mic, relative to the other. When mixed to mono, these phase discrepancies could result in variations in frequency response or the partial cancellation of various instruments or sound components in the pickup field.

The XY technique is an intensity-dependent system; it uses only the cue of amplitude to discriminate direction. With the XY coincident-pair technique, two directional microphones of the exact same type and manufacture are placed with their grills as close together as possible (without touching) and facing at a certain angle to each other. The midpoint between the two mics is faced toward the musical instrument and the mic outputs are panned left and right. Even though the two mics are placed together, the stereo imaging is excellent—often better than the spaced pair. An added advantage is having no appreciable phase problems due to the close proximity of the mics. The angles of the two mics may be changed to fit the occasion, with the preferred angles being from 90° to 135°. The generally accepted polar pattern is cardioid (see Figure 4.31a), although two crossed bidirectional mics (known as the Blumlein technique and named after the unheralded inventor, Alan Dower Blumlein) can also give excellent ambient results (see Figure 4.31b).

Stereo microphones are available that contain two diaphragms in the same microphone case—with the top diaphragm generally being rotatable by 180°—so that they can be adjusted to various coincident XY angles (see Figure 4.32).

Another coincident-pair method, M-S (or mid-side), shown in Figure 4.33, is similar to the XY technique in that it uses two diaphragms in close proximity, most often with a stereo mic like the one shown in Figure 4.32. The mid-side method differs from the previous two methods in that it requires an external transformer or active matrix that operates in conjunction with the microphones. In the classic M-S system, the capsule designated as the mid (M) microphone has a cardioid pattern that is oriented towards the sound source. The other capsule has a figure-eight pattern facing sideways—with the null side of the pattern coinciding with the main axis of the cardioid mic. This side (S) pickup collects the ambient sound, while the mid mic mostly picks up the direct sound from the instruments. The two are then mixed together by way of a matrix to reconstruct a stereo image. By varying the direct versus ambient sound patterns, the stereo image can be made to sound closer/more distant or narrower/wider without physically moving the mic.

Figure 4.31. *XY stereo miking technique.*
a. *Crossed cardioid pair.*
b. *Blumlein crossed bidirectional pair.*

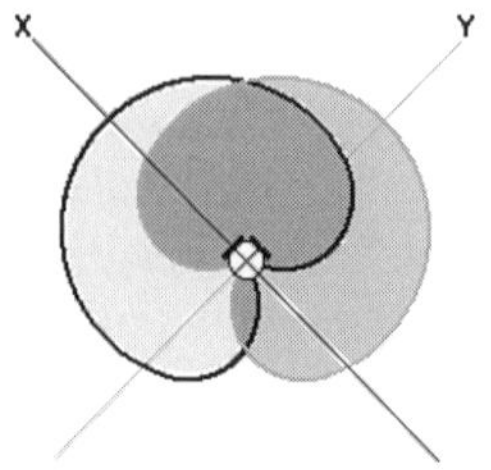

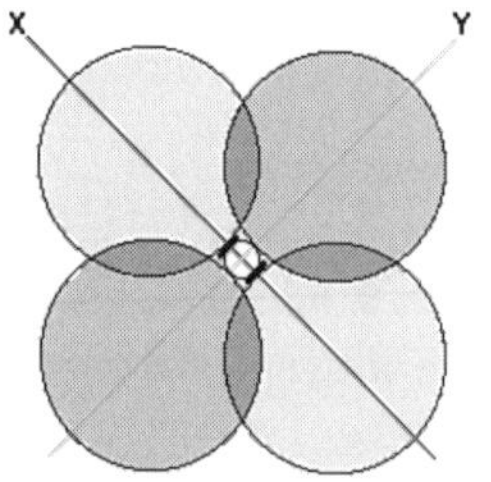

Figure 4.32. *Shure VP88 coincident stereo microphone. (Courtesy of Shure Brothers, Inc.)*

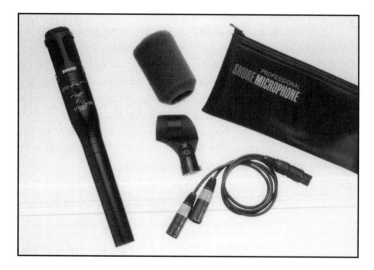

Figure 4.33. *M-S stereo miking technique.*

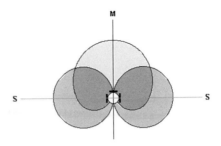

Recording Direct

The signal of an electric instrument (guitar, keyboard, and so on) can be recorded directly to tape without the use of a microphone. This option produces a cleaner, more present sound by bypassing the distorted character of an amplifier. It also reduces leakage into other mics by eliminating the amplifier.

In the project or recording studio, the *direct injection* (D.I.) box (shown in Figure 4.34) serves to interface an electric instrument to the audio console by the following means:

- Reducing an instrument's line-level output to mic level for direct insertion into the console.
- Changing an instrument's high source impedance (unbalanced line) to a low source impedance (balanced line) needed by the console's input stage.
- Electrically isolating audio signal paths (thereby reducing the potential for a ground loop hum).

Most commonly, the instrument's output is plugged directly into the D.I. box (where it is stepped down in level and impedance) and the box's output is then fed to the console or mixer. If a "dirtier" sound is desired, certain boxes allow for high-level signals to be taken directly from an amplifier's external speaker jack.

A D.I. box and microphone can be mixed to create a subtle combination of the punchy, full sound of the mic with the crispness of the direct sound. These signals can be combined onto a single tape track or recorded to the separate tracks of a multitrack recorder (allowing for more flexibility in the mixdown phase).

Figure 4.34. *The direct injection (D.I.) box. (Courtesy of Whirlwind Music Dist. Inc.)*

Microphone Placement Techniques

The following section offers a general guide to microphone placement as it relates to certain popular acoustic instruments. Keep in mind that these are only guidelines. The general application and characteristics detailed in Table 4.1 and the descriptions of various popular microphones in the "Microphone Selection" section provide further support for making mic choices.

As a general rule, the type of microphone you choose depends on the type of sound you're after. A dynamic microphone often provides a rugged or punchy sound, which can be assisted by a close proximity bass boost in most directional mics. A ribbon microphone may provide a mellow, slightly "croony" sound when used at close distances. A condenser microphone will generally give a clear, present, and full-range sound.

Table 4.1. Microphone selection chart.

Application	Microphone Characteristic/Choice
Natural, smooth tone quality	Flat frequency response
Bright, present tone quality	Rising frequency response
Extended lows	Dynamic or condenser with extended low-frequency response
Extended highs (detailed sound)	Condenser
Reduced "edge" or detail	Dynamic
Boosted bass at close working distances	Directional microphone
Flat bass response up close	Omnidirectional
Reduced pickup of leakage, feedback, and room acoustics	Directional microphone, or omnidirectional microphone at close working distances

Application	Microphone Characteristic/Choice
Enhanced pickup of room acoustics	Omnidirectional microphone, or directional microphone at greater working distances
Extra ruggedness	Moving-coil microphone
Reduced handling noise	Omnidirectional, or directional microphone with shock mount
Reduced breath popping	Omnidirectional, or directional microphone with pop filter
Distortion-free pickup of very loud sounds	Condenser with high maximum-SPL spec, or dynamic
Noise-free pickup of quiet sounds	Low self-noise, high sensitivity

Brass Instruments

The following section describes the various sound characteristics and miking techniques encountered for the brass family of instruments.

Trumpet

The fundamental frequency of a trumpet ranges from E3 to D6 (165 Hz to 1,175 Hz) and contains overtones that stretch upwards to 15 kHz. Below 500 Hz, the sound emanating from the trumpet projects uniformly in all directions; above 1,500 Hz, the projected sound becomes much more directional; and above 5 kHz, the dispersion emanates at a tight 30° angle in front of the bell.

The formants of a trumpet (the relative harmonic and resonance frequencies that give it a specific sound character) lie at around 1 to 1.5 kHz and at 2 to 3 kHz. The tone of a trumpet can be changed radically by using a mute (a cup-shaped dome that fits directly over the bell), which serves to dampen frequencies above 2.5 kHz. A conical mute (a metal mute that fits inside the bell) tends to cut back on frequencies below 1.5 kHz while encouraging the spectrum above 4 kHz.

Due to the high sound pressure levels that a trumpet can produce (up to 130 dB SPL), it's best to place a microphone slightly off center of the bell at a distance of 1' or more (see Figure 4.35). When closer placements are needed, a −10- to −20-dB pad can help prevent input overload at the mic or console preamp input. Under such close working conditions, a wind screen might help to protect the diaphragm from wind blasts.

Figure 4.35. *Typical mic placement for a single trumpet.*

Trombone

The trombone comes in a number of sizes; however, the most commonly used "bone" is the tenor that has a fundamental note range that spans from E2 to C5 (82 Hz to 520 Hz) and produces a series of complex overtones that range up to 5 kHz (when played medium loud) to 10 kHz (when overblown). The trombone's polar pattern is nearly as symmetrical as the trumpet's: frequencies below 400 Hz are distributed evenly, whereas with frequencies at 2 kHz and above, the dispersion angle is down to 45° from the bell.

The trombone most often appears in jazz and classical music. The Mass in C minor by Mozart, for example, has parts for soprano, alto, tenor, and bass trombones. This style obviously lends itself to the spacious blending that can be achieved by distant pickups within a large hall or studio. On the other hand, jazz music often calls for a closer miking distance. At 2" to 12", for example, the trombonist should play slightly to the side of the mic to reduce the chance of overload and wind blasts. In the miking of a trombone section, a single mic might be placed between two players and recorded onto a single track.

Tuba

The bass and double-bass tubas are the lowest pitched of the brass/wind instruments. Although the bass tuba's range is actually a fifth higher than that of the double-bass, it's still possible to obtain a low fundamental of B (29 Hz). A tuba's overtone structure is limited—with a top response ranging from 1.5 kHz to 2 kHz. The lower frequencies of the tuba (around 75 Hz) are dispersed evenly; as frequencies rise, however, the sonic distribution angle is reduced.

Under normal considerations, this class of instruments is not miked at close distances. A working range of 2' or more, slightly off-axis to the bell, yields the best results.

French Horn

The fundamental tones of the French horn range from B1 to B5 (65 Hz to 700 Hz). Its "oo" formant gives it the round, broad quality that can be found at about 340 Hz with other frequencies falling between 750 Hz and 3.5 kHz. French horn players often place their hands inside the bell to mute the sound and promote the formant at about 3 kHz.

The French horn player or section traditionally is placed at the rear of an ensemble, just in front of a rear, reflective stage wall. This wall serves to reflect the sound back towards the listener's position, which tends to create a fuller, more defined French horn sound. An effective pickup of this instrument can be achieved by placing an omni- or bidirectional pickup between the rear, reflecting wall and the instrument bells, thereby receiving both the direct and reflected sound. Alternatively, the microphone pickups can be placed in front of the players, thereby receiving only the sound that is reflected from the rear wall.

Guitar

The following sections describe the various sound characteristics and miking techniques encountered for the guitar.

Acoustic Guitar

The popular steel-strung guitar carries a bright, rich set of overtones (especially when played with a pick). Mic placement can vary from instrument to instrument and may require experimentation to pick up the full tonal balance.

A balanced pickup can be obtained by placing the mic at a point slightly off-axis and above or below the sound hole at a distance of between 6" to 1', as shown in Figure 4.36. A condenser mic is often preferred for its smooth, extended frequency response and excellent transient response.

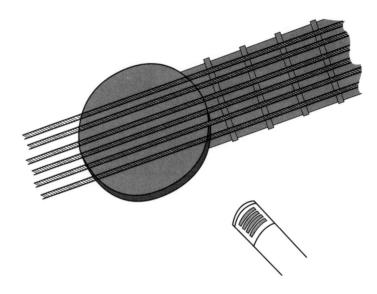

Figure 4.36. *Typical mic placement for the guitar.*

The smaller-bodied classical guitar normally is strung with nylon or gut and is played with the fingertips, giving it a warmer, more mellow sound than its steel-strung counterpart. To make sure that the instrument's full range is picked up, these reduced upper harmonics can be helped out by placing the mic closer to the center of the bridge, at a distance of between 6" and 1'.

Miking Near the Sound Hole

The sound hole located at the front face of a guitar serves as a bass port which resonates at lower frequencies (around 80 Hz to 100 Hz). Thus, a mic placed too close in front of this port tends to sound boomy and unnatural. However, a miking position close to the sound hole is popular on stage or around high acoustic levels because the guitar's output is highest at this position. To achieve a more natural pickup under these conditions, the microphone's output can be rolled off at the lower frequencies (5 dB to 10 dB at 100 Hz).

The Electric Guitar

The fundamentals of the average 22-fret guitar extend from E2 to D6 (82 Hz to 1,174 Hz), with overtones that extend much higher. All these frequencies may not be amplified, as the guitar chord tends to attenuate frequencies above 5 kHz (unless the guitar has a built-in low impedance converter or low-impedance pickups). The frequency limitations of the average guitar loudspeaker often add to this effect as their upper limit is restricted to below 5 or 6 kHz.

Miking the Guitar Amp

The most popular guitar amplifier used for recording is the small practice-type amp/speaker system. This type of amp is designed to help the suffering high end by incorporating a sharp rise in the response range at 4-5 kHz, thus promoting a clean, open sound.

By far, the most popular mic type for picking up an electric guitar amp is the cardioid dynamic. A dynamic mic tends to add a full-bodied character to the sound without picking up extraneous amplifier noises. Often the chosen mic contains a presence peak in the upper frequency range, giving an added clarity to the pickup.

For increased separation, a microphone can be placed at a working distance of between 2" to 1'. When miking at a distance of less than 4", microphone/speaker placement becomes slightly more critical (see Figure 4.37). For a brighter sound, the mic should face directly into the center of the speaker's cone. Placing it off-center to the cone produces a more mellow sound while reducing amplifier noise.

Figure 4.37. Miking the electric guitar cabinet directly in front of and off-center to the cone.

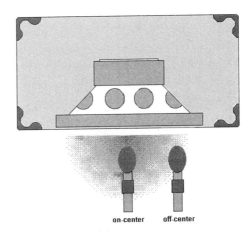

Recording Direct

A D.I. box is often used to feed the output signal of an electric guitar directly into the mic input stage of a recording console or mixer; from there it can be routed to a tape track (see Figure 4.38a). By using the direct output signal of the instrument, a cleaner, more present sound can be recorded. This technique also reduces the leakage that results from having a guitar amp in the room.

A combination of direct and miked signals often results in a sound that has the "bite" of a D.I. but the characteristic richness and fullness of a miked amp. These may be combined onto one tape track; alternatively, when a number of spare, open tracks exists, they can be routed to their own tracks, which allows for greater control during the mixdown phase (see Figure 4.38b).

Figure 4.38. *Direct recording of an electric guitar.*
a. *Direct recording.*
b. *Combined direct and miked signal.*

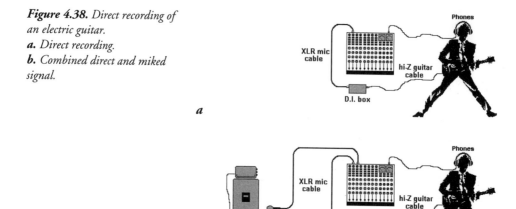

The Electric Bass Guitar

The electric bass guitar's fundamentals range from about E1 to F4 (41.2 Hz to 343.2 Hz). If it's played loudly or with a pick, the added harmonics can range upwards to 4 kHz. Playing in the "slap" style or with a pick gives a brighter, harder attack. A "fingered" style, on the other hand, produces a mellower tone.

In modern music production, the bass guitar is often recorded direct for the cleanest possible sound. As with the electric guitar, the electric bass can be either miked at the amplifier or picked up through a D.I. box. If the amp is miked, dynamic mics usually are chosen for their deep, rugged tones. The new, large-diaphragm dynamic designs tend to subdue the high-frequency transients. When combined with a boosted response at around 100 Hz, these large diaphragm dynamics give a warm, mellow tone that has added power in the lower register. Equalization of the bass signal can increase a guitar's clarity, with the fundamental being affected from 125 Hz to 400 Hz and the harmonic punch being from 1.5 kHz to 2 kHz.

One other tool for the electric and acoustic bass is the compressor. The signal output from one bass note may be weaker than the next note, causing frequency "holes" in the bass line. A compressor that is set to have a smooth input/output ratio of 4:1, a fast attack (8-20 milliseconds), and a slower release time (1/4 to 1/2 second) often smooths out these levels and yields a strong, present bass line.

Keyboard Instruments

The following section describes the various sound characteristics and miking techniques encountered for various keyboard instruments.

Grand Piano

The grand piano is an acoustically complex instrument that can be miked in a variety of ways depending on the style and preferences of the engineer or producer. The overall sound emanates from the instrument's strings, soundboard, and mechanical hammer system. Because of its large surface area, a minimum miking distance of 4' to 6' is needed for the tonal balance to fully develop and to be picked up. Due to leakage from other instruments, these miking distances aren't always practical or possible. As a result, pianos often are miked at distances that favor a certain part of the instrument's overall sound-generating surface, such as the strings and soundboard (often yielding a bright and relatively natural tone), or the hammers (often yielding a sharp, percussive tone), or the soundboard holes alone (often yielding a sharp, full-bodied sound).

In modern music production, two basic styles of grand pianos can be found in the recording studio: one that yields a traditionally rich and full-bodied tone (often used for classical music and ranging up to 9' in length) and one that is more suited for modern music production and is designed to have a sharper, more percussive edge to its tone (often about 7' in length).

Figure 4.39 shows a number of the possible microphone positions acceptable for the pickup of the grand piano. (The numbered mic positions are explained in the list following the figure.) It's important to keep in mind that these are only guidelines from which to begin. Your own personal sound can be achieved through personal mic choice and experimentation with mic placement.

Figure 4.39. *Possible miking combinations of the grand piano.*

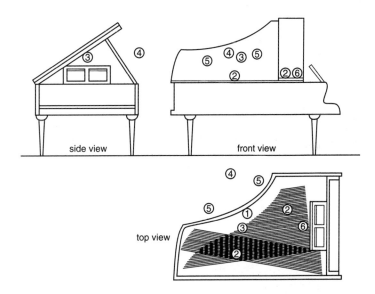

The following lists explains the various miking positions shown in the figure:

◆ *Position 1.* The microphone is attached to the partially or entirely open lid of the piano. The most appropriate pickup type is the boundary mic, which can be permanently attached or temporarily taped to the lid. This method uses the lid as a collective reflector and provides excellent pickup under restrictive conditions (such as stage and live video).

◆ *Position 2.* Two mics are placed in a spaced stereo configuration at a working distance of 6" to 1'. One mic is positioned over the low strings and one is placed over the high strings.

◆ *Position 3.* A single mic or coincident stereo pair is placed just inside the piano between the sound board and its fully or partially open lid.

◆ *Position 4.* A single mic or stereo coincident pair is placed outside the piano, facing into the open lid. This is most appropriate for the solo or accent miking of the instrument.

◆ *Position 5.* A spaced stereo pair is placed outside the lid, facing into the instrument.

◆ *Position 6.* A single mic or stereo coincident pair is placed just over the piano hammers at a working distance of 4" to 8" for a driving popular or rock sound.

A condenser or extended range dynamic mic is most often the choice when miking an acoustic grand piano, as either of these tend to accurately represent the transient and complex nature of the instrument. Should excessive leakage be a problem, a cardioid or tighter polar pattern can be used. If leakage isn't a problem, however, an omnidirectional mic may be preferred for capturing the instrument's overall tonal balance.

Separation

Separation is often a problem that's associated with the grand piano whenever it is placed next to raucous musical neighbors. Separation when miking a piano can be achieved in the following ways:

◆ Place the piano inside a separate isolation room.

◆ Place a flat (gobo) between the piano and its louder neighbor.

◆ Place the mics inside the piano and lower the lid onto its short stick. A heavy moving (or other) blanket can be placed over its lid to further reduce leakage. This technique works best with close miking techniques.

◆ Overdub the instrument at a later time. In this situation, the lid can be removed or propped up by the long stick so that the mics can be placed at a more natural-sounding distant position.

Upright Piano

You would expect the techniques for this seemingly harmless type of piano to be similar to its larger brother. This is, for the most part, true. However, because this instrument was designed for home enjoyment and not performance, the miking techniques are slightly different. Often, it's more difficult to achieve a respectable tone quality. You may want to try the following methods:

◆ *Miking over the top.* Place two mics in a spaced fashion just over and in front of the piano's open top—one over the bass strings and one over the high strings (see Figure 4.40). If isolation isn't a factor, remove or open the front face that covers the strings in order to reduce reflections and, therefore, the instrument's "boxy" quality. Also, you might want to angle the piano about 17° relative to the wall behind it to reduce resonances.

Figure 4.40. *"Over the top" mic placement of an upright piano.*

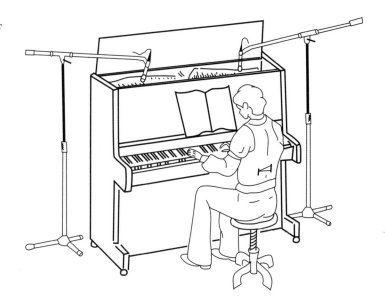

◆ *Miking the kick board area.* For a more natural sound, remove the kick board at the lower front part of the piano to expose the strings. Place a stereo spaced pair over the strings (one each at a working distance of about 8" over the bass and high strings). If only one mic is used, place the mic over the high end strings. You may want to use caution, however, as this placement may pick up too much foot-pedal noise.

◆ *Miking the upper soundboard area.* To reduce excessive hammer attack, place a microphone pair at about 8" from the soundboard, covering both the bass and high strings. In order to reduce muddiness, the soundboard should be facing into the room or be moved away from the wall.

Electronic Keyboard Instruments

The signal from most electronic instruments (such as synthesizers, samplers, and drum machines) is taken directly from the device's line level output(s) and inserted into a console—either through a D.I. box or directly into a channel's line level input. Alternatively, the keyboard's output can be plugged directly into the tape machine's line level input.

The approach to miking an electronic organ can be quite different from the techniques just mentioned. A good Hammond or another older organ can sound wonderfully dirty through miked loudspeakers. Such organs are often played through a Leslie cabinet, which adds a Doppler-based vibrato. Inside the cabinet is a set of rotating speaker baffles that spin on a horizontal axis and, in turn, produce a pitch-based vibrato as the speakers are accelerated towards and away from the listener or mic.

The upper high-frequency speakers can be miked by either one or two mics (each panned left and right), with the low-frequency driver being picked up by one mic. Motor and baffle noises can produce quite a bit of wind, possibly creating the need for a wind screen or experimentation with placement.

Percussion

The following section describes the different sound characteristics and miking techniques encountered for various percussion instruments.

Drum Set

The standard drum kit, shown in Figure 4.41, often provides the foundation of modern recorded music; it provides the "heart beat" of the basic rhythm track. Consequently, a proper drum sound is extremely important to the outcome of most music projects.

Figure 4.41. *Peter Erskine's studio drum kit. (Courtesy of Beyerdynammic, Inc.)*

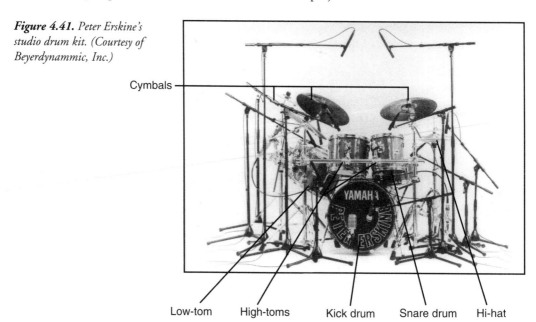

Cymbals

Low-tom   High-toms   Kick drum   Snare drum   Hi-hat

Generally, the drum kit is composed of the kick drum, snare drum, high-toms, low-tom (one or more), hi-hat, and a variety of cymbals. A full drum kit is a series of interrelated and closely-spaced percussion instruments, so it often takes real skill to translate the proper spatial and tonal balance onto tape. The larger-than-life driving sound of the acoustic rock drum set that we've all become familiar with is the result of an expert balance between playing techniques, proper tuning, and mic placement. Should any of these variables be lacking, the search for that perfect drum sound could prove to be a long and hard one. As a general rule: a poorly tuned drum will sound just as out-of-tune through a good mic as it will through a bad one. Therefore, it's important that the drum sound good to the ears, before even attempting to place the mics.

Miking the Drum Set

After the drum set has been optimized for the best sound, mics can be placed in their proper pickup positions, as shown in Figures 4.42a, 4.42b, and 4.42c. Because each part of the drum set is so different in sound and function, at this point, it probably is best to consider each part as an individual instrument.

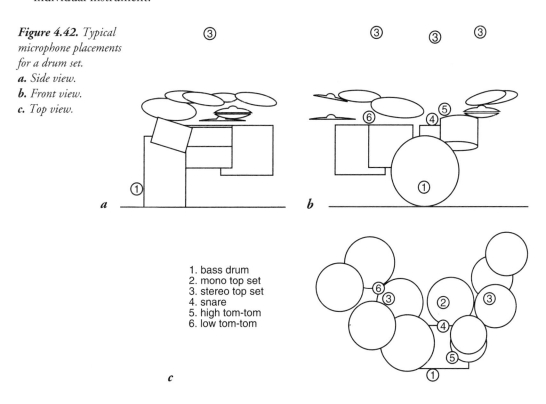

Figure 4.42. *Typical microphone placements for a drum set.*
a. *Side view.*
b. *Front view.*
c. *Top view.*

1. bass drum
2. mono top set
3. stereo top set
4. snare
5. high tom-tom
6. low tom-tom

The following mic characteristics can be used to match a mic to the various parts of a drum: frequency response, polar response, proximity effect, and transient response. Dynamic range is another important consideration when miking drums. Because a drum set is capable of generating extremes of volume and power, as well as soft, subtle sounds, a mic must be able to withstand strong peaks without distorting, yet still be able to capture the more delicate nuances of a sound.

The drum set usually is one of the loudest sound sources found in a studio setting, so often it is placed on a rigidly supported 1 1/2' riser. This reduces the amount of low-end "thud" that otherwise would be leaked through the floor to other parts of the studio. Depending on the construction, the following drum scenarios may occur:

◆ The drums are placed in their own room—isolated from other instruments.

◆ To achieve a bigger sound, the drums are placed in the large studio room while the other instruments are placed in smaller iso-rooms.

◆ To reduce leakage, the drums are placed in the room but are enclosed by 4' (or higher) divider flats.

Snare Drum

Generally, a snare mic is aimed just inside the top rim of the snare drum at a distance of about 1" (see Figure 4.43). The mic should be angled for the best possible separation from other drums and cymbals. The microphone's rejection angle should be aimed at either the hi-hat or rack-toms depending on the leakage difficulties. Most often, the mic's polar response is cardioid, although a super-cardioid response offers a tighter pickup angle.

Figure 4.43. *Mic positioning for the snare drum. (Courtesy Beyerdynamic, Inc.).*

With certain musical styles, such as jazz, you may want a crisp or "bright" sound. You can achieve this by miking the snare's bottom head in addition to the top head and then combining the two mics onto a single track. As the bottom snare head is 180° out-of-phase with the top, generally it is wise to reverse the bottom mic's phase polarity.

Hi-Hat

The hi-hat usually produces strong, sibilant energy in the high-frequency range, whereas the snare's frequencies often are more concentrated in the mid-range. Although moving the hat's mic won't change the overall sound as much as it would on the snare, you should still keep in mind the following three points:

◆ Placing the mic above the top cymbal reproduces all the nuances of the sharp stick attacks.

◆ The open and closing motion of the hi-hat often produces rushes of air. Consequently, when miking the hat's edge, angle the mic slightly above or below the point at which the cymbals meet.

◆ If only one mic is available or desired, both the snare and hi-hat can be picked up simultaneously by carefully placing the mic between the two, facing away from the rack toms as much as possible.

Rack-Toms

The upper rack-toms can be miked individually (see Figure 4.44) or with a single mic placed between the two at a short distance (see Figure 4.45). When miked individually, a "dead" sound can be achieved by placing the mic close to the drum's top head about 1" above and 1" to 2" in from the outer rim. A sound that is more "live" can be achieved by increasing the height above the head to about 3". If isolation or feedback is a consideration, a hyper-cardioid pickup pattern can be chosen.

Another way to reduce leakage and to get a deep, driving tone (with less attack) is to remove the tom's bottom heads and place the mic inside, 1" to 6" away from the top heads.

Floor-Tom

Floor-toms can be miked in a manner similar to rack-toms (see Figure 4.46). The mic can be placed 2" to 3" above the top head, or it can be placed inside, 1" to 6" from the head. Again, a single mic can be used between two floor-toms, or each tom can have its own mic, which yields a greater degree of control over panning and tonal color.

Figure 4.44. *Individual miking of a rack-tom. (Courtesy of Beyerdynamic, Inc.)*

Figure 4.45. *Single mic placement for picking up two toms.*

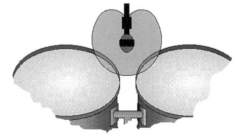

Kick Drum

The kick drum gives a low energy drive or punch to a rhythm groove. The kick drum has the capability to produce low-frequencies at high sound pressure levels, so it is necessary to use a mic that can both handle and faithfully reproduce these signals. Often the best choice for the job is a large-diaphragm dynamic mic, such as the AKG D-12E, AKG D-112, Electro-Voice RE-20, or the Beyer M-380.

Because of the extreme proximity effect (bass boost) that occurs when using a directional mic at close working distances, and because the drum's harmonics vary over its large surface area, even a minor change in placement may have a profound effect on the overall sound pickup. Moving the microphone closer to the head (see Figure 4.47) can add a degree of warmth and fullness, whereas moving it farther back often emphasizes the high-frequency "click." Placing the mic closer to the beater emphasizes the hard "thud" sound, whereas an off-center mic captures more of the drum's characteristic skin tone.

Figure 4.46. *Typical microphone placement for the floor-tom. (Courtesy of Beyerdynamic, Inc.)*

Figure 4.47. *Placing the mic at a distance just outside the kick drum head brings out the low-end and natural fullness. (Courtesy of Beyerdynamic, Inc.)*

By placing a blanket or other damping material inside the drum shell firmly against the beater head, a dull and loose kick sound can be tightened to produce a sharper, more defined transient sound. Cutting back on the kick's equalization at 300 to 600 Hz reduces the dull "cardboard" sound, whereas boosting at from 2.5 to 5 kHz adds a sharper attack, "click," or "snap."

Overheads

Overhead mics generally are used to pick up the high-frequency transients of cymbals with crisp, accurate detail, while also providing an overall blend of the entire drum kit. Because of the cymbals' transient nature, a condenser microphone is often chosen for its accurate high-end response.

Overhead mic placement can be very subjective and personal. One type of placement is the spaced pair, in which two mics are suspended above the left and right sides of the kit. These mics are equally distributed, so as to pick up their respective cymbal and overall instrument components in a balanced fashion (see Figure 4.48a). Another placement method is to suspend the mics closely together in a coincident fashion (see Figure 4.48b). This gives you a true overhead stereo image with a minimum of phase cancellations that might otherwise result from the use of spaced overhead mics.

Again, there are no rules for getting a good sound. If only one mic is available, place it at a central point over the drums; or, if you want, try using no overheads at all (the leakage spillover just might be enough to do the trick).

Figure 4.48. *Typical stereo overhead pickup positioning.*
a. *Spaced pair technique.*
b. *X-Y coincident technique.*

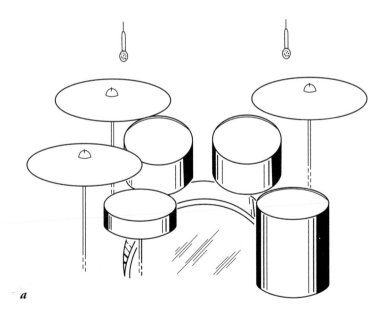

a

b

Tuned Percussion Instruments

The following section describes the various miking techniques encountered for tuned percussion instruments.

Xylophone, Vibraphone, and Marimba

The most common way to mic tuned percussion instruments is to place two high-quality condenser or extended-range dynamic pickups at distances above the playing bars and spaced at distances appropriate to the instrument's size (following the 3:1 general rule). A coincident stereo pair can help eliminate possible phase errors; however, a spaced pair often yields a wider stereo image.

General Percussion

Percussion instruments (such as the conga, timbales, and bongos) can be miked by placing a pickup just over the top head (about 1" to 3" above and 2" in from the rim). This placement provides a tight, "dead" sound with less shell resonance and more of a head attack. For a sound that is more "live," you can increase these distances above the head to 6" to 1'. If individual level control or a stereo effect is needed, a mic can be placed over each instrument; otherwise, a single mic can be shared among them.

Stringed Instruments

Of all the instrumental families, the stringed instrument is perhaps the most diverse. Ethnic music often uses instruments ranging from single-stringed to those that use highly complex and developed systems to produce rich and subtle tones. Western listeners have grown accustomed to hearing the violin, viola, cello, and double bass—both as solo instruments and in ensemble settings. Guitars abound in varieties of four-, six-, and twelve-stringed instruments. Whatever the type, design details for stringed instruments often are altered during construction to enhance or cut back on certain harmonic frequencies. These alterations are what give a particular stringed instrument its own characteristic sound.

Violin and Viola

The frequency range of the violin runs from 200 Hz to 10 kHz, so a good mic that displays a relatively flat frequency response should be used. The violin's fundamental range is from G3 to E6 (200 Hz to 1,300 Hz), and it is particularly important to use a mic that is flat within the formant frequencies of 300 Hz, 1 kHz, and 1,200 Hz. The fundamental range of the viola is tuned a fifth lower and contains fewer harmonic overtones.

In most situations, the violin or viola's mic should be placed on-axis to the instrument's front face. The distance depends on the acoustic condition and musical style. Miking at a distance generally yields a mellow, well-rounded tone, whereas a closer miking position may pick up a scratchy, more nasal quality—the amount depends on the instrument's tone quality.

For a solo instrument, a recommended miking distance is between 3' and 8'—over and slightly in front of the player—as shown in Figure 4.49. Under studio conditions, a closer mic distance—between 2' and 3'—is recommended. For a fiddle or jazz/rock playing style, the mic can be placed at a close working distance of 6" or less. The increased overtones help the instrument to cut through an ensemble.

Under PA (public address) applications, distant working conditions are likely to produce feedback. In this situation, a clip-type microphone can be attached to the instrument's tailpiece. At such close working distances, less amplification is needed, thereby reducing the possibility of feedback.

Cello

The fundamental range of the cello is from C2 to C5 (56 Hz to 520 Hz), with overtones up to 8 kHz.

If the player's line of sight is taken to be 0°, then the main direction of sound radiation lies between 10° and 45° to the right. A quality mic often is placed level with the instrument and directed towards the sound holes. The chosen microphone should have a flat response and be placed at a working distance of between 6" and 3'.

Figure 4.49. *Example of a typical mic placement for the violin.*

Double Bass

The double bass is one of the orchestra's lowest-pitched instruments. The fundamentals of the four-string type reach down to E1 (41 Hz) and up to around middle C (260 Hz). The overtone spectrum generally reaches up to 7 kHz, with an overall angle of high-frequency dispersion being ±15° from the player's line of sight. Once again, a mic can be aimed at the "f" holes at a distance of between 6" and 1 1/2'.

Voice

From a shout to a whisper, the human voice is a talented and versatile sound source that displays a dynamic and timbral range matched by few other instruments. The male bass voice may be able to extend from E2 to D4 (82 Hz to 293 Hz) with sibilant harmonics extending to 12 kHz. The upper soprano voice can range upwards to 1050 Hz with harmonics also climbing to 12 kHz.

The engineer/recordist should be aware of the following four possible traps that may be encountered when recording the human voice:

◆ *Excessive dynamic range.* This can be solved either by mic technique (physically moving away from the mic during loud passages) or by inserting a compressor into the signal path.

Some vocalists may have dynamics that range from a moderate volume to practically screaming within a single vocal passage. If you optimize your recording levels during a moderate volume passage and the singer begins to belt his or her lines, the levels will be way too hot and will distort. Conversely, if you set your recording levels for the loudest passage, the moderate volumes will barely be heard at all and will be buried in the music. The solution is to place a compressor in the mic's signal path. The compressor automatically "rides" the signal's gain and reduces excessively loud passages to a level that the system can effectively handle. (See Chapter 11, "Signal Processors," for more information about compression and dynamic range altering devices.)

◆ *Sibilance.* This occurs when sounds such as *f, s,* and *sh* are overly accentuated. This often is a result of tape saturation and distortion at high levels or slow tape speeds. Sibilance can be reduced when going to tape by inserting a frequency-selective compressor (known as a *de-esser*) into the chain.

◆ *Popping.* Explosive popping "P" sounds result from turbulent air puffs from the mouth striking the mic diaphragm. This problem can be avoided or reduced by placing a windscreen over the mic, placing a pop filter windshield between the mic and the vocalist, or by using an omnidirectional mic (which is less sensitive to popping).

◆ *Excessive bass boost due to proximity effect.* This bass build up often occurs when a directional mic is used at close working ranges. It can be reduced or compensated for in the following three ways: through equalization, by increasing the working distance between the source and the mic, or by using an omnidirectional mic (which doesn't display a proximity bass build up).

Woodwind Instruments

The flute, clarinet, oboe, saxophone, and bassoon combine to make up the woodwind class of instruments. Not all modern woodwinds are made of wood, and neither do they produce sound in the same way. The sound of the flute is generated by blowing across the hole in a tube, while others produce sound by using a reed to vibrate a column of air.

Historically, the woodwind's pitch was controlled by opening or covering finger holes along the sides of the instrument. This changed the length of the tube and, as such, the length of the vibrating air column. As instrumentation sophistication grew, the Boehm system of pads and levers further developed these instruments into their present forms.

It's a common misunderstanding that the natural sound of a woodwind instrument radiates entirely from its bell or mouthpiece. In reality, a large part of its sound often emanates from the fingerholes that span the instrument's entire length.

Clarinet

The clarinet comes in two pitches: the B clarinet with a lower limit of D3 (147 Hz), and the A clarinet with a lower limit of C3 (139 Hz). The highest fundamental is around G6 or 1,570 Hz, whereas notes an octave above middle C contain frequencies of up to 1,500 Hz when played softly. This spectrum can range upwards to 12 kHz when played loudly.

The sound of this reeded woodwind radiates almost exclusively from the finger holes at frequencies between 800 Hz and 3 kHz; however, as the pitch rises, more of the sound emanates from the bell. Often, the best mic placement occurs when the pickup is aimed towards the lower finger holes at a distance of 6" to 1' (see Figure 4.50).

Figure 4.50. Typical mic position for the clarinet.

Flute

The fundamental range of the flute extends from B3 to about C7 (247 Hz to 2,100 Hz). For medium loud tones, the upper overtone limit ranges between 3 kHz and 6 kHz. Commonly, the instrument's sound radiates along the flautist's line-of-sight for frequencies up to 3 kHz. Above this frequency, however, the radiated direction often moves outward 90° to the player's right.

When miking a flute, placement depends on the type of music and the room acoustics. When recording classical flute, the mic can be placed on-axis and slightly above the player at a distance of between 3' and 8'. When dealing with modern musical styles, the distance often ranges from

6" to 2'. In both circumstances, the microphone should be positioned at a central point between the mouthpiece and the instrument's footpiece. In this way, the instrument's overall sound and tone quality can be picked up with equal intensity (see Figure 4.51). Placing the mic directly in front of the mouthpiece generally reduces feedback and leakage; however, breath noise is accentuated without getting the full overall body sound.

Figure 4.51. *Typical mic position for the flute.*

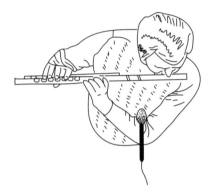

If mobility is important, a clip microphone can be secured near the mouthpiece or a specially designed contact pickup can be integrated into the instrument's headpiece.

Saxophone

Saxophones can vary greatly in size and shape. The most popular sax for rock and jazz is the S-curved B-flat tenor sax that has fundamentals spanning from B2 to F5 (117 Hz to 725 Hz), and the E-flat alto that spans from C3 to G5 (140 Hz to 784 Hz). Also within this family are the straight-tubed soprano and sopranino, and the S-shaped baritone and bass saxophones. The harmonic content of these instruments can range up to 8 kHz and can be extended by breath noises up to 13 kHz.

As with other woodwinds, the mic should be placed roughly in the middle of the instrument at the desired distance and pointed slightly toward the bell (see Figure 4.52). Keypad noises are considered to be a part of the instrument's sound; however, even these can be reduced or eliminated by aiming the microphone closer to the bell's outer rim.

Harmonica

Harmonicas come in all shapes, sizes, and keys and are divided into two basic types: the diatonic and the chromatic. The actual pitch is determined purely by the length, width, and thickness of the various vibrating metal reeds.

Figure 4.52. *Typical mic positions for the saxophone.* **a.** *Standard mic placement.* **b.** *Typical "clip-on" mic placement.*

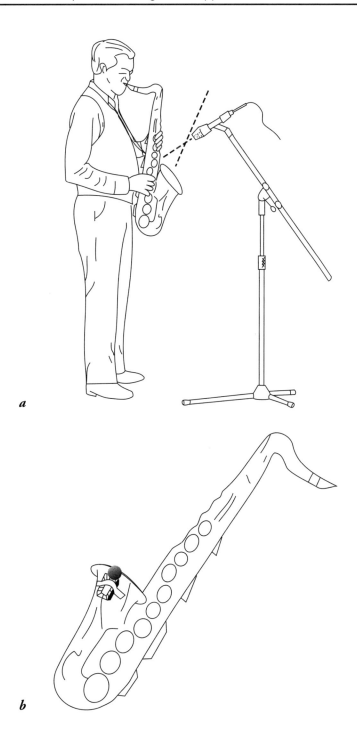

a

b

The "harp" player's habit of forming his or her hands around the instrument is a way to mold the tone by forming a resonant cavity. The tone can be deepened or a special "wahing" effect can be produced by opening and closing one hand. Often the mic is placed inside the cavity formed by the palms. Consequently, many harmonica players carry their preferred microphones with them (see Figure 4.53) rather than be rooted to one spot in front of an unfamiliar microphone and stand.

Figure 4.53. *The Shure 520D "Green Bullet" microphone is a preferred harmonica pickup for many musicians. (Courtesy of Shure Brothers, Inc.)*

Microphone Selection

The following list provides some insight into a number of professional mics used for music recording and pro-sound applications. This list is by no means complete, as literally hundreds of mics are available, each with its own particular design, sonic character, and application.

Shure SM-57

The SM-57, shown in Figure 4.54, is widely used by touring sound companies and artists' staff sound engineers for instrumental and remote recording applications. The SM-57's mid-range presence peak and good low-frequency response make it an ideal microphone for use with vocals, snare drums, toms, kick drums, electric guitars, and keyboards.

Specifications

Transducer type: Moving-coil dynamic

Polar response: Cardioid

Frequency response: 40–15,000 Hz

Equivalent noise rating: –7.75 dB (0 dB = 1V/microbar)

Figure 4.54. Shure SM-57.
(Courtesy of Shure Brothers, Inc.)

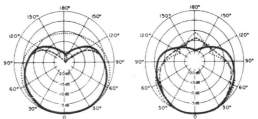

Polar response curves.

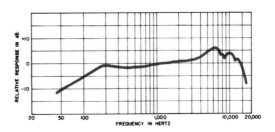

Frequency response curve.

Sennheiser MD 504

Designed primarily for instrument miking at high sound pressure levels, this mic (see Figure 4.55) is housed in a compact case with an integrated swivel mount, making it ideal for drum kit miking.

Specifications

Transducer type: Moving-coil dynamic

Polar response: Cardioid

Frequency response: 40–18,000 Hz.

Maximum SPL rating: Greater than 160 dB

Sensitivity: 1.8 mV/Pa

Figure 4.55. *Sennheiser MD 504. (Courtesy of Sennheiser Electronics Corp.)*

AKG D112

Large-diaphragm cardioid dynamic mics, such as the AKG D112 shown in Figure 4.56, are often used for picking up kick drums, bass guitar cabinets, and other low-frequency, high-output sources.

Specifications

Transducer type: Moving-coil dynamic

Polar response: Cardioid

Frequency response: 30–17,000 Hz

Sensitivity: –54 dB ±3 dB re. 1V/microbar

Figure 4.56. *AKG D112.*
(Courtesy of AKG Acoustics, Inc.)

Beyer M-160

The Beyer M-160 ribbon microphone, shown in Figure 4.57, excels in providing the transparency that often is inherent in ribbon mics. It yields a wide-frequency response/low-feedback characteristic and is capable of handling high sound-pressure levels without sustaining damage.

Specifications

Transducer type: Ribbon dynamic

Polar response: Hypercardioid

Frequency response: 40–18,000 Hz

Sensitivity: 52 dB (0 dB = 1 mW/Pa)

Equivalent noise rating: –145 dB (0dB = mW/2.10^{-5} Pa)

Output impedance: 200 Ω

Figure 4.57. *Beyer M-160.*
(Courtesy of Beyerdynamic, Inc.)

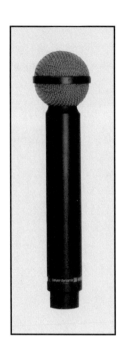

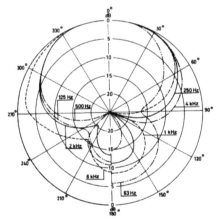

Polar response curves.

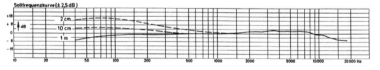

Frequency response curves.

AKG C3000

The AKG C3000 (see Figure 4.58) is a low-cost, large-diaphragm condenser mic. It includes two selectable polar patterns (cardioid and hyper-cardioid), bass roll-off switch, –10-dB pad, and an internal windscreen. The mic's dual-diaphragm capsule design is floated in an elastic suspension for an improved rejection of mechanical noise.

Specifications

Transducer type: Condenser

Polar response: Cardioid/hypercardioid

Frequency response: 20–20,000 Hz

Sensitivity: –34 dB or 20 mV/Pa (1kHz—cardioid), –36.5 dB or 15 mV/Pa (1kHz—hypercardioid)

Figure 4.58. *The AKG C3000.
(Courtesy of AKG Acoustics, Inc.)*

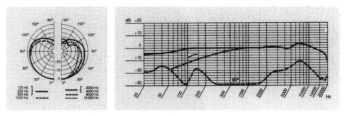

C A R D I O I D

Polar response and frequency response curves.

(continued)

HYPERCARDIOID

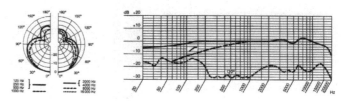

Polar response and frequency response curves.

AKG C414B/TL II

The C414B/TL II (Vintage TL) combines the re-created capsule from AKG's classic C-12 tube mic with the transformerless technology of the C414/B TL. The vintage TL, shown in Figure 4.59, is a 1" dual-diaphragm mic that has four switchable patterns, a –12-dB/octave bass roll-off switch (at 75 Hz or 150 Hz), and a switchable pad (–10 dB or –20 dB).

Specifications

Transducer type: Condenser

Polar response: Cadioid/hypercadioid/omni/figure 8

Frequency response: 20–20,000 Hz

Sensitivity: –38 dBV or 12.5 mV/Pa

Figure 4.59. *The AKG C414B/ TL II. (Courtesy of AKG Acoustics, Inc.)*

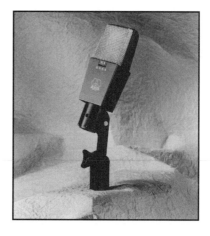

CARDIOID   OMNI-DIRECTIONAL

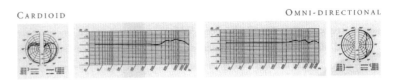

Polar response and frequency response curves.

HYPERCARDIOID   FIGURE 8

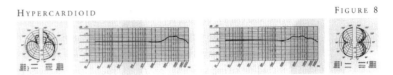

Polar response and frequency response curves.

Audio-Technica AT4050-CM5

The AT4050-CM5 (see Figure 4.60) is an improvement on the technology that has been developed for ATs popular AT4033 condenser mic. The 4050 is a transformerless, multipattern mic (cardioid, omni, and bidirectional) that sports a bass roll-off switch, –10-dB pad, and a suspension shock mount as standard equipment.

Specifications

Transducer type: Condenser

Polar response: Cadioid/omni/figure 8

Frequency response: 20–20,000 Hz

Maximum SPL rating: 149 dB SPL at 1 kHz @ 1% THD (without pad), 159 dB SPL at 1 kHz @ 1% THD (with pad)

Figure 4.60. *Audio-Technica AT4050-CM5. (Courtesy of Audio-Technica U.S.)*

(continued)

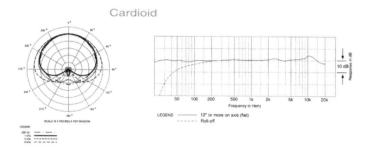

Cardioid

Polar response and frequency response curves.

Neumann TLM 193

The Neumann TLM 193, shown in Figure 4.61, is a large-diaphragm condenser mic that combines a transformerless design with the classic capsule of the U89i and TLM 170. It offers a frequency response of 20 to 20 kHz (below 10 Hz by removing an internal jumper) and has a maximum SPL rating of 140 dB, signal-to-noise ratio of 130 dB, and a low self-noise of 10 dBA.

Specifications

Transducer type: Condenser

Polar response: Cardioid

Frequency response: 20–20,000 Hz

Equivalent noise rating: 21 dB, 10 dB (A weighted)

Maximum SPL for less than 0.5% THD at 1 kHz: 140 dB

Figure 4.61. Neumann TLM 193. (Courtesy of Neumann (USA))

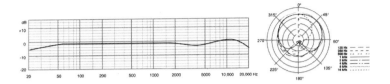

Polar response and frequency response curves.

Microtech Gefell UM92.1S

The Microtech Gefell UM92.1S, shown in Figure 4.62, is a tube version of Microtech Gefell's UM70 "Perastroika" mic. This three-pattern tube condenser mic uses a large, dual-diaphragm capsule that is similar to the one used in the original Neumann U-47. Standard accessories include an elastic shock-mount, power supply/remote pattern selection box, cable, and windscreen.

Specifications

Transducer type: Condenser

Polar response: Cadioid/omni/figure 8

Frequency response: 40–18,000 Hz

Sensitivity: 15 mV/Pa at 1 kHz

Figure 4.62. *Internal view of the Microtech Gefell UM92.1S. (Courtesy of G Prime Ltd.)*

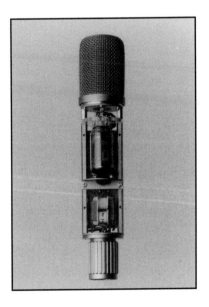

(continued)

omni

cardioid

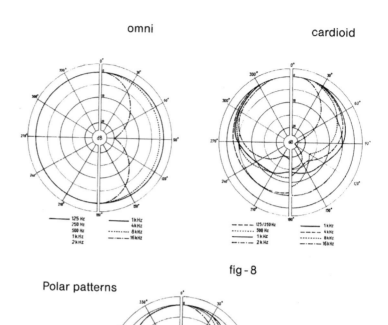

fig - 8

Polar patterns

CHAPTER

◆

The Analog Audio
Tape Recorder

In current production practices, the modern recording studio relies on magnetic media—generally in the form of magnetic tape—for storing sound. Theoretically, a sound recording device can be regarded as a block of memory that has the capacity to store audio information and, on request, to reproduce this information as faithfully as possible.

Although both tape-based and tapeless digital audio recording systems are firmly entrenched in modern recording technology, this chapter focuses on another common recording device: the analog *audio tape recorder,* or ATR.

Analog can be defined as being similar to or comparable in certain respects to something else. Calling a device an *analog ATR,* therefore, refers to its capability to transform an electrical input signal into a corresponding magnetic signal that is stored on tape in the form of magnetic remnants. The strength of these recorded remnants is directly analogous to the level of the input signal.

Magnetic Recording Media

In present day production practices, audio and other forms of information media are commonly recorded onto magnetic tape or computer disk.

Most modern magnetic storage media are composed of several layers of material, each serving a specific function (see Figure 5.1). Magnetic tape and computer floppy disks often use polyester or polyvinyl chloride (PVC) as a base material. This durable polymer is physically strong and is capable of withstanding a great deal of abuse before serious damage results. Bonded to the PVC base is a layer of magnetic oxide, which plays the operative role in the recording process. The molecules of this oxide form regions called *domains* (see Figures 5.2a and 5.2b), which comprise the smallest known permanent magnets. On an unmagnetized tape, these domains are oriented randomly over the entire surface of the tape.

Figure 5.1. *Structural layers of magnetic tape.*

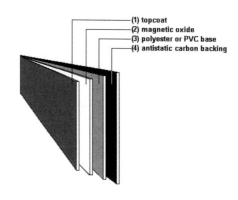

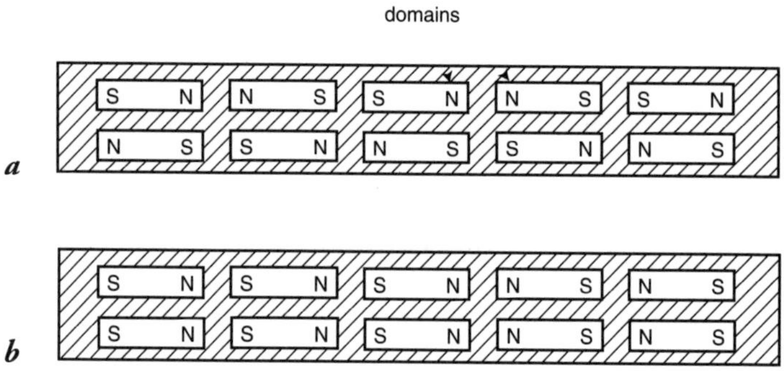

(1) topcoat
(2) magnetic oxide
(3) polyester or PVC base
(4) antistatic carbon backing

Figure 5.2. *Orientation of magnetic domains on unmagnetized and magnetized recording tape.*

a. *Random orientation on unmagnetized tape results in zero flux output.*

b. *Magnetized domains result in an average flux output at the magnetic head.*

The net result of this random magnetization is that north and south magnetic poles of each domain act to cancel each other out at the reproduce head, resulting in an average output level of zero. When a signal is recorded, however, the magnetization from the record head orients the individual domains (at varying degrees of angular direction) in such a way that their combined magnetism produces a nonzero average magnetic-flux. This alternating magnetic output can then be amplified and further processed to accurately reproduce a given signal at a later time.

The Tape Transport

The process of recording the audio bandwidth onto magnetic tape depends on the transport's capability to pass tape across a head path at a constant speed and with uniform tension. In short, this technology is used to physically pass a precise length of tape over the record head within a specific period of time (see Figure 5.3). During playback, the time relationship is kept stable by duplicating the precise speed at which the tape was recorded, thus preserving the original pitch, rhythm, and duration.

Figure 5.3. *Relationship of time to the physical length of recording tape.*

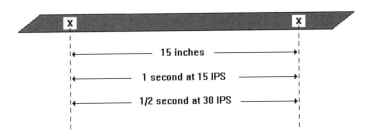

The movement of tape across the head's path at a constant speed and tension is the main function of the transport; this function is initiated when you press the Play button. You can disengage the drive at any time by pressing the Stop button, which applies the brakes simultaneously to the left and right reels. The Fast Forward and Rewind buttons cause the tape to shuttle rapidly in either direction to locate a specific point in the tape. Initiating either of these modes engages the tape lifters, which raises the tape away from the heads. After the play mode is engaged, pressing the Record button enables the audio signal to be recorded onto tape. Certain recorders require that both the Record and Play buttons be pressed simultaneously in order to begin recording; others begin recording when the play mode has been initiated and just the Record button is pressed. Figure 5.4 shows the elements of the transport deck of the Fostex R8 1/4" multitrack recorder.

On older professional transports, stopping a fast-moving tape by pressing the Stop button can stretch or ruin a master tape because a high degree of inertia is involved in the fast-wind modes. A procedure known as *rocking* the tape can prevent this damage. A tape can be rocked to its stop position by engaging the fast-wind mode in the direction opposite the current travel direction until the tape slows to a reasonable speed; then you press the Stop button.

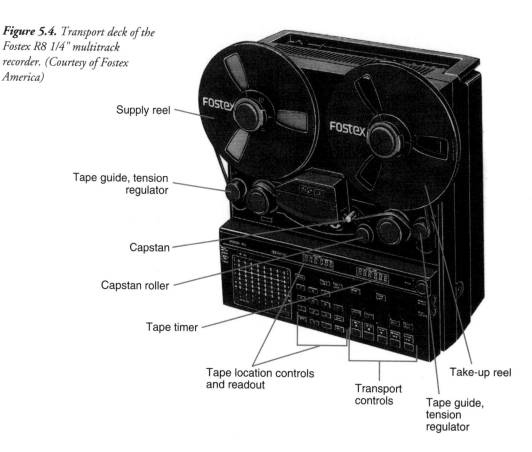

Figure 5.4. *Transport deck of the Fostex R8 1/4" multitrack recorder. (Courtesy of Fostex America)*

Supply reel

Tape guide, tension regulator

Capstan

Capstan roller

Tape timer

Tape location controls and readout

Transport controls

Take-up reel

Tape guide, tension regulator

In recent years tape transport designs have incorporated *Total Transport Logic* (TTL), which allows for complete microprocessor monitoring and control over all transport functions. This innovation has a number of distinct advantages. For example, TTL logic enables the operator to push the Play button while the tape is in either fast-wind mode without fear of tape damage. On older machines, this action could have destroyed a section of the master. With TTL, the recorder can sense that its transport isn't at a stand-still and can determine the direction the tape is traveling. The recorder can then either automatically rock the transport until the tape can safely be stopped or slow the tape to playspeed at which point it can automatically slip into play or record mode. Thus, in a limited sense, a tape machine transport equipped with TTL can be considered "intelligent." The development of this *logic-intelligence* has transformed the magnetic tape recorder into the interactive and sophisticated ATR of today.

Most modern ATRs are equipped with a *shuttle control* that enables the tape to be shuttled at various speeds in either direction. This control makes it possible to locate a cue point or to wind the tape onto its reel at a slower speed for long-term storage.

The Edit button, found on most professional machines, enables two modes of deck operation: *stop-edit* and *dump-edit*. If the Edit button is pressed while the transport is in the stop mode, the left and right tape reel brakes are released and the tape sensor is bypassed. This stop-edit mode enables the tape to be rocked back and forth so that edit points can be manually found. Often, if the Edit button is pressed while the transport is in the play mode, only the take-up turntable is disengaged and the tape sensor is bypassed again. This dump-edit mode enables the operator to run unwanted sections of a tape off the transport while listening to the material being removed.

A safety switch, incorporated in all professional ATR transports, initiates the stop mode when it senses the absence of tape along its guide path. Thus, the recorder stops automatically at the end of a reel or upon tape breakage. Such a switch may be incorporated into the tape-tension sensor or it can be a light beam that is interrupted by the presence of tape in its path.

Most professional ATRs are equipped with automatic tape counters that provide elapsed-tape readout times in hours, minutes, and seconds. Many of these recorders feature digital readout displays and can double as tape-speed indicators when in the *varispeed mode*. This mode incorporates a continually variable control to enable variation in tape speeds from the industry standard speeds. On many tape transports, this control provides for speed variations of ±20% of the standard.

Capstan Motors

The *capstan* is the most critical element in many transport systems. It is the shaft of a motor and is precisely regulated to rotate at a constant rate of speed. Two common types of capstan motors are capable of delivering a high degree of speed accuracy: the hysteresis motor and the DC servo motor.

The *hysteresis motor* derives its speed constancy from the power line's supply voltage frequency, a standard reference that usually is the stable 50-Hz or 60-Hz frequency of the main power lines. This type of motor is often found on older professional machines or newer, less expensive systems.

The more modern *DC servo motor*, on the other hand, derives its speed of rotation from the level of DC voltage supplied to it.

Essentially, the DC servo motor owes its stability and versatility to the use of motion-sensing feedback circuitry. One such feedback mechanism is constructed by mounting a notched tachometer disk, as shown in Figure 5.5, directly onto the capstan motor shaft or idler-wheel assembly. The rate of rotation is computed by counting the number of disk notches per second that pass between a light source and an optical sensor. A resolver compares the actual rate of rotation with a standard reference, giving a highly accurate and stable capstan speed. During the past decade, this latter design has become the standard for tape transports because of its greater adaptability to systems using motion-sensing and TTL.

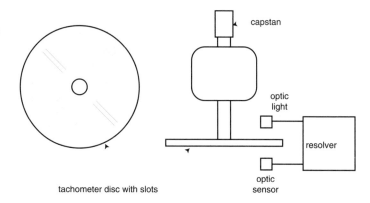

Figure 5.5. *A DC servo motor with tachometer-resolver assembly.*

capstan

optic light

resolver

optic sensor

tachometer disc with slots

Tape Transport Systems

Three methods are used on professional reel-to-reel audio recorders to transport tape across a head path with constant speed and tension: open-loop, closed-loop, and zero-loop.

The *open-loop* system, shown in Figure 5.6, is most commonly found in older professional designs and in newer, less expensive transport designs. In an open-loop system, the capstan and capstan roller (also known as a *pinch roller*) work together to move tape along the path at the proper playspeed. A small amount of torque is applied to the supply-reel motor—in the direction opposite that of tape travel—providing the required amount of tension and head-to-tape contact. A small take-up motor torque helps spool the tape passing through the capstan idler onto the take-up reel.

Figure 5.6. *Open-loop tape drive system.*

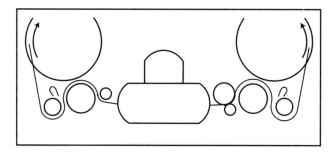

Another transport system used in the past by several manufacturers is the *closed-loop* (or *differential capstan*) drive system, shown in Figure 5.7. As the name implies, in a closed-loop system the tape guide path is isolated from the rest of the transport, with unsupported sections of tape being kept to a minimum. This tape isolation reduces many of the distortions associated with open-loop systems. In a closed-loop system, the tape is actually pulled out of the head block at a slightly faster rate than it is allowed to enter. The entrance and exit points of the head block are controlled either by two separate but synchronized capstans with slightly different diameters or by a single grooved capstan with two pressure rollers that mate with those in the capstan (which provides the same effect as two capstans with different diameters). The stretching of tape through the closed-loop system provides the necessary tension required for good tape-to-head contact. This process is possible because the tape can be stretched by as much as 5% before permanent deformation occurs—a figure well within the tolerance of closed-loop drive systems.

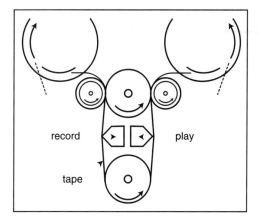

Figure 5.7. Closed-loop tape drive system.

The zero-loop system (Figure 5.8) has gained a wide degree of recent acceptance in more recent professional transport designs because it is best equipped to take full advantage of TTL computer logic control and DC servo feedback circuitry. The zero-loop system doesn't employ a capstan, as do the other methods. Instead, it makes use of motion and tape-tension sensors on both the supply and take-up sides of the head block in order to continually monitor speed and tension. When variations from the standard reference speed occur, a resolved, corrective supply voltage is applied directly to the supply or take-up DC servo motors (and possibly a capstan servo motor), thus restoring the tape's movement back to its correct speed and tension.

Figure 5.8. Zero-loop tape drive system.

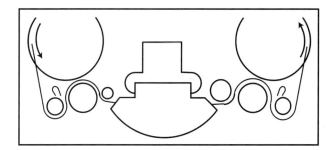

The Magnetic Tape Head

Most audio magnetic recording systems employ the use of the magnetic tape head to perform three specialized tasks: record, reproduce, and erase.

It is the function of the *record* head (see Figure 5.9) to transform the input signal into remnants on magnetic tape for permanent storage. The input current flows through coils of wire that are wrapped around the head's magnetic pole pieces, causing a magnetic force to flow through the pole pieces and across the head gap. Like electricity, magnetism flows more easily through certain media than through others. The magnetic counterpart to electric current is called *flux* and the resistance to flux is called *reluctance*. The head gap between poles creates a break in the magnetic flux, thereby creating a reluctance to the magnetic force that has been set up. The tape's magnetic oxide offers a lower reluctance path to the flux than does the nonmagnetic gap, so the flux path travels from one pole piece, through the tape, to the other pole. The actual recorded signal occurs at the trailing edge of the record head—with respect to tape motion—because the magnetic domains retain the polarity and magnetic intensity they received just before they left the head gap.

Figure 5.9. *The magnetic record head.*

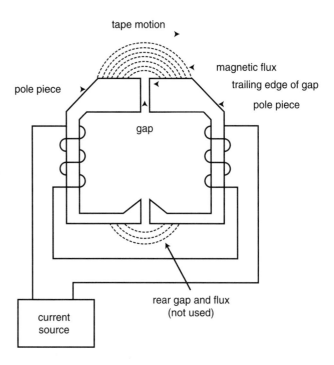

The *reproduce* (or *playback)* head, shown in Figure 5.10, operates in a similar fashion to the record head but in an inverted fashion. When a recorded flux passes across the reproduce head gap, a flux is induced into the pole pieces. This flux at the poles induces a current in the pickup coils that can either be amplified or processed.

The reproduce head is a nonlinear device because its output is proportional to both the tape's average magnitude of flux and the rate of change of this flux. The rate of change increases as a direct function of and in direct proportion to the frequency of the recorded signal. The output voltage of the reproduce head is proportional to

$$\frac{\Delta\varnothing}{\Delta t}$$

where

> $\Delta\emptyset$ is an average value change in gap flux,
> Δt is the time interval required for $\Delta\emptyset$.

As the output is directly proportional to the rate of change in flux, the output doubles for each doubling in frequency: a 6-dB increase in output for each octave (see Figure 5.11).

Figure 5.10. *The magnetic reproduce head.*

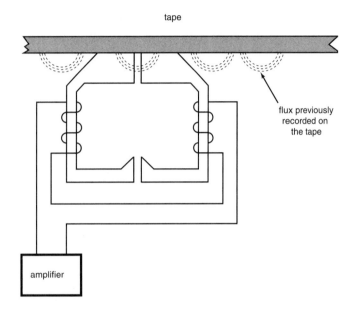

Figure 5.11. *The effects of increased frequency on the reproduced output at the magnetic head gap.*

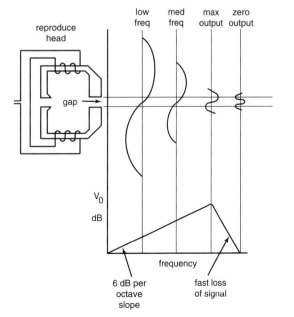

The length of the head gap and the tape speed determine the upper-frequency limit of the reproduce head, thus determining the system's overall bandwidth. The wavelength of a signal recorded on tape is equal to the speed at which tape travels past the reproduce head, divided by the frequency of the signal. Therefore, the faster the tape speed, the higher the upper-frequency limit. Similarly, the smaller the head gap, the higher the upper-frequency limit. When the wavelength of a recorded signal approaches that of twice the head-gap length, the output signal begins to decrease until the signal's wavelength is equal to that of the gap. At this point, the output will be zero and the upper-frequency limit of the system will have been reached. Making the head gap too small has the adverse effect of reducing the playback output signal, thus degrading the signal-to-noise ratio.

It should be noted that in certain multitrack ATR designs, both the record and reproduce functions are performed by a single tape head. Generally, in such a case, a compromise has been reached in the gap width that is compensated for in the equalization section of the record, playback, and sync circuitry. This design is used to reduce the cost of a system by eliminating one of the two heads (multitrack heads are generally very expensive to produce and constitute a significant portion of a recorder's price tag). Also, the need for additional sync switching and associated circuitry is eliminated because both the playback and record functions are handled by only one head.

The function of the *erase* head is to reduce the average magnetization level of a tape to zero, thereby enabling the tape to be rerecorded. After the transport is placed into the record mode, a high-frequency and high-intensity sine wave signal is fed into the erase head, resulting in a tape that is alternately saturated in both the positive- and negative-polarity directions. (Tape saturation is reached when all of the magnetic particles at the head gap are completely magnetized and an increase in magnetic force will not result in an increase in magnetism retained on tape.) This alternating saturation serves to destroy any magnetic pattern existing on the tape. As the tape moves away from the erase head, the intensity of the magnetic field decreases; the domains are left in a random orientation, with a resulting average magnetization level of zero.

The Professional Analog ATR

Professional analog ATRs are commonly configured in 2-, 4-, 8-, 16-, and 24-track formats. Each configuration is generally best suited to a specific task in production and postproduction. For example, 2- and 4-track ATRs generally are used to record the multitrack mix, whereas the 8-, 16-, and 24-track machines usually are used for multitrack recording. Some examples of the analog ATR are pictured in Figures 5.12 through 5.17.

Figure 5.12. Otari MTR-12 1/4" 2-channel or 1/2" 4-channel mastering recorder (Courtesy of Otari Corporation)

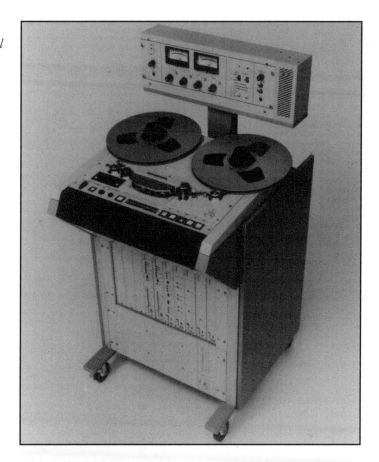

Figure 5.13. Sony APR-5003V 2-channel mastering recorder. (Courtesy of Sony Corporation of America)

Figure 5.14. *Fostex 1/2"*
16-channel multitrack recorder.
(Courtesy of Fostex Corporation of
America)

Figure 5.15. *Otari MTR-100A*
8-, 16-, and 24-channel master
recorder. (Courtesy of Otari
Corporation)

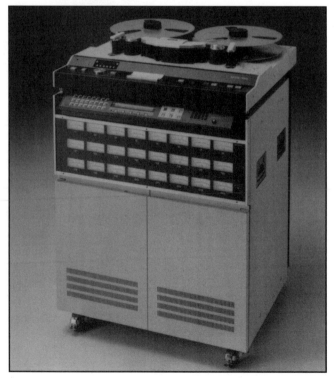

Figure 5.16. Studer A-820 24-channel master recorder. (Courtesy of Studer/Revox of America, Inc.)

Figure 5.17. Sony APR-24 analog multitrack audio recorder. (Courtesy of Sony Corporation of America)

Equalization

Equalization (EQ) is the term used to denote an intentional change in relative amplitude response at different frequencies. Because the analog recording process isn't linear, equalization is needed in order to achieve a flat frequency-response curve with magnetic tape. The 6-dB-per-octave boost inherent in the response curve of the reproduce head makes it necessary to apply a complementary equalization cut of 6 dB per octave in the playback electronics (see Figure 5.18).

Figure 5.18. *Flat frequency response curve due to complementary equalization.*

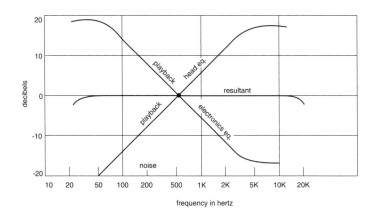

Bias Current

In addition to nonlinear changes in reproduced head output level relative to frequency, there is a discrepancy between the amount of magnetic energy applied to the tape and the amount of magnetism retained by that tape after the initial flux has been removed.

As Figure 5.19a shows, the magnetization curve of magnetic tape is linear only between points A and B, and between points C and D. Signals greater than A (in the negative direction) and D (in the positive direction) have reached the saturation level and are subject to clipping distortion. The signal falling within the B to C range is too low in flux level to adequately affect the tape's magnetic particles. Thus, some means must be employed to bring the signal at the recording head into the linear A-B and C-D range. The means generally used is the application of *bias current*, or *AC bias* (see Figure 5.19b).

Bias current is applied by mixing the incoming audio signal to be recorded with that of an ultrasonic signal (often between 75 kHz to 150 kHz). This has the effect of modulating the amplitude of a recorded signal to a higher average flux level and moves the recorded signal away from the nonlinear zero-crossover range and into the linear portion of the curve. On playback, because the reproduce head can't reproduce the high-frequency bias signal, the modulated audio signal is reproduced whereas the bias signal is not.

Figure 5.19. *The effects of bias current on recorded linearity.*

a. *Magnetization curve.*

b. *After bias.*

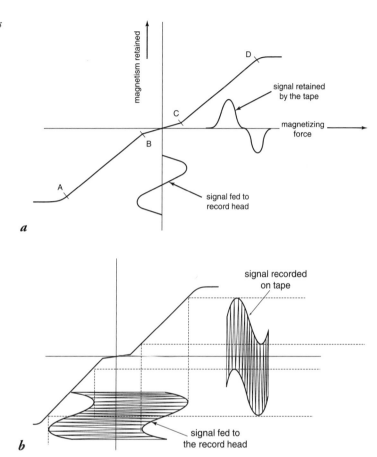

Recording Channels

Each recording channel of a modern ATR, no matter what the machine's track configuration, is designed to be electrically identical to the others; the channel circuitry is simply duplicated by the number of channels.

Just as the magnetic tape head performs three functions, the electronics of an ATR must be specialized to perform the same processes: record, reproduce, and erase. In older machines, each recording channel consists of a module in which three electronics cards are housed. Each card performs one of the three functions.

Newer ATRs, specifically multitrack machines, make use of the input/output (I/O) module, which incorporates all the adjustment controls and electronics of a recording channel on a single printed circuit board. These modules are housed in a single mainframe inside the ATR's console. Designs using I/O modules allow for greater channel interchangeability and servicing ease. These electronic modules enable adjustments to input level, output level, sync level, and equalization.

The output signal of a professional ATR may be switched between three working modes: input, reproduce, and sync.

In the *input mode*, the signal given at the selected channel output is derived from the input signal. Thus, with the ATR transport in any mode, including stop, it is possible to meter and monitor all signals present at the input of the ATR.

In the *reproduce mode,* the output signal is derived from the reproduce head, with the channel's output and metering reflecting the playback signal. The reproduce mode can be useful in two ways: it allows for the playback of a previously recorded tape for use in studio production, and it enables monitoring of material off tape while in the record mode. The latter application provides an immediate quality check of the ATR's entire record and reproduce process.

In postproduction, with the use of *time code* (a code used for transport synchronization), a multitrack ATR may need modified electronics. The modification may be necessary because control synchronizers often require a reproduce bandwidth of up to 100 kHz in order to read code at higher shuttle speeds. For standardization, the highest multitrack ATR track number is fitted with the modified time code electronics.

The *sync mode* is a required feature of the multitrack ATR because of the need to record material on one or more tracks while simultaneously listening in synchronization to the playback of previously recorded tracks (a process known as *overdubbing*). To record one or more new tracks using the record head while listening to the previously recorded tracks through the reproduce head would make the composite signal out of sync on final playback. To prevent such a time lag, the originally recorded tracks are placed in the sync mode. In this mode, the signals to be monitored are reproduced through their respective record head tracks, thereby physically aligning the reproduced and recorded tracks and maintaining absolute synchronization (see Figures 5.20a and 5.20b).

Figure 5.20. *The function of the sync mode.*

a. In the reproduce mode, the recorded signal lags behind the monitored playback signal, creating an out-of-sync condition. *a*

b. In the sync mode, the record head acts as both record and playback head, bringing the signals into sync.

b

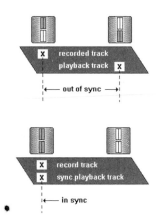

The professional ATR enables independent switching between record and playback functions and each channel. These controls may be located at the ATR's meter bridge, front panel, or remote control unit.

In addition to control over input, reproduce, and sync, the professional ATR has a function selection known as the record-enable switch. Activating this switch (usually labeled *Safe* or *Ready*) prevents the accidental erasure of a recorded track.

The Remote Control Unit and Autolocator

The remote control unit has evolved from a simple control over deck transport functions to the present-day unit that contains all the transport-function and track-status controls. These units are usually located near the audio production console, placing the operating controls close at hand to the engineer.

One feature being built into many of the newer 2- and 4-track ATR transport decks and multitrack remote controls is the *autolocator* (see Figure 5.21). The autolocator allows specific cue point locations to be stored in memory. When an engineer enters a cue point into a keypad or calls it from memory and then presses the search button, the autolocator shuttles the tape to the desired position. At that point, the transport either stops, places itself into the play mode, or automatically cycles (play-relocate-play) between two cue points. Most autolocators allow multiple reference points to be stored, providing for the location of the various cues needed in music production.

The *control synchronizer,* used in high-level postproduction, can also operate as an interactive remote control and autolocator. This form of synchronizer uses time code to exercise full control over the remote location functions of one or more transports simultaneously. In the latter case, several ATRs may be designated to chase, or follow, the tape position of a master tape or to locate to a specific time code address number.

MIDI Machine Control

Recently, a new method has been adopted that uses the MIDI specification to maintain control over the transport functions of suitably equipped tape- and nontape-based recording media.

Figure 5.21. *Example of a multitrack remote control with autolocator. (Courtesy of Studer/ Revox of America, Inc.)*

This system, known as *MIDI Machine Control* (MMC), enables a wide range of transport commands to be transmitted over standard MIDI lines from a central controller to a particular device or number of devices within a connected system. Basically, this means that transport control functions can now be executed cost-effectively from a MIDI-based central controller (such as a MIDI sequencer, mixing console, or transport controller)—either manually or under automated computer control. For example, the music production system in Figure 5.22 shows a computer-based MIDI sequencer configured to be the master transport controller for an 8-track analog tape recorder. Here is one possible scenario:

1. Clicking the *locate 1* icon on the sequencer's screen sends a message to the ATR to locate to the beginning of the song.

2. Once located, beginning the sequence sends out a *play* command. At that point, the transport begins playing and outputs SMPTE time code from track 8.

3. The SMPTE is translated to MIDI time code (MTC), which is then understood by the sequencer. The sequencer, in turn, locks to the recorder's time code track and begins playing the sequence.

4. At measure 28, track 6 is instructed to go into record so that a vocal track can be laid down.

5. At measure 43, track 6 is instructed to drop out of record.

6. At the end of the sequence a *locate 1* instruction is sent to the recorder and is followed by the *stop* command.

Figure 5.22. *Example of a music production system using MIDI machine control.*

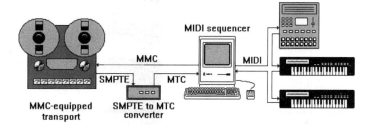

Of course, this is one of an almost limitless number of situations and system setup types that might make use of MMC. The real beauty of MMC is that because it is MIDI-based, it can be designed into software packages, controllers, audio and video recorders, mixing consoles, and even musical instruments with little or no added cost to the manufacturer. As a result, this standardized protocol has grown in popularity with both professional and electronic music manufacturers. A more detailed description of the protocol itself can be found in Chapter 7, "MIDI and Electronic Musical Instrument Technology."

Tape and Head Configurations

Professional analog ATRs are currently available in a wide number of track- and tape-width configurations. The most common track- and tape-width configurations are 2-track, 4-track, and 8-track, 1/4"; 4-track, 8-track, and 16-track, 1/4"; 8-track and 16-track, 1"; and 16- and 24-track, 2" (see Figure 5.23).

Optimal tape-to-head performance characteristics for an analog ATR are determined by several parameters: track width, head-gap width, and tape speed.

In general, the track width and head-gap width is on the order of 0.080" for the 1/4" 2-track ATR; 0.070" for the 1/2" 4-track, 1" 8-track, and 2" 16-track formats; and 0.037" for the 2" 24-track format. These widths are rather large compared to the 0.021" head-gap width of cassette tape. With a greater recorded track width, an increased amount of magnetism can be retained by the magnetic tape, which results in a higher output signal and an improved signal-to-noise ratio. The use of a wider track width also makes the recorded track less susceptible to signal-level dropouts. The guardband, a width of unrecorded tape between adjacent tracks, is present in order to prevent channel crosstalk.

Figure 5.23. *Track configurations for basic tape widths used for analog recording.*

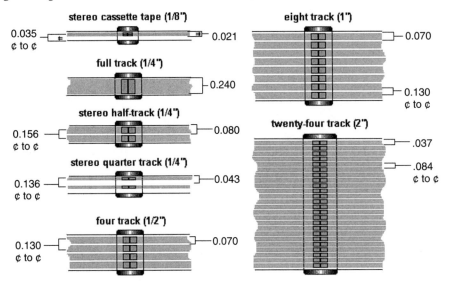

Note: ¢ to ¢ = center-to-center distance

Tape speed has a direct bearing on the performance characteristics of the ATR because it is directly related to the recorded signal's level and wavelength. At high tape speeds, the number of magnetic domains that pass over the tape head gap in a given period of time is greater than it is at slower speeds. Thus, the average magnetization received by the playback head is greater and produces a stronger output signal requiring less amplification, which results in less tape noise. Because higher tape speeds also increase the recorded signal's wavelength, less high-frequency boost is required by the record electronics, permitting increased recording levels without saturating the tape.

At faster tape speeds, the recorded bandwidth is effectively increased. This result is also due to the response characteristics of the reproduce head. As I have noted, the output of the reproduce head responds to the average value of magnetization at the head gap. As the frequency of the playback signal increases, more and more of the complete cycle falls inside the head gap at any one point in time until the signal wavelength is equal to the gap width (see Figure 5.24). At this point, the average output level is zero. This reduced output, known as *scanning loss*, determines the system's upper-frequency limit. Because the wavelength of a recorded signal increases with tape speed, the upper-frequency bandwidth limit is extended at higher tape speeds.

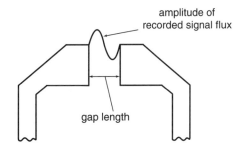

Figure 5.24. *Scanning loss occurs when the recorded wavelengths equal the width of the magnetic head gap.*

amplitude of recorded signal flux

gap length

The tape speeds generally used in postproduction are the following: 7 1/2" per second (ips), 15 ips, and 30 ips. These speeds are found on many 1/4" and 1/2" studio ATRs, although 7 1/2 ips presents too much of a compromise in signal quality for multitrack production work. In recent years, 30 ips has gained wide acceptance for analog music production. When used with low-noise, high-output tape, this speed has reduced or eliminated the need for extensive noise reduction.

Print-Through

One form of deterioration in tape quality that can occur after a recording has been completed is known as *print-through*. Print-through is the transfer of a recorded signal from one layer of magnetic tape to an adjacent layer by means of magnetic induction. This gives rise to an audible false signal or pre-echo on playback. The effects of print-through are greatest when recording levels are very high and is reduced by about 2 dB for every 1-dB reduction in signal level. The extent of this condition also depends on other factors, such as length of storage, storage temperature, and tape thickness (tapes with a thicker base material are less likely to have severe print-through problems).

Due to the directional properties of magnetic lines of flux, print-through has a greater effect on the adjacent outer layer of tape (where magnetic induction is in-phase) than the adjacent inner layer (where the magnetic induction is out-of-phase), as shown in Figure 5.25. So if a recorded tape is stored *heads-out* (with the tape rewound onto the supply reel), a "ghost" signal will transfer to the outer layer and be heard before the original signal. For this reason, the standard means of professionally storing a recorded tape is in the *tails-out* position (with the tape wound onto the take-up reel). When a tape is stored tails-out, the print-through echo follows the original signal—a condition that is masked by the natural decay of the sound and often is subconsciously perceived by the listener as reverberation.

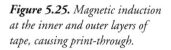

Figure 5.25. *Magnetic induction at the inner and outer layers of tape, causing print-through.*

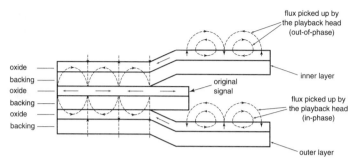

Cleanliness

It's very important to keep the magnetic recording heads and moving parts of an ATR transport deck free from dirt and *oxide shed*. Oxide shed occurs when friction causes small particles of the magnetic oxide to accumulate on surface contacts. This accumulation is most critical at the surface of the magnetic recording heads because even a minute separation between the magnetic tape and heads can cause separation loss. For example, a signal recorded at 15 ips, with an oxide shed buildup of 1 mil (0.001") at the playback head yields a playback loss of 55 dB at 15 kHz. When cleaning the transport tape heads and guides (with the exception of the machine's pinch roller and other rubber-like surfaces), use denatured (isopropyl) alcohol or an appropriate cleaning solution at least daily and always before a routine alignment.

Degaussing

Magnetic tape heads are made from a *magnetically soft* metal alloy, which means that the alloy does not readily retain magnetism but does act as an excellent conductor of flux. These heads, however, do retain small amounts of residual magnetism, which can act to partially erase the high-frequency signals on a master tape. For this reason, the degaussing of a magnetic tape head after 10 hours of operation is a recommended practice. A magnetic head degausser acts very much like an erase head in that it saturates the magnetic head with a high-level alternating signal that randomizes the residual magnetic flux. After the head has been degaussed, it is important to move the degausser away from the tape heads at a speed of less than 2" per second, so as to avoid inducing a magnetic flux in the head. Before an ATR is aligned, the magnetic tape heads should always be cleaned and degaussed in order to obtain accurate readings and to protect the alignment tape.

Head Alignment

An important factor affecting the performance of an analog recorder is that of the magnetic tape head's physical alignment. The erase, record, and playback heads have five adjustments: height, azimuth, zenith, wrap, and rack.

The *height* determines where the track's vertical positioning will be along the width of the magnetic tape (see Figure 5.26). If the track is recorded and reproduced on heads with different height settings, not all the recorded signal will be reproduced. This results in a compromised signal-to-noise ratio and increased crosstalk between multitrack channels.

Figure 5.26. The head gap must be centered on the track location.

Azimuth refers to the tilt of the head in the plane parallel to the tape (see Figure 5.27). The head gap should be perpendicular to the tape, so that all track gaps are electrically in-phase with each other.

Figure 5.27. The head gap must be perpendicular to tape travel.

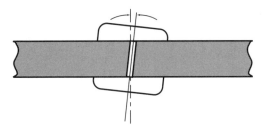

Zenith refers to the tilt of the head toward or away from the tape. The zenith must be adjusted so that the tape contacts the top and bottom of the head with the same degree of force, otherwise the tape tends to skew. Skewing occurs when tape rides up or down on a head or guide so that its edges no longer are parallel to the top plate of the transport, which causes variation in the effective height, azimuth, and tape speed.

Wrap refers to the angle at which the tape bends around the head and the location of the gap in this angle. The wrap determines the degree of tape-to-head contact and thus controls the head's sensitivity to dropouts.

Rack determines the pressure of the tape against the head. The further forward the head is, the greater the pressure.

Height can be adjusted visually or by using a test tape of the proper head configuration to give maximum output at 1 or 3 kHz. Zenith can be tested by covering the pole pieces with a white grease pencil and playing a piece of scrap tape to observe the pattern formed as the grease pencil wears off (see Figure 5.28). The edges of the wear pattern should be parallel. If they are not, use the screws on the headblock to adjust the zenith. Wrap angle can be checked at the same time as zenith by making sure that the wear pattern is centered around the gap. Rack adjustment may be needed if the wear pattern is wider on the record head than on the playback head, or vice versa.

Figure 5.28. Various tapehead wear patterns as a result of proper and improper zenith alignments.

proper alignment   adjusted too high   adjusted too low

Azimuth can be tested by deliberately skewing the tape across the heads (by pushing up and down on the edges of the tape right in front of the head), while reproducing the 15-kHz band of an older standard alignment tape. If the output increases, the azimuth needs to be adjusted. It can be adjusted by either of two methods, both of which use a full-track standard alignment tape.

The first method for checking azimuth is to play the 15-kHz section of the tape and adjust the azimuth for the highest output on all channels. This will be a compromise because various channels on a multitrack machine will rise as others fall on either side of the proper setting. The peak at the correct setting will be fairly sharp with smaller, broader peaks on either side of it. To ensure the proper peak is found, the head gap should be visually checked to be sure it is perpendicular to the tape path before the adjustment is attempted.

The second method uses the phase of a 12-kHz or 15-kHz test-tape signal to find the correct setting. After finding the peak, as in the first method, the output of the top channel is fed into the vertical input of an oscilloscope, and the bottom channel is fed into the horizontal input of the scope. The resulting pattern on the face of the scope represents the relative phase between the two channels. A straight line sloping up 45° to the right indicates that the two channels are in-phase, a circle indicates they are 90° out-of-phase, and a straight line sloping up 45° to the left indicates that the two channels are 180° out-of-phase. The azimuth should be adjusted so that the two channels are in-phase.

On a multitrack head, it's not possible to get all the gaps in-phase at one time because of *gap scatter* (variations from a vertical gap line), which occurs in manufacturing. So proper phase adjustment is made for the outer tracks, and the inner tracks are then usually less than 60° out-of-phase with one another. The record-head azimuth may be set in the same way as the playback-head azimuth: by playing the test tape while in the sync mode. Erase-head azimuth isn't critical due to its gap width. It will be correct as long as it is at the correct height and at right angles to the tape path.

Head adjustment does not have to be done frequently. Need is indicated by the deterioration of the ATR performance. The adjustment should be performed by a qualified maintenance engineer.

Electronic Calibration

Because sensitivity, output level, bias requirements, and frequency response vary considerably from one tape formulation to the next, ATRs have variable-level electronic adjustments for record/ playback level, equalization, bias current, and so on. It is extremely important that a standard be adhered to in the adjustment of these level settings so that a recording made on one ATR is compatible with and identical in playback response on another. The procedure used to set these controls to a standard level is called *electronic calibration* or *electronic alignment*.

Proper ATR alignment depends on the specific tape formulation and on the set of headblocks being used, so it's a good practice to recalibrate all production ATRs at regular or semi-regular intervals. In major music recording studios, machine alignments are routinely made first thing each morning and prior to each recording session, which ensures the standardization and performance reliability of each master tape.

Assuming that the record and reproduce heads are in proper alignment, the electronic calibration of the ATR can be performed. This procedure is carried out with reference to a standard set of equalization curves for each tape speed. Such curves have been established by the National Association of Broadcasters (NAB), in use throughout the US and much of the world; the Deutsche Institute Norme (DIN), in use throughout Europe; and the Audio Engineering Society (AES) for use at 30 ips. The NAB and DIN curves are shown in Figure 5.29.

Figure 5.29. *Pre- and post-equalization for the NAB and DIN characteristics.*

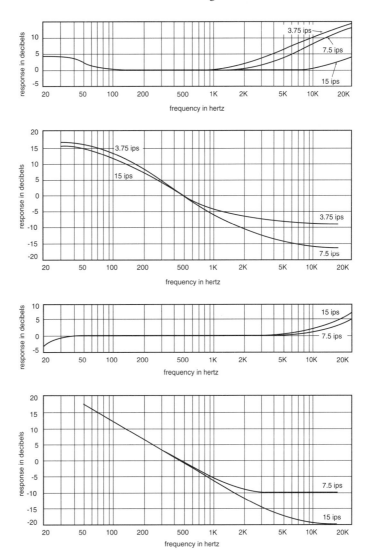

To anyone new to this process, the calibration may seem complicated; but with repetition, it becomes second nature and usually takes less than 10 minutes to perform. Electronic calibration is performed in two stages: reproduce alignment and record alignment. The step-by-step procedures for various speeds and tapes are provided in Chart 5.1.

Chart 5.1. Playback and recording alignment procedures.

Playback Alignment

Thread the playback alignment tape on the machine to be aligned. If the tape was stored tails out, rewind to the head.

A. *15-ips playback alignment for low-noise, high-output tape, using a standard full-track 15-ips alignment tape.*

1. Set repro level for 0 VU at 700 Hz.
2. Set high-frequency 15-ips playback EQ for 0 VU at 10 kHz.
3. Reset repro level for –6 VU at 700 Hz.
4. Do not adjust low-frequency playback EQ until after record adjustments.

B. *15-ips playback alignment for regular tape using a standard full-track 15-ips alignment tape.*

1. Follow steps given in A, above, but omit A3.

C. *7 1/2-ips playback alignment for low-noise, high-output tape, using a standard full-track 7 1/2-ips alignment tape.*

1. Set repro level control so that the 700-Hz tone recorded 10 dB below operating level reads 0 VU.
2. Set 7 1/2-ips high-frequency playback EQ so that the 10-kHz tone reads 0 VU.
3. Do not adjust low-frequency EQ until after record adjustments.
4. Set repro level control so that the 700-Hz tone recorded at operating level reads –6-dB VU.

D. *7 1/2-ips playback alignment for regular tape, using a standard full-track 7 1/2-ips alignment tape.*

1. Follow steps C1 through C3.
2. Set repro level control so that the 700-Hz tone recorded at operating level reads 0 VU.

Record Alignment

Thread the tape to be recorded on the machine.

A. *For all tapes at 15 ips.*

1. Set output selector to the bias position, and set the machine into the record mode on all tracks.

2. Adjust the erase peak control for maximum meter reading.

3. Feed a 1000-Hz tone into the machine inputs.

4. Set the output selector to repro and, beginning with a low bias setting, increase the amount of bias until the meter reading rises to a maximum. Continue increasing the bias level until the meter reading drops by 1 dB.

5. Feed a 700-Hz tone into all machine inputs at a 0-dB level from the console and set the level control so that the meter on the recorder reads 0 VU.

6. Set the output selector to the input position and adjust the record cal control for a 0 VU meter reading.

7. Set the output selector to repro and feed a 10-kHz tone into all machine inputs at 0-VU level from the console, and set the 15-ips record EQ so that the meter reads 0 VU (at the peak of the meter swing if the needle is unsteady).

8. Feed a 50-Hz tone into the machine inputs at 0-VU level. Adjust the 15-ips low-frequency playback EQ for a 0-VU meter reading.

B. *For all tapes at 7 1/2 ips.*

1. Follow steps A1 and A2 above.

2. Set the output selector to repro and feed a 500-Hz tone to all the machine inputs. Set all tracks into the record mode and increase the bias level to obtain maximum meter reading. Increase bias so that meter reading drops slightly below the peak reading.

3. Feed a 700-Hz tone at 0 VU to all machine inputs and set the level control for a 0-VU reading on the meter.

4. Set the output selector to input and set the record cal control for a 0-VU reading on the meter.

5. Reduce the setting of the record level control so that the meter reads −10-dB VU.

6. Set the output selector to repro and adjust the repro level control for a 0-VU reading on the meter.

7. Feed a 10-kHz tone into the machine inputs at 0 VU and adjust the 7 1/2-ips record EQ to obtain a meter reading of 0 VU.

8. Feed a 50-Hz tone into the machine inputs at 0 VU and adjust the 7 1/2-ips low-frequency playback EQ for a 0-VU meter reading.

9. Feed a 700-Hz tone into the machine inputs at 0 VU, set the output selector switch to input, and adjust the record level control for a reading of 0 VU on the meter.

continues

Chart 5.1. continued

10. Reduce the repro level control to approximately the position set with the operating level tone on the playback alignment tape. Then switch the output selector to the repro position and adjust the repro level control for a 0-VU meter reading. (Follow step 10 in the order stated to prevent pinning and possibly damaging the VU meters on the tape machine.)

The primary objective of the calibration process is to provide a standard reference for the levels and equalization of an ATR. This is accomplished through the use of a reproduce alignment tape, which is available in various tape speeds and track-width configurations and contains the following set of recorded materials:

1. Standard reference level—A 700-Hz or 1-kHz signal recorded at a standard reference flux level, 185 nWb/m-standard operating level, 250 nWb/m-elevated level, 320 nWb/m-DIN (European operating level).

2. Azimuth adjustment tone—15 kHz for a duration of 30 seconds.

3. Frequency response—Tones at 12 kHz, 10 kHz, 7.5 kHz, 5 kHz, 2.5 kHz, 1 kHz, 500 Hz, 100 Hz, 50 Hz, and 30 Hz.

In calibrating the reproduce ATR electronics, put the output switching functions on all channels in the reproduce mode. Place the test tape on the take-up turntable (it is always wise to store this tape tails-out for the purpose of an even wind tension) and rewind the tape onto the supply reel. With the level adjustment controls exposed, place the ATR in the play mode and adjust for a proper reference level and flat frequency response. The high-frequency reproduce EQ is best set such that the 10-kHz tone will read at reference level, although small deviations in response may require special tailoring.

A 7 1/2-ips reproduce alignment tape is similar to the 15-ips one, except that all tones are recorded at 10 dB below normal operating level. This is done in order to account for the increased saturation of high frequencies at lower tape speeds. The last recorded signal is a 700-Hz or 1-kHz signal at standard operating levels for proper level adjustment.

Most reproduce alignment tapes are originally recorded full-track for all track formats; that is, the recorded signal is laid down across the entire width of the tape. On an ATR with two or more tracks, the reproduction of the low-frequency equalization response cannot be properly set. This is due to an effect called *fringing* that occurs when a tape of one configuration is played back using a gap that is narrower than that used in recording the signal on tape. Signals having a longer wavelength (below 500 Hz) will pick up both the magnetic flux present at the head gap and any extraneous flux extending above or below the head's gap length, adding to the signal output at lower frequencies. For this reason, these test-tape frequencies exhibit a bass boost, ranging upward to +3 dB at 30 Hz. To avoid the inaccuracies of fringing, the low-frequency EQ settings are postponed until after the record alignment has been made.

The preceding process may now be precisely duplicated with each output of the ATR switched into the sync mode. By adjusting the appropriate sync trim pot, all level and EQ controls can likewise be aligned. After all the reproduce alignment adjustments have been made, rewind the alignment test tape again and play or slow-wind it without interruption onto the take-up reel for storage in the tails-out position.

The setup for the record alignment stage consists of placing a fresh reel of unrecorded tape on the machine. Ideally, this tape will be the same reel that is used in the actual recording session; if not, it should at least have the same manufacturer and formulation.

The first setting made in the routine adjustment of an ATR's record electronics is that of the *bias-adjustment control*. This important adjustment determines the amount of AC-bias signal that will be mixed with the recorded signal at the record head, and it has a direct relationship to the noise-to-distortion ratio of the machine. Too little bias signal results in an increased noise floor, increased distortion, and a rise at high frequencies. Too much bias results in reduced output at high frequencies. Thus, bias provides for a compromise among noise, distortion, and frequency response, with the optimum being at some intermediate point.

To set the bias control, thread the proper tape on the machine and feed a signal of 1 kHz at 15 or 30 ips or 500 Hz at 7 1/2 ips to all inputs of the ATR at operating level (0 VU on the console). With the track output selectors in the reproduce mode, switch the transport to record, using an appropriate adjustment tool to locate the individual track controls marked *Bias* or *Bias Adjust*. The ATR's output level can then be monitored by way of the VU meter bridge. Turn the bias trim counter-clockwise until the test signal drops to its lowest level. At this point, slowly turn the trim pot clockwise until the signal rises to a peak reading, after which the reading begins to fall. Continue to rotate the trim clockwise, until the VU drops 1 dB in level. The bias level will now be optimized for the specific tape head and formulation being used. Continue to adjust bias controls for the number of required channels. If the recorded signal begins to peak off the VU scale, reduce the record input-level trim pot. Do not adjust the reproduce trim after it has been set to a standard operating level.

The additional 1-dB setting is known as *overbiasing*; it has been determined to be a precise optimal compromise point. It should also be noted that a common practice in Europe is to overbias by 10 dB, using an input frequency of 10 kHz. This method claims to give a highly accurate bias reading.

After the bias-adjust controls have been set, adjust the record-level and equalization controls. With the tape rewound to the beginning of the reel, a 1-kHz signal at 0 VU is routed from the console to all inputs of the ATR. With all channel electronics switched to reproduce, place the transport in the record mode and adjust the record level for all channels until each meter reads 0 VU. Place the ATR in the stop mode and with all channel electronics switched to read the input signal, adjust each meter level control to read 0 VU. The meter and output levels between the reproduce and input modes will now be precisely matched. The final step is to adjust the record-equalization controls. In this adjustment, place the ATR back in the record mode. While monitoring the electronics, send tone signals of 100 Hz, 1 kHz, and 10 kHz to all ATR inputs at 0 VU. The individual high-equalization (10-kHz) and low-equalization (100-Hz) trims are adjusted to 0 VU in order to attain the flattest possible frequency response over the audio spectrum.

CHAPTER

Digital Audio
Technology

Over the last decade, digital audio has evolved from an infant technology, available to only a few, to its present position as a primary driving force in audio production. In fact, digital audio—along with electronic music production—has had a direct and profound impact on the art, application, and technology of almost every aspect of the audio industry.

Digital audio theory isn't difficult to understand. At its most basic level, it is a means of encoding data through the use of the binary number system. Just as humans communicate by combining any of 26 letters together into groups known as *words*, and by manipulating numbers using the decimal (base^{10}) system, the system of choice for a digital device is the binary (base^2) system. The binary system provides a fast and efficient means for manipulating and storing digital data. By translating the alphabet, base^{10} numbers, or other information types into binary form (represented as on/off, voltage/no voltage, magnetic flux/no flux, or logical 1 and 0 conditions), a digital device (such as a computer or central processor) can perform calculations and tasks that might otherwise be cumbersome, less cost-effective, or downright impossible to perform in the analog domain. After a task has been performed, the results can be changed back into a representative form that we humans can readily understand.

Before moving on, let's take another look at this concept. If you were to type the letters *C, A,* and *T* into a word processor, the computer would translate your keystrokes into a series of 8-bit digital words that would be represented as [0100 0011], [0100 0001], and [0101 0100]. These "alpha-BITS" don't have much meaning when examined individually. However, when placed together into a group, they represent a four-legged animal that's either seldom around or always under foot (see Figure 6.1). When these overall word groupings are pieced together so that they flow in a logical manner, a meaningful message is conveyed. Similarly, a digital audio system works by *sampling* (measuring the instantaneous voltages of) an analog signal and converting these samples into a series of digitally encoded words. Upon reproduction, this stream of words is converted back into a series of voltages that represent the original analog signal.

Figure 6.1. *Digital and analog equivalents for a strange four-legged animal.*

C, A & T = (0100 0011) (0100 0001) (0101 0100) =

(alpha-bits)   (digital words)   (analog equivalent)

The Basics of Digital Audio

In Chapter 2, you learned about the two most basic characteristics of sound: frequency (the component of time) and amplitude (the overall component of signal level). Digital audio can be broken down into two analogous components: sampling (time) and quantization (level).

Sampling

Analog recording technology implies the recording, storage, and reproduction of changes in signal level that are continuous in nature (see Figure 6.2). The digital recording process, on the other hand, doesn't operate in a continuous manner. Rather, digital recording takes periodic samples of a changing audio waveform (see Figure 6.3) and transforms these sampled signal levels into a representative stream of binary words that can be manipulated or stored for later reproduction.

Figure 6.2. An analog signal is continuous.

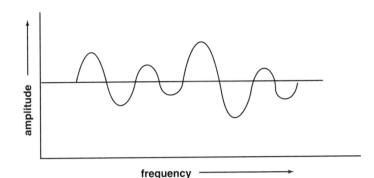

Figure 6.3. A digital signal employs periodic sampling to encode information.

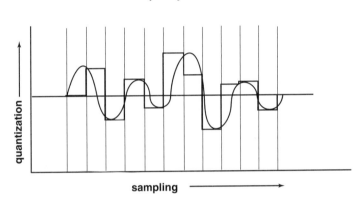

Within a digital audio system, the *sampling rate* is defined as the number of measurements (samples) that are taken of an analog signal in one second. Its reciprocal, *sampling time,* is the elapsed time between each sampling. For example, a sample rate of 48 kHz corresponds to a sample time of 1/48,000th second. Because sampling is tied directly to the component of time, the sampling rate of a system determines its overall bandwidth, with higher sampling rates yielding an increased upper-frequency limit.

During the sampling process (see Figure 6.4), an incoming analog signal is sampled at discrete time intervals (determined by the sample rate). At each interval, this analog signal is momentarily "held" for observation and therefore represents a specific, measurable voltage level. During this "sample and hold" period, a mathematical conversion process is used to generate a digital word of n-bits that represents this signal level as closely as possible at that instant in time. After this conversion has been made, this digital word may be processed or stored, at which time the system is ready to sample the next voltage level, and the process is repeated.

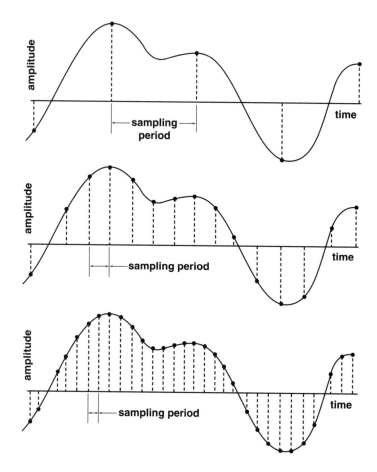

Figure 6.4. *Discrete time sampling.*

The Nyquist Theorem

According to the *Nyquist Theorem,* in order to digitally encode the desired frequency bandwidth, the selected sample rate must be at least twice as high as the highest recorded frequency (sample rate 2 × highest frequency). Thus, an audio signal with a bandwidth of 20 kHz would require a sampling rate of at least 40,000 samples/second. In addition, it's of equal importance that no audio signal greater than half the sampling frequency enter into the digitization process. If frequencies greater than one-half the sample rate are allowed to enter into the conversion process, erroneous frequencies—known as *alias frequencies* (see Figure 6.5)—would enter into the audible audio signal band as false frequencies and produce audible harmonic distortion.

In order to eliminate the effects of aliasing, a low-pass filter is placed before the analog-to-digital (A/D) conversion stage. In theory, a filter that would pass all frequencies up to the Nyquist cutoff frequency and have infinite attenuation thereafter would be ideal (see Figure 6.6a). In the real world, however, such a "brick wall" filter doesn't exist. For this reason, a slightly higher sample rate must be chosen in order to account for the attenuation slope required for the filter to be effective (see Figure 6.6b). A sample rate of 44.1 kHz, for example, has been chosen in order to accurately encode an effective bandwidth up to 20 kHz.

Figure 6.5. *Alias frequencies introduced into the digital audio chain.*

a. *Sampled frequencies above the Nyquist half-sample frequency limit.*

b. *Alias frequencies introduced into the audio band.*

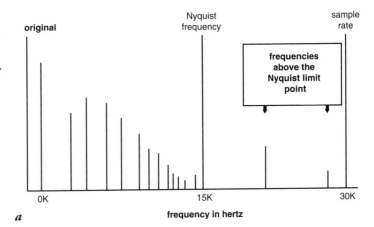

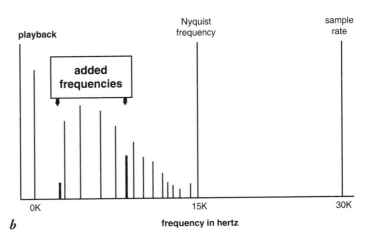

Figure 6.6. *Anti-alias filtering.*

a. *Ideal filter having infinite attenuation at the 20,000 Hz Nyquist cut-off frequency.*

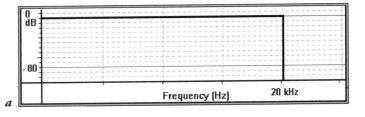

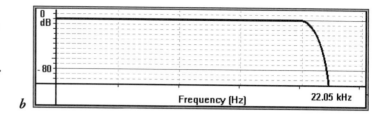

Figure 6.6. *(continued)*
b. Real-world filters require an additional bandwidth of 2.05 kHz in order to fully attenuate unwanted frequencies above the 1/2 bandwidth Nyquist limit.

Oversampling

Oversampling is a process commonly used in professional and consumer digital audio systems to improve anti-aliasing filter characteristics. Oversampling has the effect of further reducing intermodulation and other forms of distortion).

Whenever oversampling is employed, the effective sampling rate in a signal processing block is multiplied by a specified factor—commonly ranging between 12 and 128 times the original rate. This significant increase in the sample rate is accomplished by interpolating sampled level points between the original sample times. This technique, in effect, makes educated guesses as to where the sample levels would fall at the new sample time and generates an equivalent digital word for that level. This increased sample rate likewise results in a much wider frequency bandwidth—so much so that a simple, single-order filter can be used to cut off the frequencies above the Nyquist limit. By shifting the sample rate down to its original value, the bandwidth of the Nyquist filter is likewise narrowed to such a degree that it approximates a much more complex and expensive cutoff filter.

Quantization

Quantization represents the amplitude component of the digital sampling process. It is the technique of translating the instantaneous voltage levels of a continuous analog signal into discrete sets of *binary digits* (bits) for the purpose of manipulating and/or storing this information in the digital domain. The amplitude of the incoming analog signal is broken down into a series of discrete voltage steps. Each step is then assigned an analogous set of binary numbers that are arranged together to form a binary word (see Figure 6.7). This representative word encodes the signal level with as high a degree of accuracy as can be permitted by the word's bit length and the system's overall design quality.

Currently, the most common binary word length for professional audio is 16-bit (for example, 0110010100101101); however, systems having 20- and 24-bit capabilities are also currently available.

Digital signal processors (such as reverb, equalizers, and dynamics processors) often perform their internal calculations using a 24- or 32-bit wordlength structure. This keeps calculation errors to a minimum by increasing the effective processing headroom. Because errors are most likely to accumulate within the least-significant bits (the final and smallest numeric value within a digital word), these bits can be dropped at the device's output. The final result is 16-bit data that is relatively free of errors. This leads to the conclusion that greater word lengths translate directly into an increased resolution due to the added number of finite steps into which a signal can be digitally encoded. The following example details a number of encoding steps for the most commonly encountered bit lengths:

8-bit word	=	(nnnn nnnn)	=	256 steps
16-bit word	=	(nnnnnnnn nnnnnnnn)	=	65,536 steps
20-bit word	=	(nnnnnnnnnn nnnnnnnnnn)	=	1,048,576 steps
24-bit word	=	(nnnnnnnnnnnn nnnnnnnnnnnn)	=	16,777,216 steps

Figure 6.7. *The instantaneous amplitude of the incoming analog signal is broken down into a series of discrete voltage steps, which are then assigned an equivalent set of grouped binary numbers.*

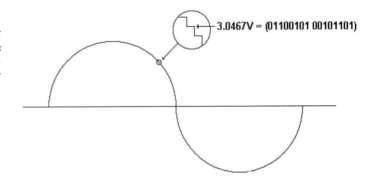

3.0467V = (01100101 00101101)

Signal-to-Error Ratio

Although an analog signal is continuous, the process of quantizing a signal into an equivalent digital word is not. Because the accuracy of the quantization process is limited by the number of discrete steps that can be encoded within a digital word, the representative digital word can only be a close approximation of the original analog signal level.

Signal-to-error ratio is closely akin, although not identical, to signal-to-noise ratio. Whereas signal-to-noise ratio is used to indicate the overall dynamic range of an analog system, the signal-to-error ratio of a digital audio device indicates the degree of accuracy used when encoding a signal's dynamic range with regard to the step-related effects of quantization. Given a properly designed system, the signal-to-error ratio for a signal coded with N bits is

$$S/E = 6N + 1.8 \text{ (dB)}$$

For a 16-bit system, this would yield an error figure of 97.8 dB—a value about 30 dB below the noise figure for most conventional analog tape recorders.

Dither

Through the addition of small amounts of *white noise* (a signal that includes the random distribution of all frequencies at equal levels across the entire audio spectrum), it is possible to further reduce signal-to-error and distortion to figures below their standard levels. By adding this noise (known as *dither*), it's possible to encode signals that are less than the least-significant bit (that is, less than a single quantization step).

Although a small amount of noise is introduced into the circuit, the result is far preferable to the increased quantization distortion that would otherwise result.

The Digital Recording/Reproduction Process

The following sections provide a basic overview of the various stages encountered within the process of encoding analog signals into equivalent digital data (see Figure 6.8a) and converting digital data back into its original analog form (see Figure 6.8b).

Figure 6.8. *The digital audio chain.*

a. *Record.*

b. *Reproduction.*

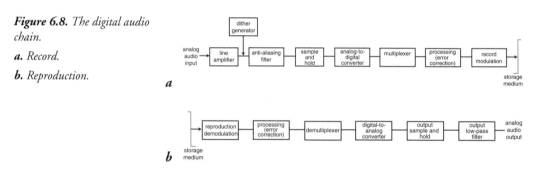

The Recording Process

In its most basic form, the digital recording chain includes a low-pass filter, a sample-and-hold circuit, an analog-to-digital converter, and the circuitry for signal coding and error correction.

At the input of a digital sampling system, the analog signal must be band-limited with a low-pass filter so as not to allow frequencies above half the sample rate frequency to pass into the A/D conversion circuitry. Such a stop-band (*anti-aliasing*) filter, generally employs a gradual roll-off slope at its high-frequency cut-off point because an immediate or "brick wall" slope would introduce severe signal distortion and phase shifts. As a result, the Nyquist Theorem requires that a sampling rate be chosen that is higher than twice the highest frequency to be recorded in order to accurately encode the full bandwidth. For example, a system with a bandwidth reaching up to 20 kHz is often sampled at a rate of 44.1K or 48K samples/second.

Following the low-pass filter, a sample-and-hold (S/H) circuit holds and measures the analog level for the duration of a sample period, until the analog-to-digital (A/D) converter is able to generate a corresponding digital word. Timing information, which determines the sample-rate frequency, is generated by a crystal reference clock. During this sample period, the analog-to-digital conversion process begins. A/D conversion is a critical component of the digitization process, in that the sampled DC voltage level at the S/H circuit must be quantized to the nearest step level.

At this point, computations must be performed to translate the sampled level into an equivalent binary word for further digital conditioning and storage—all within a single sample period (commonly less than 1/44,100th of a second).

After the signal has been converted into digital bit form, the data must be conditioned for further data processing and storage. This conditioning includes data coding, data modulation, and error correction. In general, the binary digits aren't directly stored on a recording medium. Rather, *data coding* is used to encode this data (along with synchronization and address codes) into a form that allows for the most efficient means of storing digital data onto a medium and for attaining a maximum degree of data density. The most common form of digital modulation is *pulse-code modulation,* or PCM (see Figure 6.9).

Figure 6.9. *Pulse code modulation.*

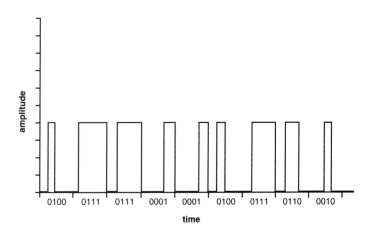

The density of the information in PCM recording and playback equipment is extremely high (about 20,000 bits per inch at 15 ips on a fixed-head PCM deck), so any imperfections, such as dust adhering to the surface of the magnetic recording medium, will readily generate error signals. To reduce these errors to within acceptable limits, a form of error correction is used. Several forms of error correction may be employed, depending on the media type. One method uses redundant data in the form of parity bits and check codes, while a second means of error correction involves interleaving techniques in which data is scattered across the digital bit stream in order to reduce the effects of dropouts (see Figure 6.10).

Figure 6.10. *An example of interleaved error correction.*

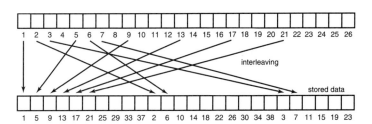

The Reproduction Process

The digital reproduction chain essentially works in a complementary manner to the standard encoding process.

Because most digital media encodes data onto magnetic media in the form of highly saturated transition states, the reproduced signal must be reconditioned so as to restore the digital bitstream back to its originally modulated binary state. Once this has been done, the data is then de-interleaved back into its original form where it can be easily restored back into PCM data. It's a fact of life that without error correction the quality of the digital audio medium would be greatly reduced or (in the case of the CD) almost useless.

After the signal has been reconstituted into its original PCM data form, the process of digital-to-analog (D/A) conversion can take place. Often, a *stepped resistance network* (sometimes called an R/2R network) is used to convert each word back into an analogous voltage. During a complementary sample-and-hold period, each bit in the word under current conversion (moving from the most-significant to the least-significant bit) is assigned to a leg in the network. Each leg is designed to pass one-half the reference voltage level that can be passed by the previous step (see Figure 6.11). The presence or absence of a logical "1" in each step determines how much of the overall voltage source will be summed together and passed on to the converter's output.

Figure 6.11. *A stepped resistance network is a common device for accomplishing D/A conversion by assigning each word bit to a series of resistors that are scaled by factors of two.*

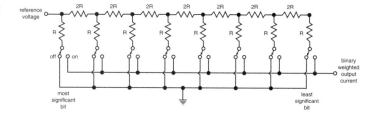

Following the conversion process, another complementary low-pass filter is inserted into the signal path. This filter smooths out non-linear steps introduced by the sampling process. The final result is a smooth waveform that (when the circuit is properly designed) will faithfully represent the originally recorded analog waveform.

Digital Transmission

In this digital age, it has become commonplace for digital audio data to be distributed from one device to another or throughout a connected production system in the digital domain. Using this medium, digital audio information can be transmitted in its original numeric form and thus, in theory, will not suffer from degradation whenever copies are made from one generation to the next.

When looking at the differences between the distribution of digital audio and analog audio, you should keep in mind that, unlike its counterpart, the transmitted bandwidth of digital audio data occurs in the megahertz range. Therefore, digital audio transmission has more in common with video signals than the lower-bandwidth analog audio range. This means that impedance must be much more closely matched and that quick-fix solutions (such as using a Y-cord to split a digital signal between two recorders) are a major no-no. Failure to follow these precautions could seriously deform the resultant digital waveform.

Because of these tight signal restrictions, several digital transmission standards have been adopted that uniformly allow for the quick and reliable transmission of digital audio between devices that support these standards. The two most commonly encountered formats are the AES/EBU and S/PDIF protocols.

AES/EBU

The AES/EBU (Audio Engineering Society and the European Broadcast Union) protocol has been adopted for the purpose of transmitting digital audio between professional digital audio devices. This standard is used to convey two channels of interleaved digital audio through a single, three-pin XLR microphone cable. This balanced configuration connects pin 1 to the signal ground, while pins 2 and 3 are used to carry signal data. AES/EBU transmission data is low impedance in nature (typically 110) with waveform amplitudes that range between 3 and 10 volts. These factors allow a maximum cable length of up to 328 feet (100 meters) to be used without encountering undue signal degradation.

Digital audio channel data and sub-code information is transmitted in blocks of 192 bits that are organized into 24, 8-bit words. Within the confines of these data blocks, two sub-frames are transmitted during each sample period that convey information and digital synchronization codes for both channels in a L-R-L-R... fashion. Because the data is transmitted as a self-clocking bi-phase code (see Figure 6.12), wire polarity can be ignored, and the receiving device will receive its reference clock timing from the digital source device.

Figure 6.12. *AES/EBU subframe format.*

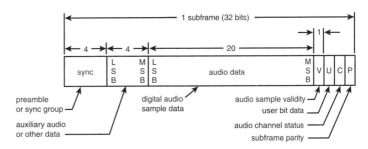

S/PDIF

The S/PDIF (Sony/Phillips Digital Interface) has been adopted for transmitting digital audio between consumer digital audio devices; it is similar in data structure to its professional counterpart.

Instead of using a balanced 3-pin XLR cable, the S/PDIF standard has adopted the single conductor, unbalanced phono (RCA) connector, which conducts nominal peak-to-peak voltage levels of 0.5 volts with an impedance of 75 ohms. In addition, transmissions via optical lines that use the Toslink optical connection cable generally use the S/PDIF data protocol.

As with the AES/EBU protocol, S/PDIF channel data and sub-code information is transmitted in blocks of 192 bits; however, they are organized into twelve 16-bit words. A portion of this information is reserved as a category code that provides the necessary setup information (sample rate, copy protection status, and so on) to the copy device. A portion of the 24 bits set aside for transmitting audio data is used to relay track indexing information, such as start ID and program ID numbers, enabling this relevant information to be transferred from the master to the copy. It should be noted that the professional AES/EBU protocol isn't capable of digitally transmitting these codes during transfer.

SCMS

Initially, the DAT medium was intended to provide consumers with a way of making high-quality digital recordings for their own personal use. Soon after its inception, however, for better or for worse, the recording industry began to see this new medium as a potential source of royalties lost due to home copying and piracy practices. As a result, the RIAA (Recording Industry Association of America) and the former CBS Technology Center set out to create a "copy inhibitor." After certain failures and long industry deliberations, the result of these efforts was a process that has come to be known as the *Serial Copy Management System* or SCMS.

SCMS (pronounced *scums*) has been implemented in many consumer digital devices to prohibit the unauthorized copying of digital audio at the 44.1 kHz sample rate (standard CD rate). It doesn't apply to the making of analog copies, to digital copies made using the AES/EBU protocol, or to sample rates other than 44.1 kHz.

So what is SCMS? Technically, it is a digital protection flag that is encoded in byte 0 (bits 6 and 7) of the S/PDIF's subcode area. This flag can have only one of three possible states:

- *Status 00:* No copy protection, allowing unlimited copying and subsequent dubbing.
- *Status 10:* No more digital copies allowed.
- *Status 11:* A single copy can be made of this product, but that copy cannot be copied.

Suppose that you have a CD player with an optical output and two consumer DAT machines equipped with SCMS. If you try to digitally copy a CD that has a 10 SCMS status, you would simply be out of luck. But suppose that you found a CD that has an 11 status flag. By definition, the bitstream data would inform the initial DAT copy machine that it's OK to record the digital signal. However, the status flag on the copy tape is then changed to a 10 flag. If at a later time you were to clone this DAT copy, the machine doing the second generation copy wouldn't allow itself to be placed into Record. At this point, you have two possible choices: you can record the signal using the analog ports (often with a minimum of signal degradation), or you can purchase a digital format converter that (among other things) enables you to strip the SCMS copy protection flags from the bitstream and continue to make multigeneration copies.

Signal Distribution

Both the AES/EBU and S/PDIF digital audio signals can be distributed from one digital audio device to another in a daisy-chain fashion (see Figure 6.13). This method works well if only a few devices are to be chained together. If a number of devices are connected together, time-base errors (known as *jitter*) may be introduced into the path, with possible side effects being added noise and a "blurred" signal image. One way to reduce potential time-based errors, is to use a digital distribution device that can route data from a single digital source to a number of individual device destinations (see Figure 6.14).

Figure 6.13. Digital audio can be distributed in a daisy-chain fashion.

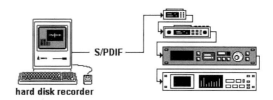

Figure 6.14. A distribution system can be used to route digital audio data to individual devices.

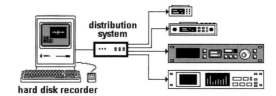

What Is Jitter?

Jitter is a controversial and widely misunderstood phenomenon. It has been explained best by Bob Katz of Digital Domain (NYC) in his article "Everything You Always Wanted to Know about Jitter but Were Afraid to Ask." The following is a brief excerpt:

> Jitter is time-base error. It is caused by varying time delays in digital audio circuit paths from component to component. The two most common causes of jitter are poorly designed phase-locked loops (PLLs) and waveform distortion due to mismatched impedances and/or reflections in the signal path.
>
> The waveform shown in [Figure 6.15a] represents a theoretically perfect digital source. Its value is 101010, occurring at equal slices of time and represented by the equal-spaced, dashed vertical lines. When the first waveform passes through long cables of incorrect impedance, or when a source impedance is incorrectly matched at the load, the square wave becomes rounded, and fast rise times become slow. Reflections in the cable can cause misinterpretation of the actual zero crossing point of the waveform.

The second waveform (see Figure 6.15b) shows some of the ways in which the first might change. Depending on the severity of the mismatch, you might see a triangle wave, a square wave with ringing, or simply rounded edges. Notice that the new transitions (measured at the zero line) in the second waveform occur at unequal slices of time. Even so, the numeric interpretation of the second waveform is still 101010! There would have to be very severe waveform distortion for the value of the new waveform to be misinterpreted, which usually shows up as audible clicks or tics in the sound. If you hear tics, you really have something to worry about.

Figure 6.15. *Example of time-base errors.*

a. *A theoretically perfect digital signal source.*

b. *The same signal with jitter errors.*

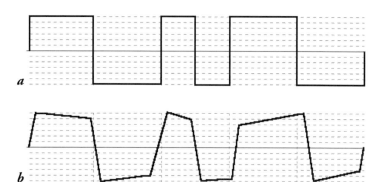

If the numeric value of the waveform is unchanged, why should you be concerned? Let's rephrase the question: "When (not why) should you become concerned?" The answer is "hardly ever." The only effect of time-base distortion is in the listening; it does not affect the dubbing of the tapes or any digital-to-digital transfer (as long as the jitter is low enough to permit the data to be read).

A typical D-to-A converter derives its system clock (the clock that controls the sample-and-hold circuit) from the incoming digital signal. If that clock is not stable, the conversion from digital to analog will not occur at the correct moments in time. The audible effect of this jitter is a possible loss of low-level resolution caused by added noise, spurious (phantom) tones, or distortion added to the signal.

A properly dithered 16-bit recording can have over 120 dB of dynamic range; jitter can deteriorate that range to 100 dB or less, depending on the severity of the jitter. We have performed listening experiments on purist, audiophile-quality musical source material recorded with a 20-bit accurate A/D converter (dithered to 16 bits within the A/D).

The sonic results of passing this signal through processors that truncate the signal at −110, −105, or −96 dB are as follows:

◆ Increased "grain" in the image

◆ Instruments lose their sharp edge and focus

◆ Apparent loss of level, causing the listener to want to turn up the monitor level (even though high-level signals are reproduced at unity gain)

Contrary to intuition, these effects are audible without having to turn up the listening volume beyond normal! Nevertheless, the effects are subtle and require the highest-grade audiophile musical material to be audible at all (as well as high resolution D-to-A converters), and some engineers [may] not deem the effects to be audible with most music recorded today. Jitter in the signal path can produce a subtle loss of resolution similar to the above.

Addendum to article:

Since the preceding article was written, evidence has been uncovered that there may be subtle accumulation of jitter on digital copies. A 20-bit digital recording has enough resolution to render the additional jitter in the copy more audible than was previously believed. This is obviously a very controversial topic. One theory proposes that the power supply powering the recorder's input PLL carries a small residual of the source jitter components to subsequent stages even as the bits are stored onto tape. Simply put, at the heart of the technological matter is the fact that good, high-speed digital circuits are just as difficult to design as good analog circuits. In high-speed digital circuits, grounds begin carrying residual high-frequency energy; if the designer isn't careful, this high-frequency energy can be passed on to other stages—a phenomenon known as *logic-induced modulation*.

Many professional engineers doubt whether this residual "stored" jitter is audible on playback or later dubs, but numerous golden-eared engineers have noted deterioration on dubbing. Careful tests, however, have proved that the bits have been copied accurately during the dubbing, so this is not an issue. A powerful jitter attenuation circuit inserted during playback of the final generation (or the pressed CD) helps to render the output virtually indistinguishable from the original generation. But to satisfy the qualms of producers and engineers who have noticed sonic differences between the CD and the master—even when played through the same D/A converter— a major record company has purchased $50,000 of jitter-analysis equipment. The company is attempting to determine the mechanisms that cause jitter on CD pressings and whether any jitter in the masters or intermediate generations actually contribute to jitter on the pressing. Stay tuned for more developments.

Fixed-Head and Rotating-Head Digital Audio Recorders

Digital audio recording systems have become increasingly popular and cost-effective in recent years. These devices are based on two distinct technologies: fixed head and rotating head systems.

The Fixed-Head Digital Audio Recorder

The fixed-head digital audio recorder is a reel-to-reel system that often emulates its analog counterpart in form and function. However, all similarities end there. These recording systems use digital audio conversion and special data encoding structures to store digital audio data on specially formulated digital audio tape in a longitudinal format. Although two-channel fixed-head recorders do exist, the majority of fixed-head systems in current operation are multitrack. Of these systems, the most commonly found recorders are 24- and 48-track systems that employ the Digital Audio Stationary Head (DASH) format, which was ratified and jointly announced by Sony, Willi Studer AG, and Matsushita Electric Industries.

The DASH Format

The DASH format was developed to help ensure standardization between different generations of digital ATR and tape manufacture.

Format Structure

The format consists of fast, medium, and slow versions (see Figure 6.16), the choice of which is determined by the tape speed of the recorder. One track is necessary to record a single audio channel for the fast version, two tracks are necessary for the medium version, and four tracks are necessary for the slow version.

Figure 6.16. DASH format tape speed and sampling rate. *(Courtesy of Sony Professional Audio)*

sampling rate	tape speed		
	fast	medium	slow
48 kHz	30 ips	15 ips	7.5 ips
44.1 kHz	27.56 ips	13.78 ips	6.89 ips

Tracking Density

A recording with double the track density, as shown in Figure 6.17, can be made with the state-of-the-art, thin-film heads, while maintaining a recording compatibility with standard track densities. Although not shown in Figure 6.17, the slow version for 1/2-inch tape is a viable choice.

Error Correction

The operation of the encoders shown in Figure 6.18 is based on the Cross Interleave Code (CIC)— with increased interleaving made between the even- and odd-numbered words. The CIC can correct errors corresponding to a maximum of three words.

Figure 6.17. *DASH format track density and channel numbers. (Courtesy of Sony Professional Audio)*

tape width	1/4 inch		1/2 inch	
track density	normal	double	normal	double
digital tracks	8	16	24	48
aux tracks	4	4	4	4
digital audio channels — fast	8	16	24	48
digital audio channels — medium	–	8	–	24
digital audio channels — slow	2	4	–	–

Figure 6.18. *The three tape speed versions of the DASH format (encoders and decoders of all versions are identical). (Courtesy of Sony Professional Audio)*

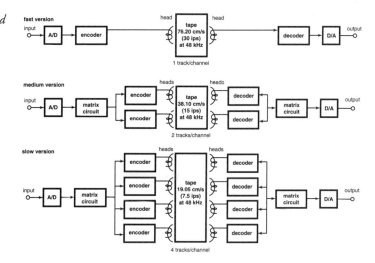

The interleaving of even and odd words allows for tape-splice edits while protecting against possible damage to the recorded tape. The correctability of *burst* (large-scale) errors is determined by the encoders and is the same for all three versions. Encoding and decoding of the error-correction codes are done independently for each track. If excessive errors appear on one of the recorded tracks (as with a dropout), the correction capabilities of the other tracks aren't affected—a feature that safeguards the recording under adverse conditions.

Editing

All functions of crossfading, punch in/out, tape-splice, and electronic editing operations are feasible under this system, which allows for a natural continuation of recorded sound.

DASH Recording Systems

Recorders using the DASH format are commonly available in 2-, 24- and 48-track configurations. With the development of new large-scale integrated circuits (LSIs), smaller and lighter recorders that are lower in both cost and power consumption than their first-generation models are currently being produced.

Both Sony and Studer have developed a series of popular multichannel digital audio recorders that use the DASH format. Both the Sony 24-track PCM-3324S (see Figure 6.19) and the Studer D 827 MCH-24 (see Figure 6.20) utilize 1/2-inch tape running at a speed of 30 ips and with a sample rate of 48 kHz. The 24 digital tracks are located across the width of the tape, with 2 outside analog tracks and an additional center control or external data track. The error-detection circuitry of both of these recorders allows splice edits to be made by creating a seamless crossfade that interpolates the data before and after the splice. These recorders employ thin-film head technology (see Figure 6.21)—a technology borrowed from integrated circuit fabrication. The headblock mechanism for the Sony 3324 is shown in Figure 6.22.

Figure 6.19. The Sony 3324S digital multitrack recorder. (Courtesy of Sony Professional Audio)

Figure 6.20. *The Studer D 827 MCH digital multitrack recorder. (Courtesy of Studer Professional Audio)*

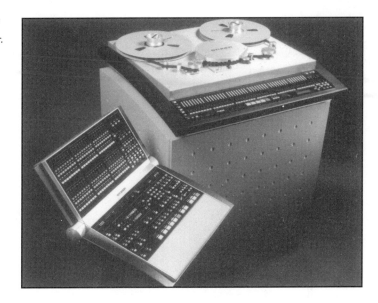

Figure 6.21. *Detail of digital thin-film head construction. (Courtesy of Sony Professional Audio)*

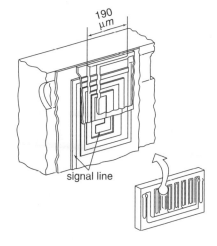

Figure 6.22. *Headblock mechanism of the Sony 3324A. (Courtesy of Sony Professional Audio)*

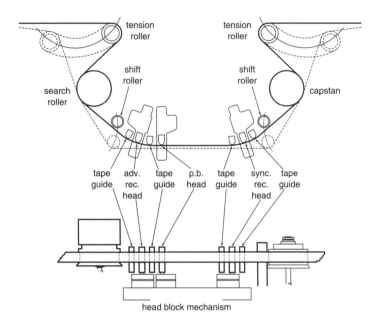

Both the Sony PCM-3348 48-track digital audio recorder (see Figure 6.23), and the Studer D 827 MCH-48 DASH recorder have the added bonus of being fully compatible with the 24-track DASH format, in that tapes previously recorded on a 24-track DASH machine can be recorded and reproduced with no signal degradation using tracks 1-24. The additional tracks (25-48) can then be recorded onto available tracks on the same tape with no compatibility problems.

Figure 6.23. *The Sony PCM-3348 digital audio multitrack recorder. (Courtesy of Sony Professional Audio)*

The Rotating-Head Digital Audio Recorder

Rotating-head digital recorders fall into three categories: digital processor/VCRs, DAT (digital audio tape) systems, and MDMs (modular digital multitrack) recording systems.

The Rotary Head

Due to the tremendous degree of data density required to record/reproduce PCM digital audio (approximately 2.77 million bits/second), the recording of data onto a linear tape track is virtually impossible. In order to operate within such a wide-frequency bandwidth, a rotating-head helical scan path is used to effectively increase the overall head-to-tape contact speed. This process enables the longitudinal tape speed to be substantially lowered, which results in a vast reduction in tape consumption. An example of a transport that employs rotary head technology is shown in Figure 6.24.

Figure 6.24. Helical scan path.
a. Tape path around drum mechanism.
b. Recording of slanted tracks.

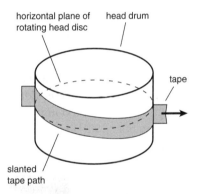

a

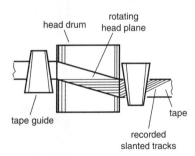

b

Digital Processor-VCR Digital Audio System

The digital processor-VCR digital audio system has (not surprisingly) two components: a PCM A/D and D/A conversion unit (see Figure 6.25) and a video cassette recorder (VCR). The role of the digital processor is to accept analog audio at its input, convert this signal into digital PCM audio, and then modulate this digital information so that it can be encoded within the video field of a VCR. On playback, the encoded data is demodulated and converted back into its original analog form.

Figure 6.25. *The Sony 1630 digital audio processor. (Courtesy of Sony Professional Audio)*

In recent years, PCM-VCR combinations have been taken off the market, and their use in audio production has fallen off drastically due to the DAT (Digital Audio Tape) recorder and computer-based hard-disk recording systems. Editing is impractical with PCM-VCR systems because of their inability to manually splice helically-scanned tape, and many of the features commonly found in systems with DAT technology—such as index search functions—are not available on these systems. The Sony 1630 processor, which is often paired with a 3/4-inch U-Matic VCR, is still a standard system for preparing compact disc final masters.

Digital Audio Tape (DAT) System

The *rotary-digital audio tape*, or R-DAT format (most commonly known as DAT), combines both the rotary head and PCM digital technologies. This combination has enabled the development of a compact, dedicated PCM digital audio recorder (see Figure 6.26) that displays a wide dynamic range, low distortion, and an immeasurable amount of wow and flutter. This recorder is designed to equal or exceed many standard digital specifications.

DAT technology employs an enclosed compact cassette that's even smaller than a compact audio cassette. Equipped with both analog and digital input/outputs, the DAT format can record and play back at three standard sampling frequencies: 32 kHz, 44.1 kHz, and 48 kHz (although sample rate capabilities and system features often vary from one DAT recorder to the next). When using a consumer DAT machine, the 44.1 kHz sampling frequency is often reserved for prerecorded

DAT tapes and is designed to discourage the unlawful copying of prerecorded program material through the implementation of the SCMS copy code system. Current DAT tapes offer running times of up to two hours when sampling at 44.1 kHz and 48 kHz and reserve three record/reproduce modes for the 38-kHz sampling rate. Option 1 provides two hours of maximum recording time with 16-bit linear quantization. Option 2 provides up to four hours of recording time with 12-bit nonlinear quantization. Option 3 enables the recording of four-channel, nonlinear 12-bit audio.

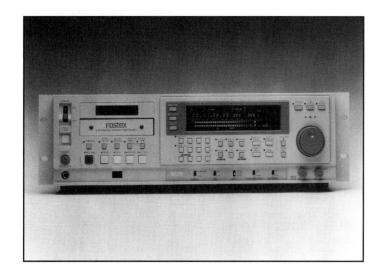

Figure 6.26. Fostex D-10 production DAT recorder. (Courtesy of Fostex Corporation of America)

DAT Tape/Transport Format

The actual track width of the DAT format's helical scan can range downward to about 1/10th the thickness of a human hair, allowing a density of 114 million bits per square inch. This is the first time such a density has been achieved using magnetic media. To assist with tape tracking and therefore to maintain the highest quality playback signal, a sophisticated tracking correction system is used to center the heads directly over the tape's scan path.

The head assembly of a DAT recorder uses a 90° half-wrap tape path. Figure 6.27 shows a two-head assembly in which each head is in direct contact with the tape 25% of the time. Such a discontinuous signal requires that the digital data be encoded within the slant track in digital "bursts," necessitating the use of a digital buffer. On playback, these bursts are again smoothed into a continuous data stream. Such encoding methods have the following advantages:

◆ Only a short length of tape is in contact with the drum at one time. This reduces tape damage and enables a high-speed search to be performed while the tape is in contact with the head.

◆ A low tape tension can be used to ensure a longer head life.

Figure 6.27. *A half-wrap tape path, showing 90° tape contact.*

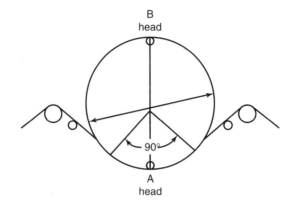

The high-speed search function (approaching 300 times normal speed) is a key feature of the DAT format. In addition to this function, this format makes provisions for nonaudio information to be written into the digital stream's subcode area. This subcode area serves as a digital identifier, just as a compact disc uses subcodes for selection and timing information. These subcodes may be written as any one of three data types: start ID, which indicates the beginning of a selection; skip ID, which indicates a selection should be skipped over; and program number. These identifiers assist the user in finding selections when searching at high speeds. Subcode data is treated independently by the DAT system and may be written or erased at any time without altering the audio program. In addition, the subcode area can be used to encode SMPTE time code for use with audio-for-video and synchronous music production.

Modular Digital Multitrack Systems (MDMs)

One of the most monumental developments in recent recording history has been the introduction of the *modular digital multitrack* system, or MDM (see figure 6.28). MDMs are small multitrack digital audio recorders that are capable of recording eight tracks of digital audio onto standard videotape cassettes that can often be bought at your favorite neighborhood drugstore.

Figure 6.28. *Studio equipped with an MDM system.*

These recorders are called *modular* devices because they can be linked together in a proprietary sync fashion; additional tracks can be added to the system (in blocks of eight), with a theoretical maximum limit of up to 128 tracks! If the capability to create and expand a digital audio recording system to suit your needs isn't enough to pique your interest, you may want to check out the price of an MDM. As of this writing, a basic 8-track digital MDM can be purchased for less than $5,000 (often significantly less). Until recently, the only available digital option was a 12-track rotary head system that cost more than $20,000. So it's easy to see why these modular, expandable systems have begun to revolutionize the music industry—both in the home and in the studio.

ADAT

The ADAT from the Alesis Corporation, shown in Figure 6.29, is an eight-track modular digital multitrack recorder that uses standard S-VHS videotape and includes the following features: 16-bit resolution; sample rates of 40.4 and 50.08, which can be varied over a wide range; two location memories, including a return-to-zero location; auto-loop/rehearse functions that work in both the play and record modes; and a choice of four digital crossfade times. When using a standard 120-minute S-VHS tape at the highest sampling rate, an ADAT yields a total recording time of 40 minutes, 44 seconds. At the time of writing, when a 160-minute tape is used, however, this time is increased to 53 minutes.

Figure 6.29. ADAT modular digital multitrack. (Courtesy of Alesis Corporation)

Analog input/output connections include both 1/4-inch unbalanced jacks, (–10 dBV) as well as ±4 dBu connections that are provided on a single, standard 56-pin Elco connector. Digital I/O connections include a proprietary fiber-optic cable link capable of carrying all eight channels. This cable link is used to connect an ADAT to a digital peripheral I/O device (such as the A1-S/PDIF, A2-AES/EBU digital interface, Digidesign's ProTools interface, Alesis Quadrasynth synthesizer, or the Quadraverb II digital effects processor), or to connect one ADAT to another to make direct digital-tape copies.

Remote functions are carried out through the use of the LRC (Little Remote Control) and BRC (Big Remote Control). The LRC is shipped with every ADAT and contains all the device's front panel transport functions on a single, palm-sized remote. The optional BRC, shown in Figure 6.30, is capable of acting as a full-featured remote control, digital editor, expanded-feature autolocator, and synchronizer on a single tabletop or freestanding surface. It is capable of controlling up to 16 remote ADAT units and can bounce data from one track to another in the digital domain (while also allowing tracks to be shifted in location). An example of track location shifting is the capability to copy a set of background vocals in a song's bridge to a point near the end of the song. By stating a track and location source, as well as a track and location destination, the BRC can handle all transport, record, and audio data requirements in carrying out the necessary functions.

Figure 6.30. Alesis BRC (Big Remote Control) unit for the ADAT. (Courtesy of Alesis Corporation)

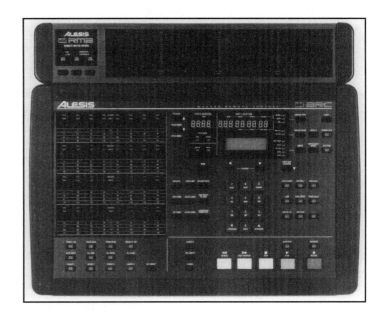

Another ADAT-compatible system is the Fostex RD-8 MDM (see Figure 6.31). The RD-8 uses the same S-VHS transport and data/track encoding technology as the ADAT; however, it operates at sample rates of 44.1 and 48 kHz. These sample rates can be varied over a wide range to match the Alesis ADAT. Other features include an on-board SMPTE synchronizer/reader/generator, Sony 9-pin RS-232 control that allows the unit to be controlled from a video editor or 9-pin compatible device, video sync and word sync I/O, as well as being able to handle pull-down rates of 44.056 and 47.952 (when translating between various film and video frame rates).

Figure 6.31. Fostex RD-8 digital
multitrack recorder. (Courtesy of
Fostex Corporation of America)

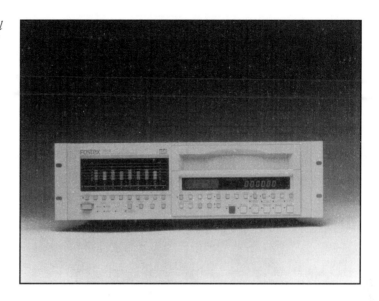

DA-88

The DA-88 from Tascam, shown in Figure 6.32, is an 8-track MDM that uses a *digital tape recording system* (DTRS) to record up to 108 minutes of digital audio onto a standard 120-minute Hi-8mm video tape. In addition to the usual transport functions, the DA-88 includes such controls as tape shuttle, sample rate select, two location points, auto-punch and rehearse modes, pitch change (±6%), and playback delay on any channel (up to 7,200 samples). Similar to the ADAT format, the DTRS format can be built up (in 8-track increments) to a system having a maximum of 128 tracks.

Figure 6.32. Tascam DA-88
digital multitrack recorder with
RC-848 remote control. (Courtesy
of Tascam)

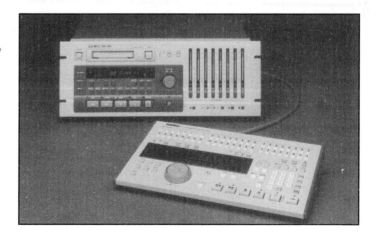

Analog I/O is made via –10 dBV, unbalanced RCA phono jacks, as well as by using balanced +4 dB connections using two 25-pin D-sub multipins. Digital I/O is made through Tascam's proprietary TDIF-1 (Tascam Digital Interface), which uses a 25-pin D-sub connector to link to external interface accessories or to another DA-88 for use with making digital clones.

External synchronization to time code can be accomplished by an optional SY-88 interface card that offers SMPTE chase-lock sync, MIDI machine control, Sony 9-pin RS-422 video editor control, and video I/O sync on standard BNC jacks.

Sony's PCM-800 (see Figure 6.33) also employs DTRS Hi-8mm technology and features both AES/EBU digital and analog XLR audio input/outputs for recording up to eight channels (again up to 16 PCM-800s can be linked together) at sampling rates of either 44.1 or 48 kHz. Full sync capabilities with SMPTE time code are also built into the system.

Figure 6.33. Sony PCM-800 digital multitrack recorder. (*Courtesy of Sony Corporation*)

Third-Party Developments and Accessories for the MDM

A number of accessories are available for both the ADAT and DTRS MDM formats. These accessories can be used to link both the digital audio and synchronization signals to an external device to carry out distribution, processing, and advanced (or cost-effective) sync functions. Examples of these accessories include the following:

◆ The dataMASTER (shown in Figure 6.34) and dataSYNC (not shown) from JLCooper Electronics (Los Angeles) take the output from the ADAT's proprietary control track and derive MIDI time code, which enables external devices, such as synchronizers, analog recorders, and hard-disk recorders, to be cost-effectively locked to the system.

◆ The Steinberg ACI from Steinberg/Jones (Northridge, CA) has been designed for use with Steinberg's Cubase sequencer to add MIDI sync and MIDI machine control to any ADAT system. The ACI works with any sequencer or device that supports MIDI machine control.

◆ The PC-connection enables IBM-compatible computers to directly control ADATs via a hardware/software interface.

◆ Digidesign offers an optical interface (see Figure 6.35) that enables direct transfer of digital audio to and from either a Pro Tools or Session-8 digital audio workstation.

Figure 6.34. dataMASTER *professional synchronizer for the Alesis ADAT. (Courtesy of JLCooper Electronics)*

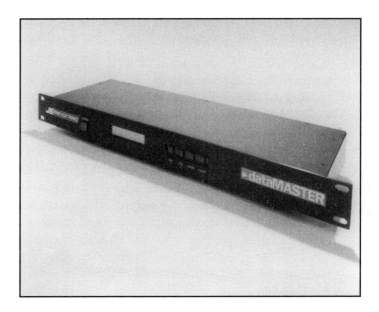

Figure 6.35. Digidesign's ADAT *optical Sync interface. (Courtesy of Digidesign)*

The preceding list offers just a few of the third-party accessories available for the MDM. Supporting third-party companies (as well as the manufacturers themselves) continue to create new accessories. The best way to keep on top of these developments is to contact the manufacturers and to keep reading the various trade magazines.

Digital Audio Sampling Systems

One of the biggest trends in recent audio production has been to merge digital audio with computer technology to create a sample- and samplefile-based approach to sound recording. The encoding of audio data into digital memory or onto a storage medium provides us with a means for storing or manipulating defined blocks of digital data. This data can be stored as a *samplefile* (data that can be imported, played, manipulated and exported to various sampling devices) or as a *soundfile* (providing recorded "tracks" that can be recorded, manipulated, and reproduced from hard disk).

Perhaps the most important difference that can be distinguished between a tape-based system (be it digital or analog) and a sample-based recording system is that the latter is random access. *Random access* production refers to the fact that digital audio data can be stored within a read-only memory (ROM), a random-access memory (RAM), or a disk-/disc-based memory medium in such a way that the data can—virtually instantaneously—be accessed, processed, or reproduced in any order and at any point in time. A random-access medium differs drastically from a tape-based medium in that the tape-based medium is linear in nature and requires time to locate to the various points where the audio data is stored. Neither medium is better than the other, as each has its strong points and inherent weaknesses. In fact, the two production styles complement each other in ways that can boost the effectiveness of a production system.

Samplers

A sampler (see Figure 6.36) is a device that's capable of recording, musically transposing, processing, and reproducing segments of digitized audio directly from RAM. Because this memory storage media often is expensive and limited in size (relative to digital audio's memory-intensive nature), these segments generally are of a limited length (ranging from only a few seconds to one or more minutes).

Assuming that sufficient memory is available, these devices often let you load any number of samples into the system and then access them at any point in time, often polyphonically (whereby more than one sound can be reproduced at any given time). A sampler's real power, however, comes from the fact that these samples can be musically transposed in real time, either up or down, and over a wide number of octave ranges. Quite simply, this musical transposition occurs by reproducing the recorded samplefiles at various sample rates that correspond to established musical intervals.

Figure 6.36. *Kurzweil K2500R sampler module. (Courtesy of Kurzweil/Young Chang America)*

Most samplers contain extensive edit capabilities, which allow sounds to be modified in much the same way as a synthesizer modifies such electronically generated characteristics as LFO (low-frequency oscillation), velocity, envelope processing, keyboard scaling, aftertouch, and full dynamic MIDI control. After a sample has been edited or modified, it can be assigned to a specific note on a MIDI keyboard or controller or mapped across a range of keys in multiple-voice fashion. A sampler with a specific number of voices—for example, 32 voices—simply means that 32 notes can be simultaneously played on a keyboard at any one time. After a set of samples has been recorded or recalled from disk memory, they can be split across a performance keyboard in such a way that individual samples can be assigned to a specific key or range of notes (see Figure 6.37).

Figure 6.37. *A split keyboard setup.*

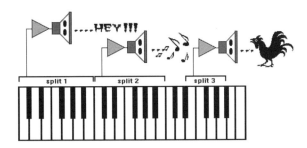

In this day and age, an enormous range of previously recorded and edited samplefiles containing instruments, effects, noises, and so on, can be purchased on disk, CD, and CD-ROM. Such commercially available sample libraries are the mainstay of both electronic musicians and visual postproduction facilities.

Commercial sample libraries need not be the only option. Professional and nonprofessional artists alike often create personal sample libraries. These libraries can be created from original acoustic or electronically generated sound sources, although it's a common practice for samples to be "lifted" from previously recorded source material (such as CD, TV, records, and videotapes).

Sample Editing

When a recorded sound is transferred into a sampling device, the original source material often contains extraneous sounds—breathing noises, fidget sounds, or other music—that occur both before and after the desired sound (see Figure 6.38a). At this point, the unwanted sampled sounds can be edited out of memory by trimming the in-points and out-points. *Trimming* is accomplished by instructing the system's microprocessor to ignore (not access or reproduce) all the samples that exist before a user-defined in-point and/or those samples following a desired out-point (see Figure 6.38b). After trimming, the final sample can be played, processed (if a fade or other function needs to be performed), and saved to floppy or hard disk for later recall.

Figure 6.38. *Sample editing.*
a. *Unedited sample.*
b. *Sample that has been edited and faded at its end.*

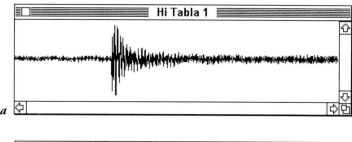

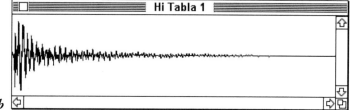

Looping

Another editing technique that is used regularly by samplers to maximize the system's available RAM and disk-based memory is a process known as *looping*. By using looping, a sample that occupies a finite memory space in RAM can be sustained for a long period of time (well past the length of the original sample), thereby preventing the sound from abruptly ending while the keys on a MIDI keyboard are still held down.

Such a loop is created by defining a segment of sound within a sample that doesn't significantly change in amplitude and composition over time and then repeatedly accessing this section from within RAM (see Figure 6.39). This loop can be created from waveform segments that are very short (consisting of only a few periods), or they can be longer in length. In general, this segment can be found in a sample's internal dynamic area within the sustain portion of a waveform.

Figure 6.39. Example of a sample with a sustain loop. (Courtesy of Sonic Foundry)

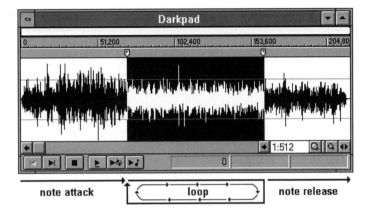

When creating a looped splice, you can often make your life easier by following this simple rule: Match the waveform shape and amplitude at the beginning of the loop with the waveform shape and amplitude at its end. This simply means that the waveform amplitude at the beginning of the loop must match the amplitude of the ending loop point. If these aren't matched, the signal levels vary and an annoying "pop" will result. Many samplers and sample-editing programs provide an automatic means for searching out the closest level match and/or an on-screen display that shows loop crossover points and enables the user to manually match amplitude levels (see Figure 6.40).

Figure 6.40. A loop waveform window enables the beginning and end levels of a loop to be manually matched. (Courtesy of Sonic Foundry)

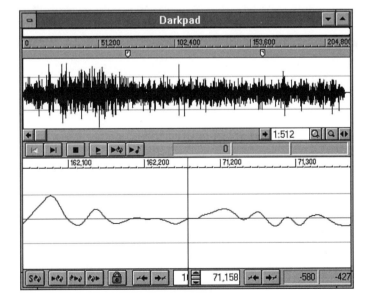

Certain sampling systems allow more than one loop to be programmed into a samplefile. This has the effect of making the sample less repetitive and more natural and adds to its overall expressiveness when played on a keyboard. In addition to having multiple sustain loops, a release loop can be programmed to decay the sample when the keyboard note is released.

Distribution of Sampled Audio

Within a sample-based MIDI setup, it's important that the distribution of sampled audio be as fast and as painless as possible. In order for most samplers and related software management programs to communicate systemwide, a number of guideline standards have been adopted for the transmission and storage of sampled digital audio.

MIDI Sample Dump Standard

The *sample dump standard* (SDS) was developed and proposed to the MIDI Manufacturers Association by Chris Meyer and Dave Rossum as a protocol for transmitting sampled digital audio and sustain loop information from one sampling device to another. This data is transmitted over regular MIDI lines as a series of MIDI system-exclusive (Sys-Ex) messages, which are unspecified in length and data structure.

Although various sampler types can be used to perform similar musical functions, the inner electronics and the way that data is internally structured often differ from device to device. As a result, most samplers communicate using their own unique system-exclusive data structure (as identified by a unique manufacturer and device ID number). In order to successfully transmit sample-based data between samplers, they must be of the same or compatible manufacture and design. If this is not the case, a computer-based program (such as a sample editor) must be used to translate from one sampler's data format and structure into that which can be understood by another make or model.

One of the greatest drawbacks of SDS is that data transfer can be a rather slow process because it conforms to the MIDI standard of transmitting serial data at the 31.25 Kbaud rate.

SCSI Sample Dump Formats

A number of computer-based digital audio systems and sampling devices are capable of transmitting and receiving sampled audio by way of a *small computer systems interface,* or SCSI (pronounced *scuzzy*). SCSI is a bidirectional communications bus that is commonly used by many PCs and digital devices to exchange data between systems at high speeds. When used in digital audio applications, SCSI provides a direct data link for transferring soundfiles at a rate of 500,000 bits/second (nearly 17 times the speed of SDS) or higher. Such an interface, when combined with a computer-based digital audio system, provides the user with a fast and straightforward means of transferring data to and from an editing program, hard disk, or CD-ROM sample library.

SMDI

The SCSI Musical Data Interchange (SMDI) was developed by Matt Issaacson for Peavey Electronics as a nondevice-specific format for transferring digitally sampled audio between SCSI-equipped samplers and computers at speeds up to 300 times faster than MIDI's transmission rate of 31.25 kbytes per second. Using this format, all you need to transfer digital audio directly from one supporting sampler to another is to connect the SMDI ports by way of a standard SCSI cable.

Although SMDI is loosely based on MIDI SDS, it has more advantages over its slower cousin than just speed. For example, SMDI can distribute stereo or multichannel samplefiles. In addition, SMIDI isn't limited to files that are less than two megawords in length and can transmit associated file information, such as filename, pitch values, and sample number range. Sound patch and device-specific setup parameters can also be transmitted and received over SMDI lines through the use of standard System Exclusive (Sys-Ex) messages.

SMDI is not, however, designed to replace MIDI. Although MIDI isn't the fastest at shuffling samples around, it's still the best way for communicating real-time performance and control data. Therefore, SMDI can simply be thought of as a fast side-chain for distributing samples and Sys-Ex data.

In the applied real world, SMDI could let you save all your samples to disk (possibly within a single database) and then quickly distribute these sounds to any supporting sampler. Sample editors wouldn't really require that your computer have an internal digital signal processing (DSP) sound card because edited samples could be quickly shuffled to and from your sampler for auditioning. Inexpensive sample players can have easy access to tons of samples, and sample dumps from a central library can be managed much more easily during a live performance.

As of this writing, the following products have implemented the SMDI protocol into their system:

> Sonic Foundry's Sound Forge 3.0 (Windows sample editor)
> Opcode MAX (Mac music programming construction kit)
> Passport Alchemy (Mac sample editor)
> Dissident MIDI Sample Wrench (Amiga sample editor)
> Dissident disSX (Amiga control program for Peavey SX)
> Kurzweil K2000, K2000R, and K2500 (sampling systems)
> Peavey DPM-SX/SXII (sampling modules)
> Peavey DPM-SP (sample player)
> Turtle Beach SampleVision (PC sample editor)
> E-mu ESI-32 (sampling module)

Sample Edit Software

In order to take full advantage of the protocols just mentioned for digitally transmitting and receiving samplefiles between various types of samplers, a number of sample editing software packages (see Figures 6.41 through 6.43) have been developed. Programs such as these enable a variety of important sample-related edit and processing tasks to be performed using the PC. Functions you can perform using these software packages include:

◆ Loading samplefiles into a computer, where they can be stored in a larger, central hard disk; arranged in a library that best suits the users needs; and transmitted to any sampling device within the system (often to samplers with different sample rates and bit resolutions).

◆ Editing samplefiles using standard computer cut-and-paste tools in order to properly arrange the sample before copying it to disk or transmitting it back to a sampler. Loops are often supported, which enables segments of a sample to be repeated and saves valuable sample memory.

◆ Digitally altering or mixing samplefiles using extensive signal processing techniques (such as gain changing, mixing, equalization, inversion, reversal, muting, fading, crossfading, and time compression).

From these tasks, you can see that the fundamental role of the sample editor is to integrate the majority of sample editing and transmission tasks within an electronic music setup.

Figure 6.41. Example of a sample editor network diagram.

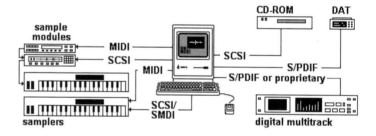

Figure 6.42. Sound Forge 3.0 sample editing software for Windows. (Courtesy of Sonic Foundry)

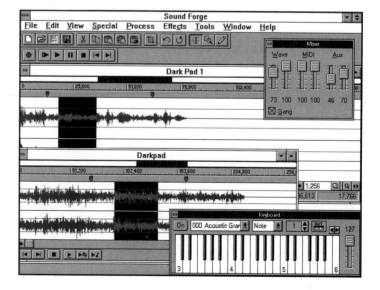

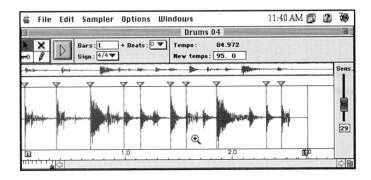

Figure 6.43. *Recycle sample-groove editing software for the Macintosh. (Courtesy of Steinberg)*

Synthesis and Sample Resynthesis

In addition to computer-based sample edit and signal processing functions, software packages exist that use various MIDI and SMDI sample dump formats to import a sample and then perform extensive sound synthesis and resynthesis on that sound (see Figure 6.44). After being processed, the resulting sound can again be transferred back to a sampling device or hard-disk-based recorder.

Figure 6.44. *Turbosynth modular synthesis and sample processing program. (Courtesy of Digidesign, Inc.)*

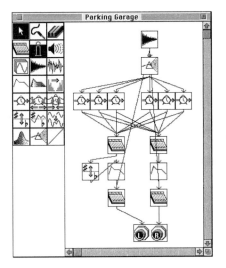

Often, synthesis and sample-processing programs combine elements of digital synthesis, sampled audio, and digital signal processing to generate a sound according to user-defined parameters or to modify existing samplefiles in ways ranging from subtle modifications to totally wild or exotic sounds. In effect, these programs are capable of turning virtually any sampler into an additive or FM digital synthesizer.

Modular synthesis works by combining various algorithmic oscillators, filters, and signal processing blocks in order to alter a sample's frequency, harmonic, delay, and time-based content. Many of these programs contain digital processing modules that don't exist in analog modular synths for the production of a wider range of sounds. After it is created, a sound patch can be saved to disk in a number of file formats for later recall or transferred to any of the currently available samplers for full polyphonic reproduction.

Hard-Disk Recording

Once developers began to design updated sample editor software versions, it was discovered that through additional processing hardware, digital audio editors were capable of recording digitized audio directly to a computer's hard disk. Thus, the concept of the *hard-disk recorder* was born. These devices (see Figures 6.45 through 6.47), sometimes known as *digital audio workstations* (DAW), serve as computer-based hardware and software packages that are intended specifically for the recording, manipulation, and reproduction of digital audio data that resides on hard disk. Commonly, such devices are designed around and controlled by a standard personal computer and associated digital audio hardware (which may be built into the computer or added on as a separate hardware device). Often, real-time audio processing systems require the use of one or more digital coprocessors. Such a coprocessor may be required if the speed and number crunching capabilities of the PC's processor alone are sometimes inadequate to perform the complex DSP and sustain-to-disk calculations that may be encountered in digital audio production. More expensive multichannel systems often use proprietary computer and processing systems that are specifically designed to perform digital audio and signal processing tasks.

There are multiple advantages to having digital audio workstations in an audio production environment. The following list describes some of these advantages:

◆ *The capability to handle longer samplefiles.* Hard-disk recording time is often limited only by the size of the hard disk itself (commonly one minute of stereo audio at 44.1 kHz occupies 10.5 MB of hard-disk memory).

◆ *Random-access editing.* As audio is recorded on a hard disk, any point within the program can be accessed at any time, regardless of the order in which it was recorded. Nondestructive editing allows audio segments (often called *regions*) to be placed in any context and/or order in a program without changing or affecting the originally recorded soundfile in any way. Once edited, these regions can be reproduced consecutively to create a single performance, or individually at a specified SMPTE time-code address.

◆ *DSP.* Digital signal processing can be performed on a segment or entire samplefile in either real time or non-real time in a nondestructive fashion.

In addition to these advantages, computer-based digital audio devices serve to integrate many of the tasks related to both digital audio and MIDI production. As with the sample editor, many hard-disk-based workstations are capable of importing, processing, and exporting samplefiles. These hard-disk recording systems offer a new degree of power to the artist who relies on sampling technology, as they can also categorize, edit, and process (and possibly import/export) samplefiles to connected samplers within the system.

Figure 6.45. *Example of a hard-disk recording system in a connected production system.*

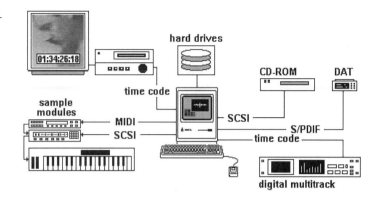

Figure 6.46. *Fostex Foundation 2000 nonlinear recorder/editor/ mixer. (Courtesy of Fostex of America)*

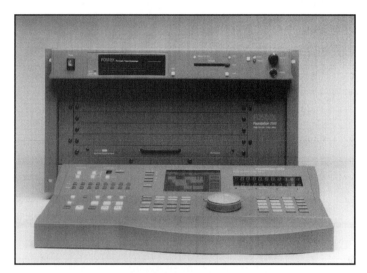

Figure 6.47. *Timeline DAW-80 digital audio workstation. (Courtesy of Timeline Vista Inc.)*

Hard-Disk Editing

In addition to being able to record and play back larger soundfiles, one of the strongest features of a hard-disk recorder is how quickly and easily it can perform an extensive edit on a soundfile compared to the time it would take to perform a similar function using an analog recorder. Furthermore, if you don't like the final results, a disk-based system easily lets you "undo" an edit. These systems also let you nondestructively save an edited program, create a number of different versions, and then choose among them.

Nondestructive Editing

Nondestructive editing refers to a disk-/disc-based recorder's capability to edit a soundfile without in any way altering the data that was originally recorded to disk. This invaluable facet of hard-disk recording technology means that any number of edits, alterations, program versions, and so on, can be performed and saved to disk without altering the original sound or performance.

The nondestructive editing process is accomplished by accessing precise segments of a recorded soundfile and outputting them according to a user-defined *editlist*. In effect, when you use your mouse to define a specific region, you are telling the program to define a block of memory that begins at a specific memory address on the hard disk/disc and continues until the ending address has been reached (see Figure 6.48). Once defined, these regions can be inserted into the list in such a way that they can be accessed and reproduced in any order, without affecting the originally-recorded soundfile. Most of the time, however, these editlists are built up in the background as a result of instructions that are graphically input with the help of a trackball or mouse.

In addition to being able to cut, copy, and paste various regions in a single soundfile (or between numerous soundfiles), it's also possible to perform extensive signal processing to a soundfile nondestructively. This task can be carried out in two ways: by processing the signal to disk in a non-real-time environment or by processing it in real time.

Figure 6.48. Example of a built-up editlist.

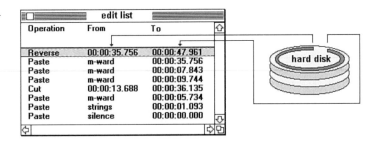

Operation	From	To
Reverse	00:00:35.756	00:00:47.961
Paste	m-ward	00:00:35.756
Paste	m-ward	00:00:07.843
Paste	m-ward	00:00:09.744
Cut	00:00:13.688	00:00:36.135
Paste	m-ward	00:00:05.734
Paste	strings	00:00:01.093
Paste	silence	00:00:00.000

hard disk

Non-Real Time

Extensive signal processing, such as changes in level, EQ, dynamics, and reverb, can be accomplished by defining the waveform segment to be processed. After being defined, the segment can be processed in a non-real-time environment, which means that the processor most likely won't be available for playback or other processing tasks until the necessary signal calculations have been performed and the resulting calculations are written to disk as a separate file that can be tagged to a soundfile or editlist.

Assume, for example, that you want to fade a sample after its initial transient without changing the original samplefile (see Figure 6.49). You can do this easily by defining a segment that ranges from an area just past the initial hit, all the way to the end of the sample. After the segment is defined, you can instruct the system to perform a *fade-out* (a simple set of calculations that reduces the amplitude over time according to defined parameters), after which the resulting processed file is automatically saved to disk. During playback, the hard-disk recorder plays back the initial transient and, at the appropriate time, switches to the processed file and continues—without interruption—to the end of the file. If a processed file were simply inserted into the body of a soundfile or editlist, upon being played out, it would again seamlessly begin playback of the desired program following this point without an audible break.

Figure 6.49. A non-real-time fade can be performed and written to disk as a separate file.

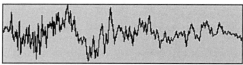

original soundfile

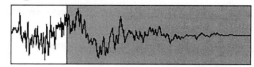

defined region is faded and the resulting data is written to disk as a separate file

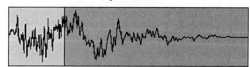

The original soundfile is reproduced until the fade is reached, at which point the tagged file containing the fade's data will begin to play.

Real-Time DSP

Real-time DSP differs from its non-real-time counterpart: instead of writing processed data to disk, these systems use additional high-speed signal coprocessors within their digital chains that are able to perform complex DSP calculations during actual playback (Figure 6.50).

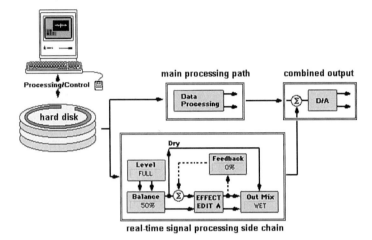

Figure 6.50. Real-time signal processing is carried out via side-chain coprocessors.

Because no calculations are written to disk in an off-line fashion, significant savings in time and disk space can be realized when you are working with productions that involve complex or long processing events. In addition, real-time processing systems often embed signal processing instructions within the editlist or playlist that allow these effects to be recalled, dynamically automated, or changed at any later time.

Destructive Editing

Destructive editing occurs whenever the recorded data is altered and rewritten to disk in such a way that it can't be recovered in its original form. Obviously, this edit form is less desirable than its nondestructive counterpart when you're working in a random-access production environment because edited data can't be easily recovered without forethought. (However, it always is possible to copy the original soundfile to DAT or to disk as a backup.) Fortunately, most systems don't work in a destructive fashion.

There are times, however, when a form of destructive editing is useful. Although the original segment and editlist files may be left intact (arguably making it nondestructive), on occasion you may want to save an edited file as a single, separate file that you can easily retrieve and reproduce without additional playlist assembly or processing. For example, you might want to save a complex series of mixed and edited files to disk as a single sound effects file and then export it to a sampler.

On-Screen Editing

Most hard-disk recording systems graphically display soundfile information in addition to letting us hear through the definitive interface: our ears. These graphic displays may take some getting used to; however, if the chosen display type is intuitive and detailed enough, you may be surprised at how quickly you come to rely on it for finding and defining soundfile segments with speed and relative ease.

Graphic Editing

One of the most common ways to display a soundfile on a computer or LCD screen is to graphically display the waveform as a series of vertical lines that represent the overall amplitude of the waveform at that point in time (see Figure 6.51). This display is WYSIWYG (*what-you-see-is-what-you-get*) in nature because it displays the waveform as a continuous data stream moving from left to right. Depending on the system type, soundfile length, and the degree of zoom, the entire waveform may be shown on the screen, or only a portion may be shown—with the waveform continuing off one or both sides of the screen.

Figure 6.51. *Example of Deck II's graphic editing screen. (Courtesy of OSC)*

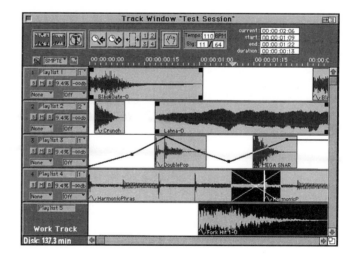

Graphic editing differs greatly from the "razor blade" approach used to cut analog tape, in that it offers both visual and audible cues as to where a precise edit point will be. Using graphic editing, waveforms that have been cut, pasted, reversed, and assembled are visually reflected on the screen. Usually these edits are nondestructive, enabling the original soundfile to remain intact.

Only when a waveform is zoomed-in to a point that begins to display the samplefile at the individual sample range is the actual soundfile shape shown (see Figure 6.52). This zoom feature lets you easily remove clicks and pops, redraw the waveform to remove potential offenders, or smooth amplitude transitions between loops or adjacent regions.

Figure 6.52. *Using most systems, when a soundfile is fully zoomed-in, actual sample data is displayed.*

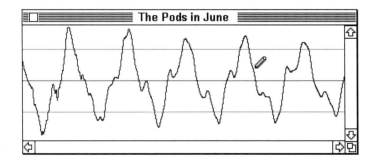

When you are working in a graphic editing environment, a defined soundfile segment is often referred to as a region. Generally, you can define a region by positioning the cursor within the waveform and then pressing and holding the mouse or trackball button and dragging the cursor to the left or right of the initial point. Usually, the selected region is highlighted for easy identification, as shown in Figure 6.53. After the region is defined, it can be edited, marked, named, or otherwise processed.

Figure 6.53. *A defined region. (Courtesy of Innovative Quality Software)*

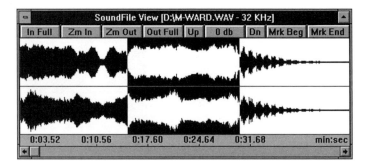

Playlist Editing

The *playlist* is an extension of a graphic editing system in that it enables regions to be graphically defined, named, and then placed into a sourcelist. After the desired regions have been placed in the sourcelist, they can be graphically dragged (via a trackball or mouse) into a destination list known as a playlist (see Figure 6.54). The playlist is then assembled (in a sequential fashion) so that the defined, named regions are played back as either a single, continuous program, or programmed to be triggered sequentially at specific time-code addresses.

A continuous, sequential playlist can be an invaluable tool for assembling individual songs into a finished album or CD project. In this context, all songs can be recorded to disk as a single soundfile or as a number of individual soundfiles. After this has been accomplished, each song can be edited and defined as a region, where it can be defined by name in the sourcelist. This enables songs to be placed into a playlist in any order. Then by recording a short file that contains silence, the desired length of silence can be placed between each song. If you want to try out a different order, all you have to do is save the original playlist (which doesn't include the soundfile data itself, but only the address information as to where the data can be found on the hard disk), and then try a new song order.

Triggering individual regions of a playlist at specific time-code addresses often is done in audio-for-video or film production. Playlists are extremely useful in this context, as any number of sound effects can be loaded into a sourcelist. By *spotting* the videotape (the process of locating points that need effects) and determining the necessary SMPTE addresses, it becomes a simple matter to search out the desired effects in the sourcelist and place them into the playlist at the appropriate times. By playing the tape, each effect is triggered automatically (possibly with automated effects, level, and other mixing parameters).

Figure 6.54. *A sourcelist (left) contains defined and named regions that can be placed into a destination playlist (right), which can then be sequentially played as a single, continuous audio program. (Courtesy of Sonic Foundry, Inc.)*

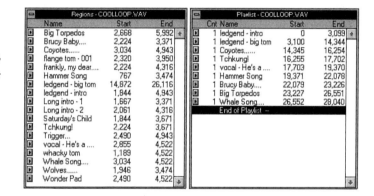

Basic Editing Tools

One of the strongest assets of a hard-disk recording system is its capability to edit segments of digital audio with speed and ease. The following sections offer a brief introduction to many of these valuable editing tools.

Cut, Copy, and Paste

The cut, copy, and paste methods used in hard-disk recording are entirely analogous to those of a word processor or almost any graphics-based program.

- ◆ *Cut.* Places the highlighted region into memory and deletes the selected data from its current position (see Figure 6.55).

- ◆ *Copy.* Places the highlighted region into memory and does not alter the selected waveform in any way (see Figure 6.56).

- ◆ *Paste.* Copies the waveform data in the system's clipboard memory to the selected waveform at the current cursor position (see Figure 6.57).

Figure 6.55. Cutting a
waveform region.

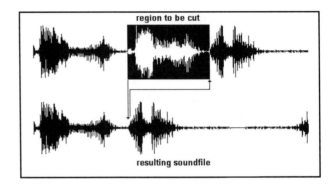

Figure 6.56. Copying a
waveform region.

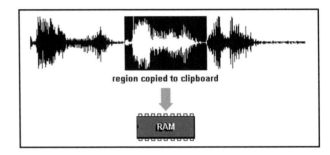

Figure 6.57. Pasting a waveform
region from the Clipboard.

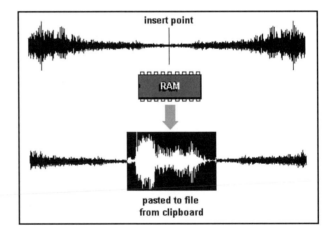

Fades and Crossfades

The *fading in* or *fading out* of a region is a DSP function accomplished by calculating the soundfile's relative amplitude over its duration. For example, fading in a file (see Figure 6.58a) proportionately increases a region's gain from infinity to full gain. Likewise, a fade-out has the opposite effect of creating a transition from full gain to infinite attenuation (see Figure 6.58b). These DSP functions have the advantage of creating a much smoother transition than would be humanly possible during a manual fade.

Figure 6.58. Examples of various fade curves.
a. Fade-in.
b. Fade-out.

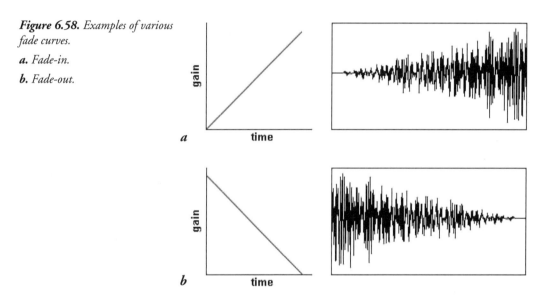

A *crossfade* (or *x-fade*) often is used to smooth the transition between two audio segments that are either sonically dissimilar in nature or don't match in amplitude at a particular edit point (a condition that would lead to an audible "click" or "pop"). A crossfade basically is a fade-in and fade-out that overlaps to create a smooth transition from one segment to the next (see Figures 6.59 and 6.60). Technically, the overall effect of this process is that the amplitude of the two signals are averaged over a user-definable length of time, thereby masking the offending edit point.

Figure 6.59. Example of a crossfaded stereo audiofile. (Courtesy of Sonic Solutions)

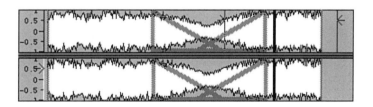

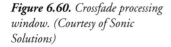

Figure 6.60. *Crossfade processing window. (Courtesy of Sonic Solutions)*

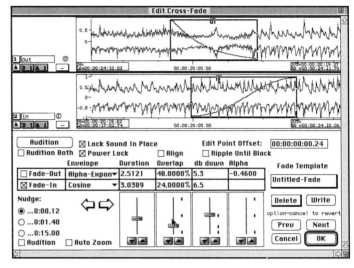

Gain Change and Normalization

Gain change is a relatively straightforward processing function that recalculates the amplitude of each sample (either upward or downward) according to a user-defined ratio.

Normalization is a specialized gain-related process that optimizes a digital system's dynamic range by automatically determining the amount of gain that would be required to increase the highest amplitude signal (within a sample or samplefile) to the system's full-scale value. Once this amount has been calculated, the amplitude of the entire file is increased by this gain ratio (see Figure 6.61).

Figure 6.61. *Original signal and normalized (full-gain) signal level.*

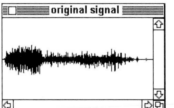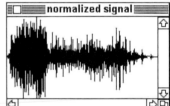

Advanced DSP Editing Tools

Most hard-disk recording systems offer additional editing functions beyond the basic cut and paste commands. However, the number of processing functions and the degree of complexity and flexibility often vary between systems. Depending on the system's capabilities, these functions may be performed in either real time or non-real time. Further explanation of the signal processing functions themselves can be found in Chapter 11, "Signal Processors."

Equalization

Digital EQ has become a common implementation that often varies in form and degree of flexibility among systems (see Figure 6.62). Most systems provide full parametric control over the entire range, including provisions for variable bandwidth (Q).

Figure 6.62. Pro Tools III EQ screen. (Courtesy of Digidesign)

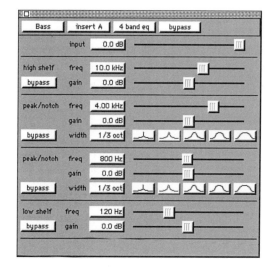

Software-controlled digital equalization often enables the user to equalize a single region, single track, or the entire program with complete automation and repeatability. In addition, equalization can easily range from being subtle and broadband to tightly controlled and severe (for example, a notch filter).

Dynamic Range Processing

Dynamic range processing has become more popular in recent systems for the same reasons that digital EQ has increased in popularity: a portion, single track, or entire program can be compressed, limited, or expanded with total flexibility and repeatability—often under a system's automation.

Pitch and Time Changing

Pitch change enables the relative pitch of a defined region or an entire soundfile to be shifted upward or downward by a specific percentage ratio or musical interval. Less expensive systems often pitch shift by determining a ratio between the present and the desired pitch and then either adding (lowering pitch) or dropping (raising pitch) samples from the existing region or soundfile (see Figures 6.63a and 6.63b). One potential drawback of this process is that the raised or lowered pitch will also be changed in duration. For example, the duration of a soundfile that has been raised in pitch will be reduced when played back at the same, initial sample rate. Likewise, a soundfile that has been lowered in pitch will be increased in duration. The same theory applies to musical transpositions that are processed in this fashion.

Figure 6.63. Basic disk-based pitch-shifting techniques.

a. Digital audio can be shifted downward by interpolating and adding data to the original segment and then reproducing this data at its original sample rate.

b. Digital audio can be shifted upward by dropping samples from the segment and then reproducing this data at its original sample rate.

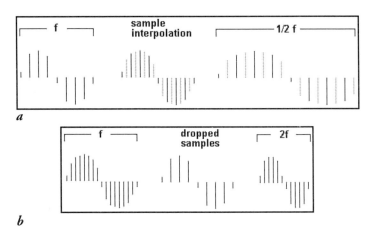

More sophisticated systems can combine variable sample rate and pitch shift techniques to alter the duration of a region or soundfile, in addition to being able to raise or lower the relative pitch. Such a system is capable of a number of pitch- and time-related combinations, including the following:

- ◆ *Pitch/duration shift.* A program's pitch can be changed while having a corresponding change in length.
- ◆ *Pitch shift only.* A program's pitch can be changed while recalculating the file so that its length remains the same.
- ◆ *Duration shift only.* A program's length can be changed while shifting the pitch so that it matches that of the original program.

DSP Plug-Ins

One of the latest and most exciting additions to the concept of hard-disk recording as a "studio-in-a-box" is that of the *plug-in*. These software-based applications have come about as the result of the extreme popularity of specific hard-disk recording platforms (most notably using Digidesign's

Sound Designer II, Pro Tools, and Sound Tools for the Macintosh and the Windows operating system for the IBM-compatible PC). As a result of these popular systems, third-party developers have begun to design and program specific applications that can perform almost any task under the sun. Usually (but not always), these tasks are beyond the manufacturing and development scope of the original platform developer; on many occasions, new small businesses have sprouted to create products based on market demand and good ideas. A small example of such plug-ins are detailed in Figures 6.64 through 6.69. Keep in mind, however, that the list of new developers is growing—almost on a monthly basis.

Figure 6.64. *The L1 - Ultra-maximizer Peak-limiting plug-in is also capable of dithering (noise-shaping) and accurately requantizing between 20-, 16-, 12-, and 8-bit signals. (Courtesy of Waves)*

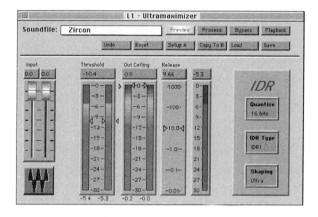

Figure 6.65. *Lexicon's NuVerb hardware/software signal coprocessor for the Digidesign NuBus system offers extensive real-time signal processing for the Macintosh PC. (Courtesy of Lexicon)*

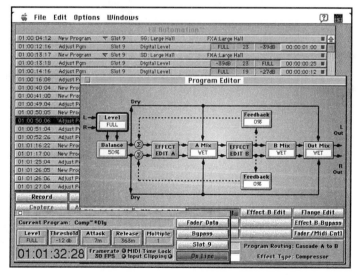

Figure 6.66. *Hyperprism offers 25 professional-quality processing functions for most Macintosh/ Digidesign systems, including time stretch, pitch shift, delays, doppler, ring modulation, and spatializers. (Courtesy of Arboretum Systems)*

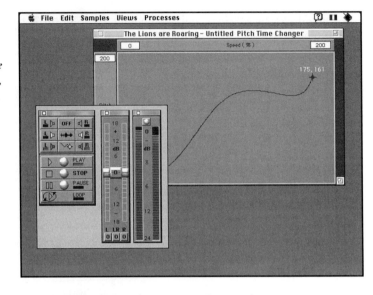

Figure 6.67. *The MDT (Multiband Dynamics Tool) offers extensive dynamic soundfile processing for the Macintosh. (Courtesy of Jupiter Systems)*

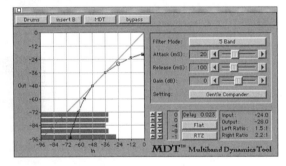

Figure 6.68. *The JVP (Jupiter Voice Processor) offers extensive dynamic, EQ, and delay processing for the Macintosh. (Courtesy of Jupiter Systems)*

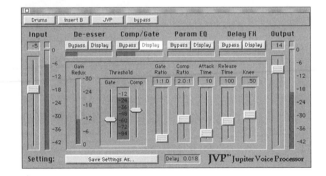

Figure 6.69. *One of a number of effects plug-in modules for the Software Audio Workshop program for Windows. (Courtesy of Innovative Quality Software)*

Two-Channel Hard-Disk Recording Systems

Two-channel hard-disk systems are commonly used for sound effects, music editing, broadcast, and the recording of digital audio within MIDI project studios. In fact, these hardware-/software-based systems can be found in almost any production environment imaginable because they are often very cost-effective. Although a number of dedicated two-channel hard-disk recording systems exist, the most commonly found system is a hardware card add-on that can be plugged into an existing personal computer (see Figures 6.70a, 6.70b, and 6.71). Because these hardware units can be integrated into most popular PC operating environments, it often follows that MIDI sequencing capabilities can also be easily integrated into the system, thereby creating a flexible and powerful production environment.

Figure 6.70. *The Card Plus two-channel editing system. (Courtesy of Digital Audio Labs)*

a. *The CardD Plus hardware card.*

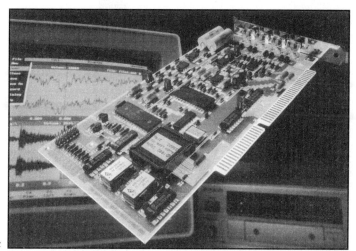

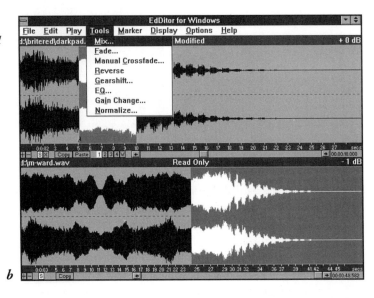

Figure 6.70. *(continued)*

b. *The EdDitor Plus two-channel editing software is a professional two-channel system for Windows.*

b

Figure 6.71. *The Pro-Master 20-bit audio interface and Sound Designer II software for the Macintosh. (Courtesy of Digidesign)*

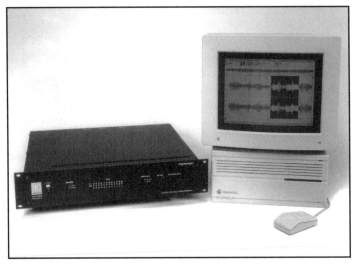

A number of two-channel systems are currently on the market and range from hardware card/ software systems with prices beginning under $100 (the Soundblaster and similar consumer multimedia clones) to high-quality, sophisticated systems priced well into the thousands of dollars. When buying such a system to integrate into your own production environment, it's always wise to stop and think about your personal needs. Is the quality sufficient? Will it integrate well into your system? What are the benefits/compromises and will it grow with you as future system expansion occurs? When researched thoroughly, a quality two-channel hard-disk system can be a tremendously flexible, creative production tool.

The Virtual Track

As signal processing power increases—due to increased hard disk/disc speeds, faster processors, improved software, and so on—it has become increasingly easier for a PC-based hard-disk system to have additional processing overhead in reserve when performing a task, compared to its older, slower counterparts. This increased processing and disk access overhead has made it possible for more than one channel (or stereo pairs of channels) to be accessed from a single two-channel or multichannel hard disk system at any one point in time. This technological fact has brought about new technology in the form of the *virtual track system*.

In plain English, a virtual track system can access multiple audio channels (or stereo pairs) and then mix them, in real time, to a single output channel or pair of channels (see Figures 6.72 and 6.73). Such a system can give you access to 8 or more independent channels on a single 2-channel hard-disk system, or it can let you mix and assign 16 or more channels to a 4-output system. At this point you might ask, "Why have all these channels if they are just going to be mixed down?" Well, since the audio data can be defined into regions that can be assigned to separate virtual "tracks," each sound can be separately processed, panned, mixed (in real time), and muted in a number of ways without altering the original soundfile data on disk. Such nondestructive systems often are fully automated so that all the necessary mix functions are automatically saved to disk, whereby new mixes can be created any number of times until the best balance is achieved.

Figure 6.72. *Software Audio Workshop 8-channel virtual digital audio software for Windows. (Courtesy of Innovative Quality Software)*

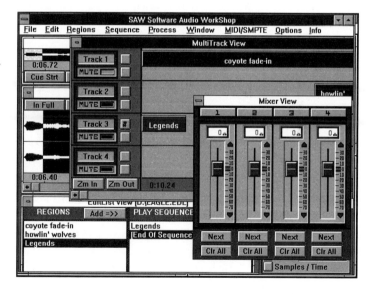

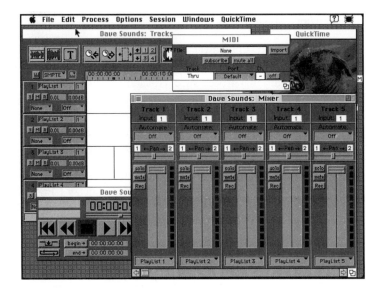

Figure 6.73. *Deck II 6- and 8-channel virtual digital audio software for the Macintosh. (Courtesy of OSC)*

Often, a virtual system is locked to an external time-code device—such as the computer's own MIDI sequencer, external audio, or video recorder—so it can provide a cost-effective way to add multitrack capabilities to your present system. Because the tracks are locked to SMPTE, each individual track could be transferred to a multitrack recorder at a later time, thereby providing total track isolation during mixdown.

Multichannel Hard-Disk Recording Systems

Multichannel hard-disk recording systems work in much the same way as their two-channel counterparts, except that any number of soundfiles or defined regions can be pasted into a multitude of on-screen "tracks." Each segment or group of segments can then be assigned to one or more of the system's designated multichannel outputs (see Figure 6.74).

Such a flexible, multichannel system is often extremely useful in MIDI production because it allows vocals, acoustic instruments, effects, and other continuous soundfiles to be integrally locked to a MIDI sequence (see Figure 6.75). Each soundfile or region can be edited, processed, looped and slipped in time to easily match on-screen action or dialog. It can also be used in a broadcast setting where station IDs, spots, and various effects can easily be built up. The system's individual on-screen track structure and multiple outputs can provide extensive flexibility over signal processing and mixing control within these and other audio production environments.

Figure 6.74. *Defined regions can be placed into discrete channels where they may be independently processed, mixed, and assigned to individual outputs. (Courtesy of Timeline Vista, Inc.)*

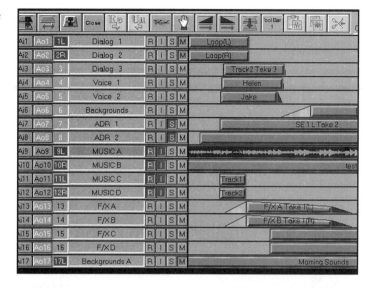

Figure 6.75. *Sun Studio's (Copenhagen, Denmark) SSL Scenaria OmniMix digital surround sound audio/video system. (Courtesy of Solid State Logic)*

Although multichannel hard-disk systems commonly offer 4-, 8-, 16-, or 24-individual outputs, it isn't uncommon for these systems to have only two or four inputs. This economic design is often implemented because most hard-disk-based productions are assembled from a number of mono or stereo sound sources that are most often recorded into the system from DAT, CD, or other 1- or 2-channel sources. A few exceptions in which multiple inputs may be necessary would be during a "direct-to-disk" recording or vocal session, when recording a number of artists who have been called on to replace on-screen dialog, or while recording talent for a radio or TV commercial. In such situations, individual takes, voice cues, intonations, and so on can be edited, processed, and assembled into the final product with a reasonable degree of speed, cost-effectiveness, and automated repeatability.

You may have noticed that this discussion hasn't included much about the concept of *track*— and for good reason. In the world of random-access audio production, the concept of the continuous track is severely limiting. As a medium, random-access audio excels at dealing with short-term sam-ples that can be repeatedly triggered, or in reproducing soundfiles that are finite in length and/or can be edited into shorter regions. These strategies save on memory—either on disk or in RAM. Unlike tape-based media, when sample or soundfile data is loaded into a system and defined regions are played from a sampler or hard-disk recorder, each event can be played back at any time—just when it is needed. During those times when a specific event isn't needed, no demands are made on disk memory or RAM.

When using tape-based technology (either analog or digital), the concept of the track is relatively simple and straightforward. Multitrack recorders use stationary or rotary heads to record and playback magnetically encoded information in a linear fashion. Once recorded, the magnetic information is physically tied to the medium (tape) and can't be moved to another track or slipped in time without physically being rerecorded. In effect, the memory medium of tape is designed to always be in active use. If a hard-disk recorder's memory medium was placed under such a constant demand, the amount of usable time would be severely limited. For example, if an eight-channel system were to continuously record data to a hard-disk system that has 1 GB (1,000 MB) of available memory at a sample rate of 44.1 kHz, each track would have a maximum recording time of almost 12 minutes (given that approximately 10.5 MB of data per minute is used at 44.1 kHz). Although this figure is still impressive, the same drive might provide track-hours that would far exceed this linear time limitation.

The Digital Audio Workstation

In recent years, the term *digital audio workstation* (DAW) increasingly has come to signify either a dedicated or computer-based hard-disk recording system that incorporates both basic and advanced editing and signal processing features (see Figures 6.76 and 6.77). Although a workstation can independently perform a wide range of audio-related functions, one of its greatest advantages is the distinct capability to integrate with other devices, production systems, and functional software that relate to audio, video, and music production. In effect, these systems are able to integrate systems to create a single, multifunctional environment that can freely communicate data and perform tasks relating to MIDI sequencing, sample/playlist editing, sampling, hard-disk recording, digital signal processing, synthesis/resynthesis, and music printing.

Figure 6.76. The Sonic System with CD Printer. (Courtesy of Sonic Solutions)

Figure 6.77. Pro Tools Digital Audio Workstation. (Courtesy of Digidesign)

Throughout music and audio production history, we have become used to the idea that certain devices were only meant to perform a single task: a recorder records and plays back, a limiter limits, and mixers mix. In response to this, I always liken digital audio workstations to a chameleon that is able to change its functional "color" to match the task at hand. In effect, a digital audio workstation isn't so much a device as a systems concept that can perform a wide range of audio production tasks with ease and speed. Some of the characteristics that are often (or should be) displayed by such systems include the following:

- *Integration.* One of the major functions of a dedicated or computer-based workstation is to provide centralized control over digital audio recording, editing, processing, and signal routing, as well as to allow both transport and/or time-based control over MIDI/ electronic music systems, external tape machines, and video tape recorders.

- *Communication.* A workstation should be able to communicate and distribute digital audio data (such as AES/EBU, S/P-DIF, SCSI, SMDI, and/or MIDI sample dump standards) throughout the system. Timing and synchronization signals (such as SMPTE time code, MTC, and MIDI sync) should also be provided for.

- *Speed and flexibility.* Speed and flexibility are probably a workstation's greatest assets. After you become familiar with a particular system, most production tasks can be tackled in far less time than would be required using similar analog equipment. Many of the extensive signal processing features would simply be next to impossible to accomplish in the analog domain.

- *Expandability.* The ideal workstation should be able to integrate new hardware and/or software components into the system with little difficulty. Of course, the operative word here is *ideal.*

- *User friendly.* An important element of a digital audio workstation is its central interface unit—humans. The operation of a workstation should be relatively intuitive and should forego any attempt at obstructing the creative process by speaking "computerese."

When choosing a system for yourself or your facility, be sure to take all these considerations into account. Each system has its own set of strengths and weaknesses. When in doubt, it's always a good idea to research the system as much as possible before you commit to it. Feel free to contact your local dealer for a test drive. Like a new car, purchasing a digital audio workstation can be a potentially expensive proposition that you probably will have to live with for a while. Once you have made the right choice, you can get down to the business at hand. Just put the top down, lean back, and have fun.

Triggered Event Digital Audio

Before finishing out this chapter, one more important subject should be discussed that relates to audio production using both samplefiles and soundfiles: *triggered event digital audio.* Triggered event production is a general description for the automated search and firing on command of specific files or segments of digital data. One popular example of a triggered-event system is a drum machine, in which digitized audio is output from read-only memory (ROM) by way of the manual trigger buttons or under the control of a MIDI sequencer. In this way, a finite set of digital segments can be triggered for a specific purpose (in this case, to build a beat or rhythm section). Another example might include the triggering of various vocal passages from the hard disk at specific points within a song being played from a MIDI sequencer. Another example might be the synchronized triggering of various sound effects (from either a sampler or hard disk) to an edited video or film track. Indeed, the number of examples and production applications are far too many to list here and are limited only by your imagination.

The structure of a triggered-event system may also make use of the CD format, as might occur in the automated management of a comprehensive CD sound effects or music library under computer control. Recent advances in the storage and retrieval of sound effects and music using the compact disc format have made this medium very flexible and desirable for many postproduction and broadcast audio applications. In addition to superior audio quality, the storage of audio and of information on CD offers the following features:

◆ *Event triggering.* Precision cueing to a time-encoded production device, allowing for a high degree of accuracy and repeatability.

◆ *Looping.* Accomplished easily and inaudibly for use with background presence and effects.

◆ *Indexing, accessing, and triggering.* These automated functions may be accomplished through the use of a computer system.

Professional compact disc players, as shown in Figure 6.78, allow for straightforward and programmable disc management. These systems often include a multiple disc- or jukebox-style transport, which literally enables hundreds of CDs to be loaded into the system. A central controller or computer system can be programmed to categorize, access, and play any number of songs, sound effects—even CD-ROM files—in any desired order and at any time.

Figure 6.78. Denon 1200C and 1200F professional CD jukebox system. (Courtesy of Denon Electronics)

Sound and sound effects library software systems make it possible for a PC to be connected to a CD jukebox system in such a way that extensive music and sound effects data banks can be built up. The applications for such a system can range from the automated triggering of sound effects for video and film to such uses as full automation for on-the-air broadcast programming.

CHAPTER 7

MIDI and Electronic Musical Instrument Technology

In present day sound production, MIDI systems (see Figure 7.1) are being used by professional and nonprofessional musicians alike. Likewise, these systems are relied on to perform an ever-expanding range of production tasks, including music production, audio-for-video and film postproduction, multimedia and stage production.

This industry acceptance can, in large part, be attributed to the cost-effectiveness and general speed of MIDI production. Currently, a huge variety of affordable MIDI instruments and devices are available on the market. Once purchased, there's often less of a need (if any at all) to hire outside musicians to help with a project. In addition, MIDI's multichannel environment lets a musician compose, edit, and arrange a piece with a high degree of flexibility, without the expressed need for recording and overdubbing these sounds onto multitrack tape. This affordability, potential for future expansion, and increased control capability over an integrated production system, have spawned the growth of an industry that often is personal in nature. For the first time in modern music history, it's possible for an individual to realize a full-scale sound production in a cost- and time-effective manner. Because MIDI is a real-time performance medium, it is possible to fully audition and edit a production at every stage of its development—all within the comfort of your own home or personal production studio.

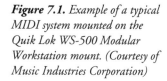
Figure 7.1. *Example of a typical MIDI system mounted on the Quik Lok WS-500 Modular Workstation mount. (Courtesy of Music Industries Corporation)*

In artistic terms, this "digital language" lets the artist express him- or herself with a degree of expression and control that was, until recently, not possible on an individual level. Through the use of this digital performance language, a musician can create and develop a song or composition in a practical, flexible, and affordable production environment. In addition to composing and performing a song, the musician can act as a technoconductor, having complete control over a wide range of sounds, their timbre (sound and tonal quality), and their blend (level, panning, and other real-time controls) during a performance.

What Is MIDI?

Simply stated, the Musical Instrument Digital Interface, or MIDI, is a digital communications language and compatible hardware specification that enables multiple electronic instruments, performance controllers, computers, and other related devices to communicate with one another within a connected network (see Figure 7.2). MIDI is used to translate performance- or control-related actions (such as playing a keyboard, selecting a patch number, or varying a modulation wheel) into equivalent digital messages. It then transmits these messages to other MIDI devices where they can be used to control their sound generation or control parameters in a performance setting. Alternatively, MIDI data can be recorded into a digital device (known as a *sequencer*) that can be used to record, edit, and playback MIDI performance data.

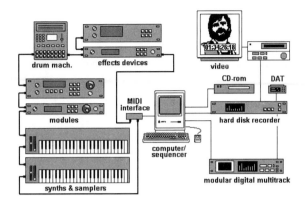

Figure 7.2. *MIDI enables electronic instruments, effects devices, and other ancillary equipment to communicate performance-related data within an audio production environment.*

System Interconnection

MIDI enables 16 channels of performance, controller, and timing data to be transmitted—in one direction—over a single data line. Consequently, it's possible for a number of devices to be connected within a network through a single data chain for communicating MIDI messages.

A MIDI cable consists of a shielded, twisted pair of conductor wires that has a male 5-pin DIN plug located at each end (see Figure 7.3a). The MIDI specification presently uses only three of the possible five pins (see Figure 7.3b), with pins 4 and 5 being used as conductors for MIDI data, and pin 2 being used as a ground connection. Pins 1 and 3 currently are not in use but are reserved for possible changes in future MIDI applications.

Figure 7.3. *The MIDI cable.*
a. Cutaway view.
b. Cable diagram.

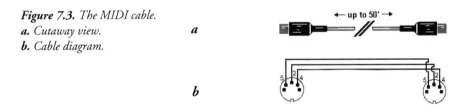

Prefabricated MIDI cables in lengths of 2, 6, 10, 20, and 50 feet often can be obtained from music stores that specialize in MIDI equipment. Fifty feet, however, is the maximum length specified by the MIDI specification in order to reduce the effects of signal degradation and external interference that tend to occur over extended cable runs.

MIDI In, Out, and Thru Ports

Three types of MIDI ports are used to connect MIDI devices in a network: MIDI in, MIDI out, and MIDI thru.

◆ *MIDI in port.* Receives MIDI messages from an external source and communicates this performance, control, and timing data to the device's internal microprocessor.

♦ *MIDI out port.* Transmits MIDI messages from a single source device to the microprocessor of another MIDI instrument or device.

♦ *MIDI thru port.* Provides an exact copy of the incoming data at the MIDI in port and
transmits this data out to another MIDI instrument or device that follows in the MIDI
data chain.

Certain MIDI devices don't include a MIDI thru port. These devices, however, may offer a
software-based transmission function that can select between a MIDI out port and a MIDI echo
port. As with the MIDI thru port, the selectable MIDI echo function is used to provide an exact
copy of any information received at the MIDI in port and then routes this data to the MIDI out/
echo port.

The Daisy Chain

One of the simplest and most commonly used methods for distributing data within a system is the
MIDI daisy chain. This method is used to distribute a single MIDI data line to every device within
a system by transmitting data to the first device and subsequently passing an exact copy of this data
through to each device in the chain (see Figure 7.4). The MIDI out data is sent from the source
device (controller, sequencer, and so on) to the MIDI in of the second device. By connecting the
MIDI thru port of the second device to the MIDI in of a third device, this last device receives an
exact copy of the original source data at its input port. This process continues throughout a basic
MIDI system until the final device is reached.

*Figure 7.4. Example of a
connected MIDI system using a
daisy chain.*

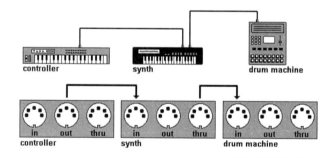

MIDI Channels

Just as it's possible for you to single out and talk to one individual in a crowd, MIDI messages can
be communicated through a common data cable to a specific device or range of devices instructed
to respond to these messages. This communication is accomplished by including a nibble (four
bits) in the status/channel number byte that instructs all receiving devices as to the MIDI channel
on which a specific message is being transmitted. The channel nibble is 4-bits wide, so up to 16
channels can be transmitted through a single MIDI cable, as shown in Figure 7.5.

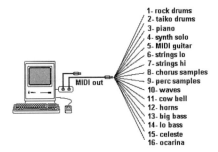

Figure 7.5. *Up to 16 channels can be transmitted through a single MIDI cable.*

1- rock drums
2- taiko drums
3- piano
4- synth solo
5- MIDI guitar
6- strings lo
7- strings hi
8- chorus samples
9- perc samples
10- waves
11- cow bell
12- horns
13- big bass
14- lo bass
15- celeste
16- ocarina

Whenever a MIDI device is instructed to respond to a specific channel number, it ignores all performance messages transmitted on any other channel. Suppose, for example, that you have a MIDI keyboard controller and two synthesizers linked together in a MIDI chain (see Figure 7.6). Assume that you have set synth A to receive data on MIDI channel 4 and synth B to receive on channel 8. By instructing your controller to transmit on channel 4, synth A receives the data and begins to play while synth B ignores the data. Likewise, setting the controller to channel 8 outputs sound from synth B while synth A remains silent. Splitting the controller so that the lower octaves transmit on channel 4 while the upper octaves are transmitted over channel 8 enables each of the synths to play their respective musical parts.

Figure 7.6. *System showing a set of MIDI channel assignments.*

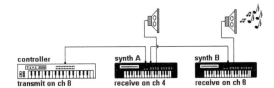

controller   synth A   synth B
transmit on ch 8   receive on ch 4   receive on ch 8

The MIDI Message

MIDI data is communicated digitally throughout a production system as a string of MIDI messages transmitted (in a serial fashion) through a single MIDI line at a speed of 31.25 Kbaud. These messages are made up of a group of related 8-bit words (known as *bytes*) that are used to convey a series of instructions to one or all MIDI devices within the system. The following table shows how a 3-byte MIDI Note On message (which is used to signal the beginning of a MIDI note) can be used to communicate these instructions in binary form.

	Status Byte	*Data Byte 1*	*Data Byte 2*
Description	Status/channel #	Note #	Attack velocity
Binary data	(1001 0100)	(0100 0000)	(0101 1001)
Numeric value	(Note On/Ch. #4)	(64)	(89)

A 3-byte Note On message of (10010100)(01000000)(01011001), therefore, would transmit instructions that would read like the following:

Transmitting a Note On message over MIDI channel #4, for note #64, with an attack velocity (volume level of a note) of 89.

Only two types of bytes are defined in the MIDI specification: the status byte and the data byte. *Status bytes* are used in the MIDI message as an identifier for instructing the receiving device as to the particular MIDI function and channel being addressed. The *data byte* is used to encode the actual numeric value to be attached to the accompanying status byte. Although a byte is made up of 8 bits, the most significant bit, or MSB (the left-most binary bit in a digital word), is used solely to identify the byte type. The MSB of a status byte is always 1, whereas the MSB of a data byte is always 0 (see Figure 7.7).

Figure 7.7. *The most significant bit of a MIDI data byte is used to determine whether a byte is a status byte (1) or a data byte (0).*

MSB of a status byte is always 1

↓

(1SSS SSSS)

MSB of a data byte is always 0

↓

(0DDD DDDD)

MIDI messages also are divided by the MIDI specification into two types: channel messages (messages assigned to a specific MIDI channel) and system messages (messages addressing all devices in a system without regard to channel assignment). These two message types are described in the following sections.

Channel Messages

Channel messages are used to transmit real-time performance data throughout a connected MIDI system. Channel messages are generated whenever the controller of a MIDI instrument is played, selected, or varied by the performer. Examples of such control changes would be playing a keyboard, pressing program selection buttons, or moving modulation or pitch wheels. Each channel message contains a MIDI channel number in its status byte and therefore can be addressed by a device assigned to that channel number. There are seven Channel Voice message types: Note On, Note Off, Polyphonic Key Pressure, Channel Pressure, Program Change, Control Change, and Pitch Bend Change. These message types are explained in the following list.

◆ *Note On.* Indicates the beginning of a MIDI note. This message is generated each time a note is triggered on a keyboard, drum machine, or other MIDI instrument (by pressing a key, striking a drum pad, and so on). A Note On message consists of three bytes of information: a MIDI channel number, a MIDI pitch number, and an attack velocity value (messages that are used to transmit the individual volume levels [0-127] of each note as they are played).

◆ *Note Off.* Indicates the release (end) of a MIDI note. Each note played through a Note On message is sustained until a corresponding Note Off message is received. A Note Off message doesn't cut off a sound; it merely stops playing it. If the patch being played has a release (or final decay) stage, it begins that stage upon receiving this message.

◆ *Polyphonic Key Pressure.* Transmitted by instruments that can respond to the pressure changes applied to the individual keys of a keyboard. A Polyphonic Key Pressure message consists of three bytes of information: a MIDI channel number, a MIDI pitch number, and a pressure value.

◆ *Channel Pressure* (or *Aftertouch*). Transmitted and received by instruments that respond to a single, overall pressure applied to the keys. In this way, additional pressure on the keys can be assigned to control such variables as pitch bend, modulation, and panning.

◆ *Program Change.* Used to change the active program or preset number in a MIDI instrument or device. Using this message format, up to 128 presets (a user- or factory-defined number that activates a specific sound-generating patch or system setup) can be selected. A Program Change message consists of two bytes of information: a MIDI channel number (1-16) and a program ID number (0-127).

◆ *Control Change.* Used to transmit information relating to real-time control over the performance parameters of a MIDI instrument (such as modulation, main volume, balance, and panning). Three types of real-time controls can be communicated through control-change messages: continuous controllers, shown in Figure 7.8 (which communicate a continuous range of control settings, generally in value ranging from 0-127), switches (which function in an ON or OFF state with no intermediate settings), and data controllers (which enter data either through numerical keypads or through stepped up/down entry buttons). A full listing of control-change parameters and their associated numbers can be found in Figure 7.9.

◆ *Pitch Bend.* Transmitted by an instrument whenever its pitch bend wheel is moved either in the positive (raise pitch) or negative (lower pitch) direction from its central (no pitch bend) position.

Figure 7.8. *Continuous controller data value ranges.*

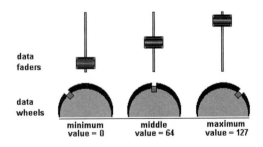

data faders

data wheels

minimum value = 0   middle value = 64   maximum value = 127

14-BIT CONTROLLER MOST SIGNIFICANT BIT				7 BIT CONTROLLERS (continued)		
Controller Hex	Number Decimal	Description		Controller Hex	Number Decimal	Description
00H	0	Undefined		.	.	.
01H	1	Modulation Controller		4FH	79	Undefined
02H	2	Breath Controller		50H	80	General Purpose Controller #5
03H	3	Undefined		51H	81	General Purpose Controller #6
04H	4	Foot Controller		52H	82	General Purpose Controller #7
05H	5	Portamento Time		53H	83	General Purpose Controller #8
06H	6	Data Entry MSB		54H	84	Undefined
07H	7	Main Volume		.	.	.
08H	8	Balance Controller		.	.	.
09H	9	Undefined		5AH	90	Undefined
0AH	10	Pan Controller		5BH	91	External Effects Depth
0BH	11	Expression Controller		5CH	92	Tremolo Depth
0CH	12	Undefined		5DH	93	Chorus Depth
.	.	.		5EH	94	Celeste (Detune) Depth
.	.	.		5FH	95	Phaser Depth
0FH	15	Undefined		**PARAMETER VALUE**		
10H	16	General Purpose Controller #1		Controller Hex	Number Decimal	Description
11H	17	General Purpose Controller #2		60H	96	Data Increment
12H	18	General Purpose Controller #3		61H	97	Data Decrement
13H	19	General Purpose Controller #4		**PARAMETER SELECTION**		
14H	20	Undefined		Controller Hex	Number Decimal	Description
.	.	.		62H	98	Non-Registered Parameter Number LSB
.	.	.		63H	99	Non-Registered Parameter Number MSB
1FH	31	Undefined		64H	100	Registered Parameter Number LSB
14-BIT CONTROLLER LEAST SIGNIFICANT BIT				65H	101	Registered Parameter Number MSB
Controller Hex	Number Decimal	Description		**UNDEFINED CONTROLLERS**		
20H	32	LSB Value for Controller 0		Controller Hex	Number Decimal	Description
21H	33	LSB Value for Controller 1		66H	102	Undefined
22H	34	LSB Value for Controller 2		.	.	.
.	.	.		.	.	.
.	.	.		78H	120	Undefined
3EH	62	LSB Value for Controller 30		**RESERVED FOR CHANNEL MODE MESSAGES**		
3FH	63	LSB Value for Controller 31		Controller Hex	Number Decimal	Description
7-BIT CONTROLLERS				79H	121	Reset All Controllers
Controller Hex	Number Decimal	Description		7AH	122	Local Control On/Off
40 H	64	Damper Pedal (sustain)		7BH	123	All Notes Off
41H	65	Portamento On/Off		7CH	124	Omni Mode Off
42H	66	Sostenuto On/Off		7DH	125	Omni Mode On
43H	67	Soft Pedal		7EH	126	Mono Mode On (Poly Mode Off)
44H	68	Undefined		7FH	127	Poly Mode On (Mono Mode Off)
45H	69	Hold 2 On/Off				
46H	70	Undefined				
.	.	. *continues*				

Figure 7.9. *Listing of Controller ID numbers, outlining both the defined format and conventional controller assignments.*

System Messages

As the name implies, system messages are globally transmitted to every MIDI device in the MIDI chain. This is possible because MIDI channel numbers aren't addressed in the byte structure of a system message. Consequently, any device will respond to these messages, regardless of what MIDI channel or channels the device is assigned to.

System Common messages are used to transmit MIDI Time Code, Song Position Pointer, Song Select, Tune Request, and System-Exclusive data throughout the MIDI system or the 16 channels of a specified MIDI port. The following list describes most of the existing System Common messages:

◆ *MIDI Time Code.* Provides a cost-effective and easily implemented means for translating SMPTE time code into an equivalent code that conforms to the MIDI 1.0 specification. The MIDI Time Code Message allows for time-based code and commands to be distributed throughout the MIDI chain. MTC Quarter-Frame messages are transmitted and recognized by MIDI devices capable of understanding and executing MTC commands.

◆ *Song Position Pointer (SPP).* Enables a sequencer or drum machine to be synchronized to an external source (such as a tape machine) from any measure position within a song. The Song Position Pointer message is used to reference a location point within a MIDI sequence (in measures) to a matching location on an external device (such as a drum machine or tape recorder). This message provides a timing reference that increments once for every six MIDI clock messages, with respect to the beginning of a song.

◆ *Song Select Message.* Used to request a specific song from the internal sequence memory of a drum machine or sequencer (as identified by its song ID number). After being selected, the song responds to MIDI Start, Stop, and Continue messages.

◆ *Tune Request.* Used to request that an equipped MIDI instrument initiate its internal tuning routine.

◆ *End of Exclusive (EOX) Message.* Indicates the end of a System-Exclusive message.

System-Exclusive Messages

The System-Exclusive (Sys-Ex) message enables MIDI manufacturers, programmers, and designers to communicate customized MIDI messages between MIDI devices. These messages give manufacturers, programmers, and designers the freedom to communicate, as they see fit, any device-specific data of an unrestricted length. Commonly, Sys-Ex data is used for the bulk transmission and reception of patch data (the stored bits and bytes that tell your instrument how to play back its internal sounds), instrument setup parameters, and sample dump data (a process that allows sampled data to be digitally transmitted via the MIDI protocol).

Unlike standard MIDI messages that communicate note, channel, volume, and controller information to any or all devices in the MIDI chain, Sys-Ex messages contain a unique identifier number that, in most cases, is used to single out and exclusively communicate with one device in the MIDI chain. Quite simply, if a device receives a Sys-Ex message that isn't coded with its manufacturer's ID number, the message that follows is ignored. If the manufacturer and device type match up, however, the receiving device begins accepting an undefined stream of Sys-Ex instructions until an end-of-exclusive (EOX) byte is received (see Figure 7.10).

Figure 7.10. *System-Exclusive data (one ID byte format).*

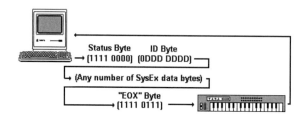

In addition to transmitting files between devices and sequencers, Sys-Ex controller or patch data can be communicated through MIDI in real time. One way to do this is to use a patch-editing program to transmit continuous patch and setup data by varying parameters directly from a computer screen. Or you can use a hardware controller to adjust volume, pan, and other settings by physically moving assignable data sliders. Both of these controller types ease the job of experimenting with parameter values or changing mix moves by avoiding instrument controls that are often cursor and LCD screen hell.

Electronic Musical Instruments

Since its inception in the early '80s, the MIDI-based electronic instrument has become a central creative force in music technology and production. These devices, along with the multitrack home recording market, have changed the face of music production by bringing cost-effective semi-professional and professional production gear into the reach of most musicians. As a result, a strong, new facet of the industry has been created: the personal project music studio.

Although electronic instruments and their associated control devices often vary in form and function, each include all or most of the following standard building block components (see Figure 7.11):

◆ *Central processing unit (CPU).* A dedicated processing chip or computer system expressly designed to handle all or most of the music performance- and control-related messages or commands that must be processed in real time.

◆ *Control panel.* The data-entry controls and display that enable sound patch and output functions to be selected or altered by the user.

◆ *Performance controllers.* Devices such as keyboards, drum pads, and wind controllers for generating real-time performance data and for transmitting this data directly to the instrument's sound generators (through the CPU) or to other instruments within a system via MIDI.

◆ *Voice circuitry.* Systems that can either internally generate sounds or reproduce digital samples and then route these sounds to the instrument's output.

◆ *Memory.* Used for storing sound- and setup-related data (such as patch information, setup configurations, and digital waveform data). This storage media can exist in a system as either read-only memory (ROM)—memory on a factory-encoded memory chip, cartridge, or CD-ROM that can be retrieved only; or as random-access memory (RAM)—memory that can be written to as well as retrieved from a memory chip, cartridge, hard disk, or optical disk.

◆ *Auxiliary controllers.* External controlling devices used in conjunction with a main instrument or controller, such as sustain pedals, remote switch pedals, pitch bend wheel, and modulation wheels.

◆ *MIDI ports.* Allow the transmission and/or reception of MIDI data.

The following sections provide brief descriptions and examples of the most commonly found MIDI-based electronic instruments. The instrument categories include: keyboard, percussion, guitar, and woodwind, as well as controlling devices.

Figure 7.11. The standard block components designed into most electronic musical instruments.

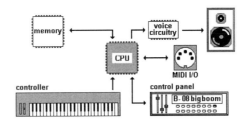

Keyboard Instruments

Keyboard instruments are the most commonly encountered electronic musical instruments in MIDI production. In part, this is because they were the first devices to be developed, and MIDI was initially developed to record and control their internal and performance parameters.

Keyboard-based devices include a central microprocessor, control panel, memory, and general provisions for auxiliary controllers. These devices may or may not incorporate a performance keyboard into their designs (with the latter device being known as a sound *module*), as too many keyboards can be both redundant and space consuming.

The two fundamental keyboard instruments are the synthesizer and the digital sampler. Synthesizers and samplers vary in the number of sounds (called *voices*) that can be produced at one time by their internal generating circuitry. For example, certain instruments can produce only one note at a time, while others, known as *polyphonic instruments,* can generate numerous notes at one time. Polyphonic instruments enable the artist to play chords and/or more than one musical line on a single instrument. In addition, an instrument can be *multitimbral* in nature, which means that it is capable of generating more than one sound *patch* at a time. A patch refers to the system configuration that's required to produce a sound or series of sounds having a particular sonic character. The word *patch* itself is a direct reference to the earlier need for using patch chords to connect the various sound-generating and sound-processing modules on analog synthesizers.

The Synthesizer

A synthesizer (*synth*), shown in Figures 7.12 and 7.13, is an electronic musical instrument that uses multiple sound generators to create complex waveforms. When combined, these waveforms "synthesize" a unique sound character. Each tone generator can be controlled with respect to frequency, amplitude, timbre (sound character), and envelope. Modern synths provide digital control over these analog parameters or, more commonly, generate these waveforms directly within the digital domain.

Figure 7.12. Alesis Quadrasynth 64-voice/76-key master keyboard. (Courtesy of Alesis Corp.)

Figure 7.13. Roland JV-90 expandable synthesizer. (Courtesy of Roland Corporation U.S.)

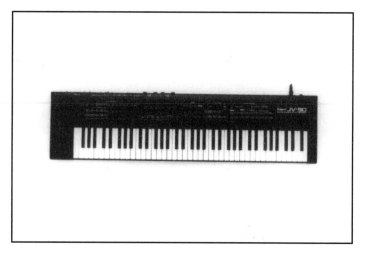

A wide range of synthesizer types currently on the market employ numerous processes for generating complex waveforms. Although analog synths are commonly sought for their simple, hands-on control and "retro" sounds, the vast majority of newer systems use complex digital algorithms (program tables) to manipulate prerecorded waveform samples from ROM to create both new and realistic sounds.

Guitar Synthesizers

Developments in pickup and microprocessor technology have made it possible for the notes and minute inflections of guitar strings to be accurately translated into MIDI data. With this innovation, many of the capabilities that MIDI has to offer (including synthesis, sampling, percussion, controlled effects, and sequencing) are now available to the electric guitarist (see Figure 7.14).

Figure 7.14. The Korg Z3 guitar synthesizer and ZD3 guitar synthesizer driver system. *(Courtesy of Korg USA, Inc.)*

Guitar players often work at stretching the vocabulary of their instruments beyond the traditional norm by using distortion, phasing, echo, feedback, and so on. Now, with the implementation of MIDI, a guitar performance can offer traditional guitar sounds coupled with the advantages of synthesis and MIDI effects technology.

Samplers

A *sampler* (see Figures 7.15 and 7.16) is a device that can convert audio signals into a digitized form, store this digital data in its internal RAM, and reproduce this data (often polyphonically) in an audio production or musical environment. Most sampling systems offer extensive edit and signal processing capabilities, allowing a sample to be modified by the user and saved to computer disk for archival purposes. Samplers also use many of the digital amplification, oscillation, and filtering stages found in digital and analog synthesizers. These filtering stages let the user modify a sample's overall waveshape and envelope.

Figure 7.15. Ensoniq ASR-10 sampler/synthesizer module and keyboard systems. (Courtesy of Ensoniq)

Figure 7.16. ESI-32 professional digital sampler. (Courtesy of E-mu Systems, Inc.)

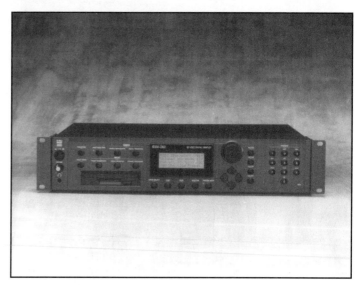

Keyboard-based samplers use performance keyboard controllers to trigger and articulate sampled audio through standard control modifiers, such as velocity, aftertouch, and modulation, as well as the mapping and layering of sounds over a keyboard. After a set of samples has been recorded or recalled from disk memory, each sample within a multiple-voice system can be split across a performance keyboard. In this way, individual sounds can be assigned to a specific key or range of keys. The assignment of a sample over a range of notes enables a sample or multiple samples to be musically played at various sample rates that correspond to a key's proper musical interval.

Newer sampling systems often offer a range of advanced features including stereo sampling, multiple outputs (allowing isolated channel outputs for added mixing and signal processing power or for recording individual samples to a multitrack tape recorder), and CD-ROM drives (allowing easy access to virtually thousands of samples). In addition to storing data on a hard disk, floppy disk, or CD-ROM, sample data can be distributed through standard MIDI lines to another sampler or computer-based sample editor via the MIDI sample dump standard or by way of a high-speed SCSI port.

Percussion

One of the first applications in sample technology was the triggering of prerecorded drum and percussion sounds. This made it possible for electronic musicians (most notably keyboard players) to add samples of actual drum sounds to their compositions without the expense of hiring drummers or percussionists. From this initial application has sprung an important class of sample and synthesis technology that enables artists to create and play back drum patterns and percussion directly from a synthesizer, drum machine, or sampler. The sampler enables an artist to create his or her own sampled drum and percussion sounds that range from traditional to the imaginatively unexpected.

In short, MIDI has placed sampled percussion into the hands of nearly every electronic musician— from those with only fundamental rhythmic talent to the professional percussionist/programmer who can use his/her skills to build complex drum patterns.

The Drum Machine

The drum machine, shown in Figure 7.17, is a sample-based device that has been expressly designed to create and reproduce high-quality drum sounds from its internal, factory-programmed read-only memory. These sounds have been carefully recorded and edited with sounds that can be triggered individually from a series of velocity-sensitive button "pads" located on the drum machine's top panel or that can be triggered from an external controller, such as a keyboard, drum pad controller, or performance trigger kit (see Figure 7.18).

After being selected, these drum voices can be edited using control parameters such as tuning, level, output assignment, and panning position. In addition, most drum machines provide multiple outputs that enable an individual voice or groups of voices to be routed to a specific output. This feature lets isolated voices be processed individually at a mixer or recording console or recorded onto separate tracks of a multitrack tape recorder.

Selected drum sounds can be arranged into rhythmic groups, known as *drum patterns*. These patterns are repeated measures that exist as basic variations on a playing style that has been programmed by the user or taken from an existing library of many of the popular playing styles, such as rock, country, and jazz.

Figure 7.17. *Alesis SR-16 drum machine. (Courtesy of Alesis Corp.)*

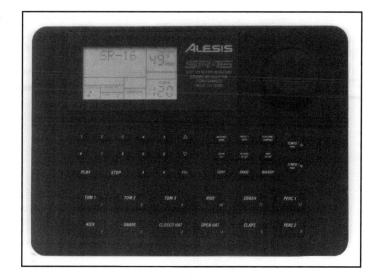

Figure 7.18. *Roland TD-5K Compact Drum System. (Courtesy of Roland Corporation US.)*

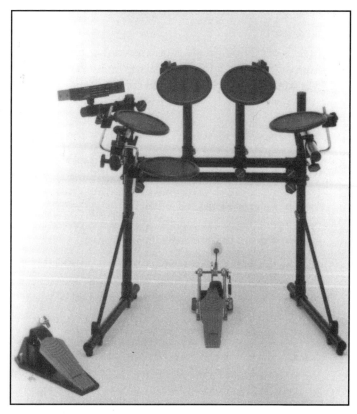

Many drum machines provide a chain function that allows patterns to be linked together into a continuous song. After a song is assembled, it can be played back using an internal MIDI clock source, or it can be synchronously driven from another device using an external MIDI clock source. In the majority of cases, however, the individual voices of a drum machine are triggered from a MIDI sequencer (explained in the following section), which enables the musician to take full advantage of a sequencer's real-time performance and editing capabilities.

Sequencers

One of the most important devices in MIDI production is the MIDI sequencer. A sequencer (see Figure 7.19) is a digitally based device or a computer program that is used to record, edit, and output performance-related MIDI data in a sequential fashion. The recorded MIDI-related channel and system messages commonly represent real-time or non-real-time performance events such as note on/off, velocity, modulation, aftertouch, and continuous controller messages. After a performance has been "recorded" into a sequencer's or a computer's internal memory, the data can be edited and saved to hard or floppy disk. When the sequence is played back, the device outputs these MIDI messages to the various connected MIDI devices within the system to re-create the performance. Unlike a recorded performance in which the instrument's sounds are produced under the direct control of a live player, a sequencer communicates real-time performance data to various electronic instruments, which in turn produce the performed sound.

Figure 7.19. *Basic functional diagram of a sequencer.*

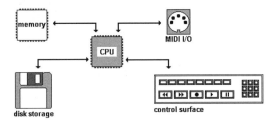

Most sequencers have a design similarity to their distant cousin, the multitrack tape recorder, in that MIDI data can be recorded onto separate "tracks" that contain isolated, yet related, performance material that is synchronous in time. Each of these tracks can be assigned to any MIDI channel and may contain any number of performance- and control-related messages (within the memory constraints of the device). When played back, the instruments and devices in the system that are assigned to a specific MIDI channel (0-16) respond only to the track (or tracks) transmitting on that particular channel.

The number of individual tracks offered varies widely from one manufacturer and model type to the next and ranges from 8 to over 500 tracks. Almost every system is capable of transmitting and receiving data over all 16 MIDI channels, although most professional sequencers can communicate data over two or more independent MIDI data lines, which enables them to address 32 or more separate MIDI channels.

Another important feature offered by most sequencers is the capability to edit MIDI data in the digital domain. Standard cut-and-paste editing techniques generally are offered, which enable segments of sequenced data to be cut, copied, or reinserted at any point in a track or to any other track. Complex algorithms for performing such tasks as velocity changes, modulation and pitch bend, transposition, and humanizing (the controlled randomization of performance data to approximate human timing errors that are generally present in a live performance), as well as control over program or continuous controller messages, can also be inserted and changed.

Hardware-Based Sequencers

Hardware-based sequencers (see Figure 7.20) are stand-alone devices designed for the sole purpose of recording, editing, and playing back MIDI sequences. These systems include a dedicated operating system, microprocessing system, and memory integrated with controls designed for performing sequence-specific functions. These sequencers often vary in the number of available tracks, as well as in the way data can be accessed and displayed.

Figure 7.20. Roland MC-50 MK II MIDI sequencer. (Courtesy of Roland Corp.)

Hardware sequencers commonly emulate the basic function of tape transport (record, play, start/ stop, pause, and so on), in addition to having fast-wind and locate commands for moving quickly to a specified measure in a song. These locate commands are often used in conjunction with manual and automatic punch in/out for overdubbing tracks at various points. In addition, extensive editing features, such as real- and step-time editing (single, note-at-a-time data entry into

a sequenced track), note, velocity, program change, copy and track merging capabilities, and tempo changes are generally offered. Programming, track, and edit information usually is displayed on a liquid crystal display (LCD) that often is limited in size and resolution and generally is limited to information relating to one parameter or track at a time.

Keyboard Workstations

Newer keyboard synthesizer and sampler systems which include internal sequencers that are integrated with the instrument's circuitry are often known as *keyboard workstations* (see Figure 7.21). Such sequencers can record performance data without the need for external MIDI peripherals and can range from having only simple track selection and transport controls to offering a full range of edit and processing capabilities. Often an internal floppy disk is designed into the keyboard to enable sequences (and patch configurations) to be saved for archival purposes.

Figure 7.21. Korg 01/Wfd music workstation. (Courtesy of Korg USA, Inc.)

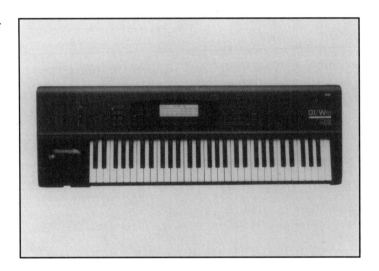

One of the main advantages to such a workstation is ease-of-use and portability. These workstations eliminate the need for carrying around a computer and MIDI accessories should you want to take a sequence to a studio, perform simple Sys-Ex dumps at a friend's house, or play sequenced songs on stage.

Computer-Based Sequencers

Sequencers also are available as software packages that use the personal computer for performing central processing, memory, and I/O (input/output) functions (see Figures 7.22 and 7.23). These systems are often powerful and extremely versatile in their speed, digital signal processing capabilities, memory management, and their capability to perform a diverse range of tasks under software control.

Figure 7.22. Performer 5.0
MIDI sequencer software.
(Courtesy of Mark of the Unicorn,
Inc.)

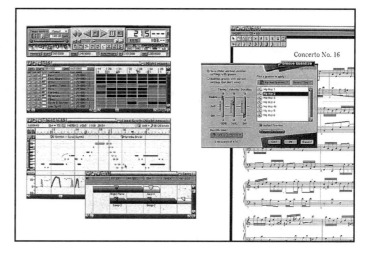

Figure 7.23. Master Tracks
Pro 4 MIDI sequencer software.
(Courtesy of Passport Designs,
Inc.)

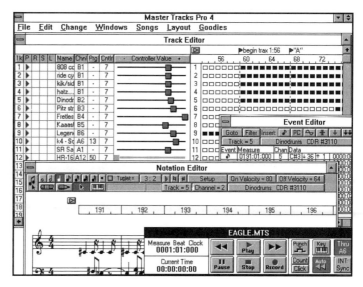

As you might expect, sequencing software is available for most Apple, IBM/compatible, Atari, and Commodore PCs. The majority of these computers require an external MIDI interface that is used for receiving and distributing MIDI data (with the exception of Atari and Commodore PCs, which have a built-in interface).

Computer-based sequencers have several advantages over their hardware-based counterparts. One of the strongest advantages is easy visibility and access to both basic and advanced editing functions, resulting from the PC's extensive DSP and graphics capabilities. Using standard cut-and-paste methods it becomes a simple matter to move a musical segment from one track to another, cut a musical passage from a song and save it to clipboard memory for later use, or copy

a passage in a track. In addition, the large screen and established graphics interface style make it much easier to perform a complex function. Graphics pattern editing also lets the user quickly and easily change the pitch, start, and duration times of a note as it appears on the screen (in a style known as *piano roll editing*), often through the simple movement of a mouse (see Figure 7.24).

Figure 7.24. Piano roll editing capabilities in OSC's METRO MIDI sequencer. (Courtesy of OSC)

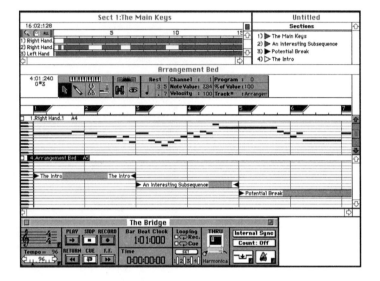

Because computer-based sequencers make use of the PC's memory management capabilities, sequenced files can be easily stored onto either hard or floppy disks, while note capacity is usually restricted only by the PC's amount of internal RAM (which often is user-expandable).

The MIDI Interface

Although both the MIDI protocol and the personal computer communicate through digital data, a digital hardware device known as a *MIDI interface* (see Figures 7.25 and 7.26) must be used to translate MIDI's serial message data into a structure that can be understood by and communicated to the computer's internal operating system.

Figure 7.25. Opcode Studio-5 240 channel MIDI interface/ patchbay with SMPTE capabilities for the Macintosh family of Computers. (Courtesy of Opcode Systems, Inc.)

Figure 7.26. *Music Quest MQX-32M MIDI interface with SMPTE capabilities for the IBM-compatible family of Computers. (Courtesy of Music Quest, Inc.)*

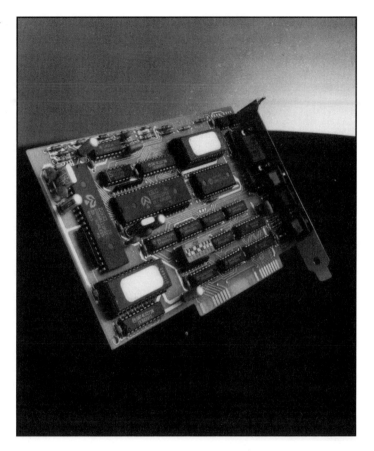

A large number of interface designs for almost every computer type can be found in today's market. They range in design from simple passive systems that essentially provide only external MIDI ports and rely on the computer to do the data exchange and conversion, to intelligent systems that incorporate internal processors for performing mundane calculations and function commands that otherwise would be left to the computer's internal processor. With the advent of multimedia, MIDI interfaces have been placed into the hands of the masses due to the fact that they are designed into almost every multimedia hardware card and many of the newer-generation General MIDI synthesizers.

High-end MIDI interface systems frequently can synchronize a multitrack tape machine and/or VCR to a running MIDI sequence. Such tape-to-data interfacing can be accomplished using sync methods that are either relative with reference to the beginning of a sequence (such as with frequency shift keying (FSK) and Song Position Pointer, both discussed in the following paragraphs) or time-code-based (such as SMPTE time code and MIDI time code).

A relative system for maintaining sync operates by keeping track of its position relative to the beginning of a song. The simplest of these systems is known as *frequency shift keying* (FSK), whereby a modulated frequency is recorded onto a dedicated audio tape track either during or before a recording. When the tape is played back, this recorded code is used to drive the MIDI timing clock, thereby keeping the sequence in sync with the tape tracks. Using this system, the tape and the sequence must be cued up to the song's beginning point in order to work. When the tape begins, the interface transmits a MIDI start command and begins transmitting the sync code. Beginning the tape or sequence from any other location results in a loss of sync. Another method, known as *Song Position Pointer*, can be used to locate a position in the song by calculating the number of measures that have passed since the beginning of the song. Using this system, a tape can be played from any point in the song; however, to alter the tempo and measures, you must spend time entering complex tempo map calculations into a suitably equipped synchronizer.

The answer to most of the problems associated with relative sync can be found in the form of an absolute timing reference known as *time code*. Time code has become a household word in almost all forms of media production. Because of the size and importance of the subject, a fully detailed explanation of time code and MIDI time code can be found in Chapter 8, "Synchronization."

Other Software Packages

In addition to sequencing packages designed to offer most of the production tools a musician might encounter in the day-to-day world, other types of software exist that offer tools for carrying out specific tasks. A few of these packages include: drum pattern editors, algorithmic composition programs, patch editors, and music printing programs.

Drum Pattern Editors

Computer-based drum pattern editors (see Figure 7.27) are designed specifically for programming drum patterns using a straightforward, on-screen system. These mouse-driven programs draw a grid pattern on the computer screen that organizes the various drum sounds along the vertical axis while time is represented on the horizontal axis. You can click the various drum sounds at specific times within a measure to build up individual drum patterns. After one or more patterns have been built up, they can be chained together to form a song. If you want changes in the pattern order, it's a simple matter to alter the pattern sequence in a way that best fits the song.

Drum pattern editors usually offer a number of features, such as the capability to change drum sounds (by changing the assigned MIDI notes), to adjust note and pattern velocities, and to save a sequence as a MIDI file for export to an external sequencer.

Figure 7.27. *Power Chords Pro's*
drum pattern editor. (Courtesy of
Howling Dog Systems)

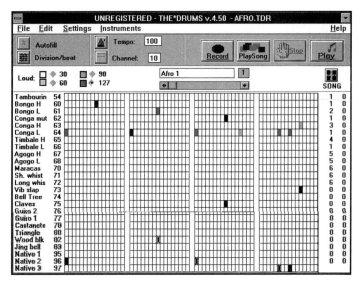

Algorithmic Composition Programs

Algorithmic composition programs (see Figure 7.28) are computer-based interactive sequencers that directly interface with MIDI controllers and standard MIDI files to internally generate MIDI performance data according to a set of user-programmed parameters. The resulting data can be used to gain new ideas for a song or for automatic accompaniment, improvisational exercises, special performances, or just for fun.

Figure 7.28. *The SuperJAM!*
algorithmic composition program.
(Courtesy of Blue Ribbon
SoundWorks, Ltd.)

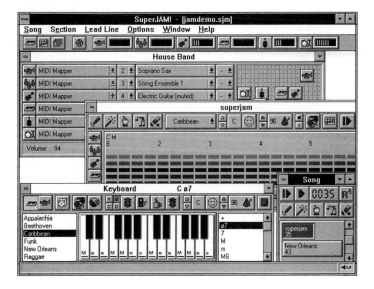

Program parameters can be entered by the user to control the performance according to musical key, notes to be generated, basic order, chords, tempo, and so on. These parameters can be varied in real time either from the computer keyboard, through mouse movements, or by incoming MIDI data.

The Patch Editor

A popular way to gain real-time control over the parameters of a specific instrument or wide range of MIDI devices is through the use of a patch editor. A patch editor is software that can control a MIDI instrument or device by varying its tone and system parameters from a PC. Direct communication between the computer and the device's microprocessor is accomplished through the transmission and reception of system-exclusive messages.

Music Printing Programs

Music printing programs (see Figure 7.29) enable the musician to enter musical data into a computerized score using a number of input methods. Most commonly, music printing and notation programs convert your performance into notes on the PC screen or on a printout. These programs provide the user with a straightforward means for manually or automatically (via MIDI) entering music notation into a score.

Figure 7.29. The Encore music composition and printing program. (Courtesy of Passport Designs, Inc.)

Notation programs vary widely in their capabilities, speed of operation, and number of features offered to the user. Features include such things as the number of staves that can be entered into a single score, the selection of musical symbols available, and the capability to enter and edit text in a composition. After the score has been edited into its final form, a hard copy printout can be produced with a high-quality printer.

Mixing in the MIDI Environment

Over the years, electronic music has had a strong impact on the system requirements and operational approach to mixing hardware. Although traditional console design hasn't changed significantly in recent years, electronic music production has placed new demands on these devices. This increase in system demand is mostly due to the fact that a large number of physical inputs, outputs, and effects are commonly encountered within a MIDI production or project studio. For example, a drum machine with six outputs and a sampler with eight outputs may dominate a smaller mixer and leave room for little else. In such a case, a MIDI system can easily outgrow a console's capabilities and leave you with the unpleasant choice of either upgrading or dealing with your present mixer as best you can. Thus, when you buy a console or mixer, it is always wise to try to anticipate your future expansion needs.

One popular way to handle the increased number of inputs is to use an outboard line mixer, as shown in Figure 7.30. These rack-mountable mic/line mixers (sometimes called *submixers*) can often handle up to 16 or 24 line-level inputs that can then be mixed down to two channels. These channels can then be routed to either two inputs or auxiliary returns on your main mixing system. On some mixer designs, special "multi" inputs enable multiple mixers to be directly tied together in a chain fashion without using up additional line-level input strips.

Figure 7.30. Mackie CR-1604 16-channel mixing system. (Courtesy of Mackie Designs, Inc.)

Mixing through MIDI

Although standard mixing practices still dominate in most multitrack recording studios, in many MIDI production studios, MIDI itself can provide the artist with the capability to mix audio by controlling how a MIDI instrument or device outputs the generated signals. This mixing capability is accomplished in a completely automated environment by altering MIDI channel

messages and transmitting them to instruments or devices in the system. As these messages can be transmitted over individual MIDI channels, they can be assigned to a specific instrument or voices within an instrument—effectively creating a powerful system that offers extensive automated mixing capabilities.

Because the vast majority of electronic instruments allow dynamic MIDI control, a continuous value ranging from 0 (minimum) to 127 (maximum) or a switched on/off value can be assigned to such performance and system parameters as velocity or controller messages.

◆ *Velocity Messages.* These messages are used to transmit the individual volume levels of each note as they are played. This parameter gives expression to a composition. Most sequencers can adjust the velocity levels by selecting the note or region to be affected and instructing the program as to the newly desired velocity and method of gain change.

◆ *Continuous Controller Messages.* These messages are generally used to control dynamic mixing events such as main volume (an instrument's overall output volume) or such other controller messages as panning, pitch bend, modulation, and aftertouch. (A full listing of these controllers can be found in Figure 7.9.)

In order to take better advantage of the full range of controller messages that are available, MIDI sequencers often provide an on-screen interface for directly controlling and automating mix-based functions. The layout and capabilities of these mix screens vary among software packages and range from a relatively straightforward screen, as shown in Figure 7.31, to the more comprehensive, user-programmable layouts found in top-of-the-line professional programs, like the one shown in Figure 7.32.

Figure 7.31. *Control over MIDI velocity and pan settings within Cakewalk Pro for Windows. (Courtesy of Twelve Tone Systems)*

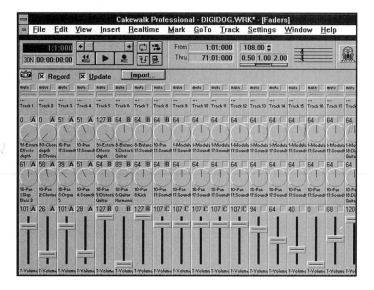

Figure 7.32. An example of Performer 5.0's MIDI mixing capabilities. (Courtesy of Mark of the Unicorn, Inc.)

Hardware MIDI Controllers

Although many sequencing programs allow for on-screen control over dynamic mixing functions, sometimes not even this feature can compare with the hands-on feel of having actual mixing faders. One possible solution is to mix using a MIDI command controller (see Figure 7.33).

Figure 7.33. Peavey PC 1600 MIDI controller. (Courtesy of Peavey Electronics Corp.)

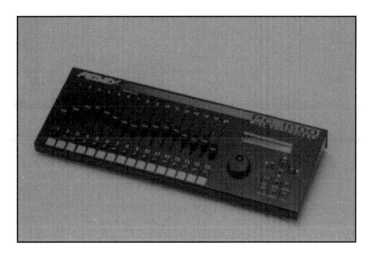

The advantage of using a command controller is that you can vary mix, patch, and system parameters in real-time by simply moving a sliding data fader. Such controllers can be programmed to effect continuous gain change, pan messages, and so on, over a number of channels or groups of channels. Controllers generally are designed with groupings of eight or sixteen faders and often provide buttons for controlling mute, solo, and other such on/off events.

MIDI Implementation in Mixers and Consoles

A growing number of audio mixers and consoles offer a degree of MIDI implementation that enables them to be remotely controlled from a MIDI controller or sequencer (see Figure 7.34). Such systems can be either analog or digital and often vary in the degree of automation over which MIDI has control. Such automation systems can range from having the simple capability to control basic mute functions to creating static level and mute "snapshots" for reconfiguring system settings. The latter often is accomplished by assigning a mix snapshot to a specific program change number. By transmitting the appropriate program change number from a controller or sequencer, the system instantly configures itself to the appropriate settings. In conclusion, complete dynamic control over automated mixing functions via MIDI is generally reserved for the growing number of fully digital mixing systems and digitally controlled analog mixers.

Figure 7.34. OTTO MIDI-based automation for the Mackie CR-1604 mixing system. (Courtesy of Mackie Designs, Inc.)

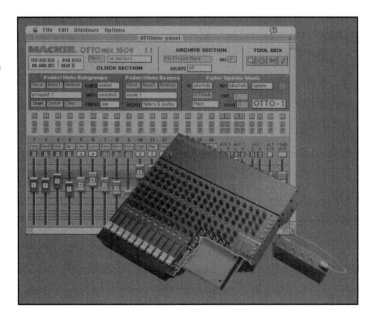

CHAPTER 8

Synchronization

The professional audio and visual markets increasingly incorporate audio, video, film, and electronic music media. In video postproduction techniques, for example, audio transports, video transports, and electronic musical instruments are routinely operated in computer-controlled tandem—allowing the operator to create and refine a video soundtrack to perfection (see Figure 8.1). The method that enables multiple audio and visual media to maintain a direct time relationship is known as *synchronization,* or *sync.*

Synchronization is the occurrence of two or more events at precisely the same time. With respect to analog audio and video systems, synchronization is achieved by interlocking the transport speeds of two or more machines. In computer-related systems, such as digital audio and MIDI, internal sync or external sync between compatible devices is often maintained by using a clocking pulse that's directly imbedded within the communicated digital word structure. Frequently, it is necessary to maintain synchronization between both analog- and digital-based devices; as a result, some rather ingenious forms of systems communication and data translation have been developed.

This chapter examines the various forms of synchronization for both analog and digital devices as well as current methods for maintaining sync between the two media.

Figure 8.1. *Example of an integrated audio production system.*

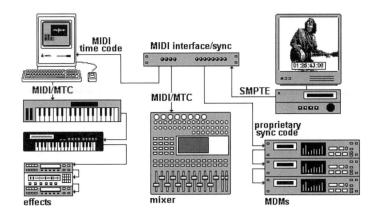

Synchronization between Analog Transports

Maintaining relative synchronization between analog tape transports doesn't require that all transport speeds involved in the process be constant; however, the transports must maintain the same "relative" speed at all points in time.

It's a fact of life that analog tape devices are unable to maintain a perfectly constant reproduction speed. For this reason, synchronization between two or more machines without some form of synchronous "locking" would be impossible over any reasonable program length. Synchronization would soon be lost as a result of such factors as voltage fluctuations and tape slippage. It therefore becomes clear that if production is to utilize multiple media, a means of interlocking these devices in synchronous time is essential.

Time Code

The standard method of interlocking audio, video, and film transports makes use of the Society of Motion Picture and Television Engineers (SMPTE) *time code.* The use of time code enables identification of an exact position on a magnetic tape by assigning a digital address to each specified length. This address code cannot slip and always retains its original location, which allows for continuous monitoring of tape position to an accuracy of between 1/24th and 1/30th of a second (depending on the media type and frame standards being used). The specified tape segments are called *frames*, a term taken from film production. Each audio or video frame is tagged with a unique identifying number, known as a *time code address.* This eight-digit address is displayed in the form 00:00:00:00, in which the successive pairs of digits represent hours:minutes:seconds:frames (see Figure 8.2).

Figure 8.2. *Readout of a SMPTE time code address.*

hours   minutes   seconds   frames

The recorded time code address is used to locate a position on magnetic tape in much the same way that a letter carrier uses an address to deliver the mail to the correct address (see Figure 8.3a). Suppose that a time-encoded videotape begins at time 00:01:00:00 and ends at 00:28:19:05 and contains a specific cue point, such as a glass shattering at 00:12:53:18 (see Figure 8.3b). By monitoring the time code (in a fast shuttle mode), it's a simple matter to locate the position corresponding to the cue point on the tape and perform whatever function is necessary, such as adding an effect to the sound track at that specific point ... CRASH.

Figure 8.3. *Location of relative addresses.*

a. A postal address.

b. Time code addresses and a cue point on longitudinal tape.

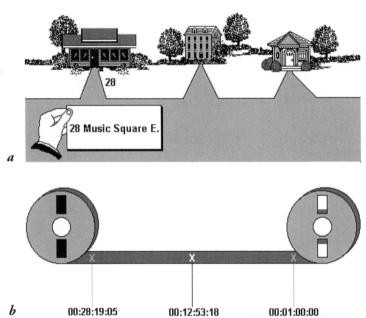

The standard means of encoding time code in audio production is to record (*stripe*) an open audio channel (usually the highest available track) with longitudinal time code (LTC) that can then be read directly from the recorded track in either direction and at a wide range of tape speeds.

The Time Code Word

The total of all time-encoded information recorded into each audio or video frame is known as a *time code word.* Each word is divided into 80 equal segments called *bits,* and these bits are numbered consecutively from 0 to 79. One word covers an entire audio or video frame, so that for every frame there is a corresponding time code address. Address information is contained in the digital word as a series of bits made up of binary ones and zeros. These bits are electronically encoded within the signal in the form of a modulated square wave. This method of encoding information is known as *biphase modulation.* With biphase, a voltage transition in the middle of a half-cycle of a square wave equals a bit value of 1, as shown in Figure 8.4. No transition within this same period signifies a bit value of 0. The primary feature about this encoding method is that detection relies on shifts within the pulse and not on the pulse's polarity. Consequently, time code can be read in either the forward or reverse direction and at fast or slow shuttle speeds.

Figure 8.4. *Biphase modulation encoding.*

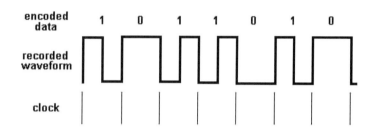

The 80-bit time code word is subdivided into groups of 4 bits (see Figure 8.5), with each grouping representing a specific coded piece of information. Within each of these 4-bit segments is the encoded representation of a decimal number ranging from 0 to 9 written in *binary-coded decimal* (BCD) notation. When a time code reader detects the pattern of ones and zeros in a 4-bit group, it interprets the information as a single decimal number. Within the time code word, eight of these 4-bit groupings constitute an address in hours, minutes, seconds, and frames.

User-Information Data

The 26 digital bits that make up the time code address are joined by an additional 32 bits called *user* bits. This additional set of encoded information, which is also represented in the form of an eight-digit number, has been set aside for the time code users to enter their own ID information. The SMPTE standards committee has placed no restrictions on the use of this "slate code," which may contain such information as date of shooting, shot or take identification, reel number, and so on.

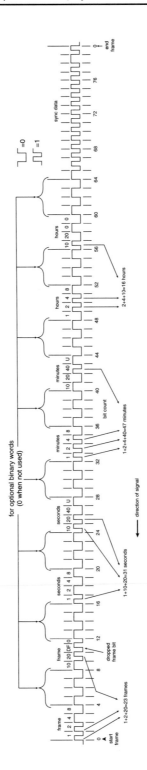

Figure 8.5. Biphase representation of the SMPTE time code word.

Sync-Information Data

Another form of information encoded within the time code word is *sync data*. The sync data—found in 16 bits at the end of the time code word—defines the end of each frame. Because time code can be read in either direction, the sync-data bits also signal the controlling device as to which direction the tape or digital device is moving.

Time Code Frame Standards

In productions using time code, it's important that the readout display be directly related to the actual elapsed time of a program, particularly when dealing with the exacting time requirements of broadcasting. In the case of a black and white (monochrome) video signal, a rate of exactly 30 frames per second (fr/second) is used. This monochrome rate is known as *nondrop code*. If this time code is read, the time code display, program length, and actual clock-on-the-wall time would all be in agreement.

This simplicity was broken, however, when the National Television Standards Committee set the frame rate for the color video signal at 29.97 fr/second. This would mean that if a time code reader set up to read the monochrome rate of 30 fr/second is used to read a color program, the time code readout would pick up an extra .03 frame for every second passed. In the duration of an hour, the readout will differ from the actual tape address by a total of 108 frames (or 3.6 seconds).

In order to correct for this discrepancy and to regain an agreement between the time code readout (for color video) and the actual elapsed time, a means of frame adjustment was introduced into the code. Because the object is to drop 108 seconds over the course of an hour, the code used for color has come to be known as *drop-frame code*. Two frame counts for every minute of operation are omitted, with the exception of minutes 00, 10, 20, 30, 40, and 50. This has the effect of adjusting the frame count to agree with the actual elapsed program duration.

In addition to the color 29.97 drop-frame code, a 29.97 non-drop-frame color standard is also commonly used in video production. When using this nondrop time code, the frame count always advances one count per frame, without any drops; however, this mode results in a disagreement between the frame count and the actual clock-on-the-wall time over the course of the program. Nondrop mode has the distinct advantage of easing the time calculations often required in the video editing process because no frame compensations have to be taken into account for dropped frames.

Another frame rate format used throughout Europe is the European Broadcast Union (EBU) time code. EBU utilizes the SMPTE 80-bit code word but differs from it in that EBU uses a 25-fr/second frame rate. Because both monochrome and color video EBU signals run at exactly 25-fr/second, an EBU drop-frame code is not necessary.

The film media makes use of a standardized 24-fr/second SMPTE format. Many newer synchronization and digital audio devices offer film sync and offset calculation capabilities.

LTC and VITC Time Code

Currently, two major methods for encoding time code onto magnetic tape for broadcast and production use exist: LTC and VITC.

Time code recorded onto an audio or video cue track is known as *longitudinal time code* (LTC). LTC encodes the biphase time code signal onto an analog audio or cue track as a modulated square wave signal with a bit rate of 2400 bits/second.

The recording of a perfect square wave onto a magnetic audio track is, under the best of conditions, difficult. For this reason the SMPTE standard has set forth an allowable risetime of 25 ±5 microseconds for the recording and reproduction of code. This tolerance requires a signal bandwidth of 15 kHz—well within the range of most professional audio recording devices.

Variable-speed time-code readers often are able to decode time code information at shuttle rates ranging from 1/10th to 100 times normal playspeed. This is effective for most audio applications; however, within video postproduction, it often is necessary to monitor a video tape at slow or still speeds. As LTC can't be read at speeds slower than 1/10th to 1/20th normal playspeed, two methods can be used for reading time code. In the first method, the time code address can be "burned" into the video image of a copy worktape in which a character generator is used to superimpose the corresponding address in an on-screen window (see Figure 8.6). This *window dub* allows the time code to be easily identified, even at very slow or still picture shuttle speeds.

The second method used by major video production houses is to stripe the picture with *vertical interval time code* (VITC). VITC makes use of the same SMPTE address and user code structure as LTC, but it is encoded onto video tape in an entirely different signal form. The VITC method actually encodes the time code information in the video signal itself—inside a field (known as the *vertical blanking interval*) that is located outside the visible picture scan area. Because the time code information is encoded in the video signal itself, it's possible for professional helical scan video recorders to read time code at slower speeds and still frame. Because time code can be accurately read at all speeds, this added convenience opens up an additional track on a video recorder for audio or cue information and eliminates the need for a window dub.

Figure 8.6. *Video image showing "burned-in" time code window.*

In most situations, LTC code is preferred for audio, electronic music, and standard video production as it is a more accessible and cost-effective protocol.

Jam Sync/Restriping Time Code

Longitudinal time code operates by recording a series of square-wave pulses onto magnetic tape. Unfortunately, it's somewhat difficult to record a square waveform onto analog magnetic tape without suffering moderate to severe waveform distortion (see Figures 8.7a and 8.7b). Although the binary-based time code reader is designed to be relatively tolerant of waveform amplitude fluctuations, such distortion is severely compounded when code is directly copied by one or more generations. For this reason, a feature known as *jam sync* has been incorporated into most time code synchronizers. Jam sync basically regenerates fresh code in order to match old time code address numbers identically during the dubbing stage or to reconstruct defective sections of code. In the jam sync mode, the output of the generator is slaved to an external time code source. After reading this incoming signal, the generator outputs an undistorted signal identical to that of the original time code address.

Figure 8.7. *Representation of the recorded biphase signal.*

a. *Original biphase signal.*

b. *Reproduced biphase signal.*

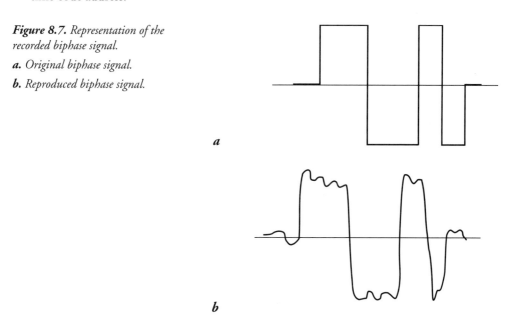

a

b

Two forms of jam sync are currently in use: one-time jam sync and continuous jam sync. Using one-time jam sync, the receipt of valid time code causes the generator's output to be initialized to the first valid address number. The generator then begins to count in an ascending order on its own, in a "freewheeling" fashion. It ignores any deterioration or discontinuity in code and produces fresh, uninterrupted address numbers.

Continuous jam sync is used in cases in which the original address numbers need to remain intact and shouldn't be regenerated into a contiguous address count. After the reader has been activated, the generator updates the address count for each frame in accordance with the incoming address.

Synchronization Using Time Code

To achieve a frame-for-frame time code lock among multiple audio, video, and film transports, it is necessary to employ a device known as a *synchronizer* (see Figure 8.8). The basic function of a synchronizer is to control one or more tape or film transports (designated as "slave" machines), whose position(s) and tape speed(s) are made to accurately follow one specific transport (designated as the "master").

Figure 8.8. *The synchronizer in time code production.*

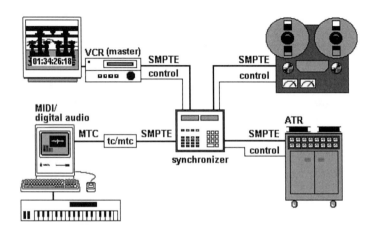

The most common synchronizer types found in modern postproduction are the chase synchronizer, the control synchronizer, and the edit decision list (EDL) synchronizer.

The *chase system* of synchronization requires that a slave transport chase the master under all conditions. This system is a bit more sophisticated than that of a playspeed-only system in that the synchronizer has additional control over the slave transport's operating functions and is often able to read the system time code for locating areas in the fast shuttle/wind mode. This enhancement enables the slave transport to automatically switch from the play mode back into the cue (search) mode, in order to chase the master under all conditions and resync to it when the master is again placed into the play mode.

The *control synchronizer* (see Figure 8.9) emphasizes control over all transport functions found in a synchronized system. Operating from a central keyboard, the control synchronizer provides control options that include the following:

◆ *Machine selection.* Enables the selection of machine(s) to be involved in the synchronization process, as well as allowing for the selection of a designated master.

- ◆ *Transport control.* Enables conventional remote control of functions over any or all machines in the system.

- ◆ *Locate.* A transport command that causes all selected machines to locate automatically to a selected time code address.

- ◆ *Looping.* Enters a continuous repeat cycle (play, rewind, play again) between any two address cue points stored in memory.

- ◆ *Offset.* Permits the correction of any difference in time code that exists between program material (that is, to adjust relative frame rates by ± X frames to achieve or improve sync).

- ◆ *Event points.* A series of time code cue points entered into memory for use in triggering a series of function commands (for example, start slave or mastering machine, record in/out, insert effects device or any externally triggered device).

- ◆ *Record in/out.* An event function that allows the synchronizer to take control over transport record/edit functions, enabling tight record in/out points to be repeated with frame accuracy.

Another entry into the field of audio and video synchronization is the *edit decision list* (EDL) controller/synchronizer. This synchronizer has evolved from the use of the EDL in the online video editing process and is most commonly found in the video and audio-for-video postproduction suite.

The EDL is a series of edit commands that can be entered, stored and retrieved as a computer file. After the information has been entered or loaded, the system electronically controls, synchronizes, and switches all the associated studio transports, and then exerts control over edit in/out points, tape positions, time code offset instructions, and so on, with complete time code accuracy and repeatability.

Figure 8.9. *TimeLine's Micro Lynx synchronizer system. (Courtesy of TimeLine Vista, Inc.)*

Production Setup for Using Time Code

In audio production, the only connection usually required between the master machine and the synchronizer is the time code playback signal (see Figure 8.10). If a control synchronizer is used, a number of connections are required between the slave transports and synchronizer. These include provisions for the time code signal, full logic transport control, and a DC control voltage (for driving the slave machine capstan).

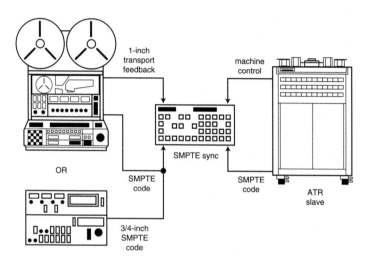

Figure 8.10. *System interconnections for synchronous audio production.*

Distribution of LTC Signal Lines

Longitudinal time code may be distributed throughout the production and postproduction system as any other audio signal is distributed. It can be routed through audio distribution amplifiers and patched through audio switching systems via normal two-conductor shielded cables. Because the time code signal is biphase or symmetrical, it's immune to problems of polarity.

Time Code Levels

One problem that can plague systems using time code is *crosstalk*, which arises from a high-level time code signal interfering with adjacent audio signals or recorded tape tracks. Currently, no industry standard levels exist for the recording of time code on magnetic tape. The levels shown in Table 8.1, however, have proven to provide the best results.

Table 8.1. Optimum time code recording levels.

Tape Format	Track Format	Optimum Recording Level
ATR	Edge track (highest number)	–5 VU to –10 VU
3/4-inch VTR	Audio 1 (L) track or time	–5 VU to 0 VU code track
1-inch VTR	Cue track or audio 3	–5 VU to –10 VU
Modular digital multitrack	Highest number track	–20 dB

Note: If the VTR to be used is equipped with Automatic Gain Compensation (AGC), override the AGC and adjust the signal gain controls manually.

Synchronization within Electronic Music Production

Just as synchronization is routinely used in audio and video production, the wide acceptance of MIDI and digital audio within the various media has created the need for devices such as MIDI sequencers, digital audio editors, effects devices, and automated mixing systems to be integrated into a synchronous production environment. The following sections discuss many forms of synchronization that can be encountered in a pre-MIDI or MIDI-based electronic music environment.

Non-MIDI Synchronization

Several types of synchronization are used by older electronic instruments and devices designed before the MIDI specification was implemented. Although sync between these non-MIDI and MIDI instruments can be a source of mild to major aggravation, many of these older devices can be found in MIDI setups because of their distinctive and great sounds.

Click Sync

Click sync or *click track* refers to the metronomic audio clicks generated by electronic devices to communicate tempo. These are produced once per beat or once per several beats (as occurs in cut-time or compound meters).

A click can be designed into an instrument or MIDI interface, so as to produce an audible metronomic tone (or series of tone pitches) that can be heard and followed as a tempo guide. Such devices often offer an unbalanced output jack that can be fed into a mixer and blended into the monitor mix so that musicians can keep in tempo with a sequenced composition. This click can also be recorded onto an unused track of a multitrack recorder for use with projects that involve both sequenced and live music.

Certain sync boxes and older drum machines are capable of syncing a sequence to a live or recorded click track. Such devices can determine the tempo based on the frequency of the clicks and can then output a MIDI start message (once a sufficient number of click pulses have been received for tempo calculation). A MIDI stop message can be transmitted whenever more than two clicks have been missed at or below the slowest possible tempo (generally 30 BPM).

This sync method doesn't work well with rapid tempo changes because its chase resolution is limited to one click per beat (1/24th the resolution of MIDI clock). Therefore, it's best to use a click source that is relatively constant in tempo.

TTL and DIN Sync

Before the adoption of MIDI, one of the most common methods of synchronization between early sequencers, drum machines, and instruments was TTL 5-volt sync. This method uses 5-volt clock pulses, whereby a swing from 0 to 5 volts represents one clock.

In this system, a musical beat is divided into a specific number of clock pulses per quarter note (that is, 24, 48, 96, and 384 pulses per quarter note—ppq), which varies from device to device. For example, DIN sync, a form of TTL sync named after the now famous DIN connector, transmitted at 24 ppq.

TTL may be transmitted in either of two ways. The first and simplest method uses a single conductor through which a 5-volt clock signal is sent. Quite simply, after the clock pulses are received by a slave device, the device starts and synchronizes to the incoming clock rate. Should these pulses stop, the devices also stop and wait for the clock to resume. The second method uses two conductors, both of which transmit 5-volt transitions. Within the system, however, one line is used to constantly transmit timing information, while the other is used for start/stop information.

MIDI-Based Synchronization

In current MIDI production, the most commonly found form of synchronization uses the MIDI protocol itself for the transmission of sync messages. These messages are transmitted along with other MIDI data over standard MIDI cables and require no additional or specialized connections.

MIDI Sync

The most fundamental way to lock together the various instruments and devices in every MIDI system is known as *MIDI sync*. This sync protocol is primarily used for locking together the precise timing elements of MIDI devices in an electronic music system. This protocol operates through the transmission of real-time MIDI messages over standard MIDI cables.

As with all forms of synchronization, one MIDI device must be designated to be the master device in order to provide the timing information to which all other slaved devices are locked.

MIDI Real-Time Messages

MIDI real-time messages consist of four basic types, with each being one byte in length: timing clock, start, stop, and continue messages.

- ◆ *Timing clock.* Generally transmitted to all devices in the MIDI system at a rate of 24 times per quarter note (24 ppq). Recently, a few manufacturers have begun to develop devices that generate and respond to 24 clock signals per metronomic beat. This method is used to improve the system's timing resolution and simplify timing when working in nonstandard meters (that is, 3/8, 5/16, 5/32, and so on).

- ◆ *Start.* Upon receipt of a timing clock message, the start command instructs all connected devices to begin playing from the beginning of their internal sequences. Should a program be in mid-sequence, the start command repositions the sequence back to its beginning, at which point it begins to play.

- ◆ *Stop.* Upon the transmission of a MIDI stop command, all devices in the system stop at their current positions and wait for a message to follow.

- ◆ *Continue.* Following the receipt of a MIDI stop command, a MIDI continue message instructs all sequencers and/or drum machines to resume playing from the precise point at which the sequence was stopped. Certain older MIDI devices (most notably drum machines) aren't capable of sending or responding to continue commands. In such a case, the user must either restart the sequence from its beginning or manually position the device to the correct measure.

Song Position Pointer

In addition to MIDI real-time messages, the song position pointer (SPP) is a MIDI system-common message that acts as a relative gauge of musical time (as expressed in measures) that has passed since the beginning of the sequence. The song position pointer is expressed as multiples of six timing clock messages and is therefore equal to the value of a 16th note.

Song position pointers enable a compatible sequencer or drum machine to be synchronized to an external source from any position in a song (with a duration of 1024 measures or less). Using SPP, therefore, it's possible for a sequencer to chase and lock to a multitrack tape from any specific measure within a song.

Within such a MIDI/tape setup, as shown in Figure 8.11, a specialized sync tone is transmitted by a device that encodes the sequencer's SPP messages and timing data directly onto tape as a modulated signal. Unlike SMPTE time code, the means by which manufacturers encode this data onto tape is not standardized. This lack of a standard format could prevent SPP data written by one device from being decoded by another device that uses an incompatible proprietary sync format.

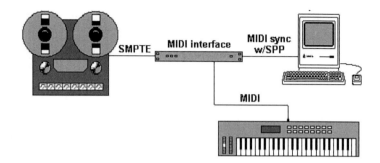

Figure 8.11. *A tape-to-MIDI synchronizer with SPP is used as a timing interface in studio production.*

Many SMPTE-to-MIDI synchronizers also can be used to instruct a slaved sequencer, drum machine, or other device (or devices) to locate to a specific position in the sequence (as defined by the number of 16th notes from the beginning of a song). After the device, or devices, have located to the correct position, the system stops and waits until a continue message and the timing clocks that follow have been received.

In order to vary tempo while maintaining sync between the sequencer and SMPTE control track, many SMPTE-to-MIDI synchronizers can be preprogrammed to create a "tempo map" that provides for tempo changes at specific SMPTE times. After the SPP control track has been committed to tape, however, the tape and sequence are locked to this predetermined tempo or tempo map.

SPP messages usually are transmitted only while the MIDI system is in the stop mode. This is because a short period of time is needed for the slaved device to locate to the correct measure before playback can begin.

Certain devices, such as earlier sequencers and drum machines, don't respond to SPPs. In order to take advantage of the sync benefits, it is best that these devices be slaved to a master timing device that does respond to these pointers.

FSK

In the pre-MIDI days of electronic music, musicians discovered that it was possible to synchronize instruments that were based on such methods as TTL 5-volt sync to a multitrack tape recorder. This was done by recording onto tape a square wave signal (see Figure 8.12a) that could serve as a master sync pulse. As the most common pulses per quarter in use were 24 and 48 ppq, the recorded square wave consisted of a 24-Hz or 48-Hz signal.

Although this system worked, it wasn't without difficulties. These difficulties were due to the fact that the devices being synced to the pulse relied on the integrity of the square wave's sharp transition edges to provide the clock. Because tape is notoriously bad at reproducing a square wave (see Figure 8.12b), not to mention its poor frequency response and reduced reliability at low frequencies, a better system for syncing MIDI to tape had to be found. Initially, the answer was found in *frequency shift keying*, or FSK.

FSK works in much the same way as the TTL sync track. Instead of recording a low frequency square wave onto tape, however, FSK uses two, high-frequency square wave pitches to mark clock transitions (see Figure 8.12c). In the case of the MPU-401/compatible interface, these two frequencies are 1.25 kHz and 2.5 kHz. The rate at which these pitches alternate determines the master timing clock to which all slaved devices are synced. These slave devices are able to detect a change in modulation, convert these into a clock pulse, and then advance their own clocks accordingly.

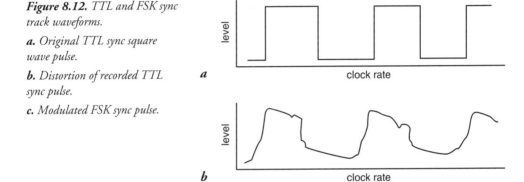

Figure 8.12. TTL and FSK sync track waveforms.

a. Original TTL sync square wave pulse.

b. Distortion of recorded TTL sync pulse.

c. Modulated FSK sync pulse.

Because FSK uses a higher frequency range than its earlier counterpart, it's far more resistant to signal deformation and low-frequency rolloff. When reproducing FSK from tape, however, the tape recorder's speed should be kept within reasonable tolerances. A wide shift upward or downward may make it difficult for the slaved device to recognize the frequency shift as a valid clock transition. For this reason, tempo can be varied only within reasonable limits when varying the speed of the tape recorder.

The level at which FSK can be recorded onto tape is also a consideration. Too low a level can make it difficult for the slaved device(s) to accurately decipher the signal. Signals that are too loud may cause problems with distortion or crosstalk. The most commonly accepted VU levels are −3 dB for semiprofessional recorders and −10 dB for professional equipment. If possible, turn off noise reduction on the FSK track.

MIDI Time Code

For decades, SMPTE time code has been the standard timing reference in audio and video production because it is an absolute timing reference that remains constant throughout a program. On the other hand, both MIDI clock and the song position pointer are relative timing references that vary with both tempo and tempo changes. As most studio-based operations are referenced to SMPTE time code addresses, as opposed to the beats in a musical bar, it would be extremely tedious for a studio engineer or musician to convert between the two timing systems when cueing or triggering a specific event.

In order for MIDI-based devices to operate on an absolute timing reference independent of tempo, *MIDI time code* (MTC) was developed. Created by Chris Meyer and Evan Brooks of Digidesign, MIDI time code provides a cost-effective and easily implemented means for translating SMPTE time code into MIDI messages. It also enables time-based code and commands to be distributed throughout the MIDI chain to those devices or instruments capable of understanding and executing MTC commands.

MTC doesn't replace, but rather is an extension of, MIDI 1.0, in that it makes use of existing message types that previously were either undefined or were being used for other, nonconflicting purposes, such as the sample dump standard. Most existing MIDI devices don't need and will never directly use MTC; however, newer devices and time-related program packages that read and write MTC are currently being developed.

MTC uses a reasonably small percentage of the available MIDI bandwidth (about 7.68% at 30-fr/second). Although MTC can travel the same signal path as conventional MIDI data, if at all possible, the MTC signal path should be kept separate from the MIDI performance path to prevent data overloading or delay.

MIDI Time Code Control Structure

The MIDI time code format can be broken into two parts: time code and MIDI cueing. The time code capabilities of MTC are relatively straightforward and allow both MIDI and non-MIDI devices (through triggered switch outputs available on many synchronizer/MIDI interface systems) to attain synchronous lock or to be triggered through SMPTE time code. MIDI cueing is a format that informs MIDI devices of events to be performed at a specific time (such as load, play, stop, punch in/out, reset). This means of communication envisions the use of intelligent MIDI devices that can prepare for a specific event in advance and then execute a command on cue.

MIDI Time Code Commands

MIDI time code uses three message types: quarter frame messages, full messages, and MIDI cueing messages.

Quarter Frame Messages

Quarter frame messages are transmitted only while the system is running in real time or varispeed time and in either the forward or reverse directions. In addition to providing the system with its basic timing pulse, four frames are generated for every SMPTE time code field. This means that should you decide to use drop-frame code (30 frames per second), the system would transmit 120 quarter frame messages per second.

You can think of quarter frame messages as groups of eight messages that encode the SMPTE time in hours, minutes, seconds, and frames. Because eight quarter frames are required for a complete time code message, the complete SMPTE time is updated every two frames. Each quarter frame message contains two bytes. The first byte is "F1," the Quarter Frame Common header, while the second byte contains a nibble (four bits) that represents the message number (0 through 7) and a nibble for each of the digits of a time field (hours, minutes, seconds, and frames).

Full Messages

Quarter frame messages are not sent while in the fast-forward, rewind, or locate modes because this would unnecessarily clog or outrun the MIDI data lines. When in any of these shuttle modes, a full message is used that encodes the complete time code address in a single message.

After a fast shuttle mode is entered, the system generates a full message and then places itself in a pause mode until the time-encoded device has autolocated to its destination. After the device has resumed playing, MTC again begins sending quarter frame messages.

MIDI Cueing Messages

MIDI cueing messages are designed to address individual devices or programs within a system. These 13-bit messages can be used to compile a cue or an edit decision list, which in turn instructs one or more devices to play, punch in, load, stop, and so on at a specific time. Each instruction within a cueing message contains a unique number, time, name, type, and space for additional information. At the present time, only a small percentage of the possible 128 cueing event types have been defined.

Direct Time Lock

Direct Time Lock (DTL), is a synchronization standard that allows Mark of the Unicorn's Mac-based sequencer, Performer, to lock to SMPTE through a converter that supports this standard. The following is a detailed technical specification.

Two messages are associated with Direct Time Lock: tape position and frame advance.

◆ *Tape position.* The tape position message is transmitted when the time code source (such as a tape machine) is started, whereby the converter achieves lock. This message is implemented as a system-exclusive message and specifies the tape's SMPTE position in

hours, minutes, seconds, and frames (HH:MM:SS:FF). The tape position message is a starting reference for the frame advance messages; it doesn't have to be sent in sync with anything. A converter can send the message a few frames before the specified frame is reached.

◆ *Frame advance.* The frame advance message is transmitted once each frame. The first frame advance sent after the tape position message corresponds to the beginning of the frame specified in the tape position message. Successive frame advances correspond to successive frames. The frame advance message is the same as the MIDI clock message used in normal MIDI sync. This is a real-time message as defined by MIDI and can be inserted in the middle of a normal MIDI message for minimum timing delay. Optionally, the converter can send periodic tape messages between frame advance while the tape is running. It is recommended by Mark of the Unicorn that periodic tape position messages be sent every half second to every second.

After a tape position message has been received, the sequence chases to that point and playback is readied. Once frame messages are received, Performer advances in sync with the master code. If more than eight frames of time pass without a frame advance message, Performer assumes that playback has stopped; it then stops the sequence and waits for a new position message.

If another position message is received, Performer compares it to the position of its last position message. If they are the same, Performer immediately continues playing. Otherwise, Performer chases to the desired sequence location. If the tape position message is not close to the current location, Performer immediately stops and chases to the new location, where it readies playback and begins looking for frame advance messages.

Problems may be encountered at slow tape motion (such as tape rocking), and it is not recommended that frame advance messages be sent at high speeds (that is, when the tape is rapidly cueing). At these non-playspeed times, frame advances are stopped and a new tape position message is sent once the tape has returned to normal playback speed.

Advanced Direct Time Lock

An advanced form of Direct Time Lock, known as *Enhanced Direct Time Lock* (DTLe), has been incorporated into Performer 3.4 and later versions. DTLe is used to synchronize Performer to the MIDI Time Piece MIDI interface by way of SMPTE time code.

DTLe differs from standard DTL in that it transmits four frame advance messages per SMPTE frame instead of one. In addition, DTL's tape position (full frame) message has been expanded to include SMPTE frame count and an identifier as to which device (within the MIDI Time Piece network) is transmitting DTLe.

DTLe offers distinct advantages over its predecessor because it enables Performer 3.4 to establish synchronous lock with tape machines without stopping the transport. Therefore, when Performer's Play button is pressed, the program jumps into sync while the audio or video machine is running. In addition, the MIDI Time Piece transmits a tape position message approximately once every second, so the user is less likely to encounter problems such as drop-outs or drift.

Performer Version 3.4 or higher is required to lock with a MIDI Time Piece through DTLe. Although the Time Piece won't support old DTL, Performer 3.4 supports it through the use of a standard DTL converter.

SMPTE-to-MIDI Conversion Systems

A SMPTE-to-MIDI converter is used to read SMPTE time code and convert it into such MIDI-based sync protocols as MIDI time code, Direct Time Lock, or song position pointer. A SMPTE-to-MIDI converter can be designed as a self-contained, stand-alone device, or the capabilities can be integrated into a synchronizer or MIDI interface/patchbay/synchronizer system (see Figure 8.13). Certain analog and digital multitrack systems also are available with built-in SMPTE-to-MIDI conversion capabilities, which means that a MIDI cable transmitting MTC or other MIDI sync form can be directly plugged from the recorder into a MIDI interface/sequencer combination without any additional hardware (see Figure 8.14).

Figure 8.13. *MOTU's MIDI Time Piece interface/patchbay/ synchronizer provides full SMPTE-to-MIDI conversion between external tape-based recorders and a MIDI production setup. (Courtesy of Mark of the Unicorn, Inc.)*

Figure 8.14. *An example of an analog multitrack recorder with a built in SMPTE-to-MIDI port for locking to an external MIDI system.*

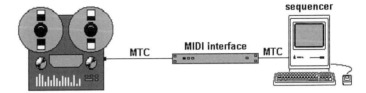

Digital Audio Synchronization

Had it been written before the onset of the digital age, this chapter would have focused on how to sync various analog and/or video machines together. MIDI would have been a mere gleam in some designer's eye, and the need to discuss sync between various digital audio systems and their analog counterparts just wouldn't exist. Present day coverage, however, would be incomplete without discussing synchronization in the context of digital and hard-disk-based audio systems. Because digital audio is an important component in modern day audio and audio-for-visual production, an understanding of digital sync becomes important when you are working in an environment in which digital audio devices are to be synchronized with each other or with other video and analog media.

The Need for Stable Timing Reference

The process of maintaining a synchronous lock between digital audio devices or between digital and analog systems differs fundamentally from the process used to maintain relative speed between analog transports. This difference is due to the fact that a digital system can achieve synchronous lock by adjusting its playback sample rate (and thus its speed and pitch ratio) so as to precisely match the relative playback speed of the master transport.

Whenever a digital system is synchronized to a time-encoded master source, the need for a stable timing source is extremely important. Such an accurate timing reference may be required to keep jitter (in this case, an increased distortion due to rapid pitch shifts) to a minimum. In other words, the program speed of the source should vary as little as possible over time to prevent any adverse effects in the quality of the reproduced digital signal. For example, all analog tape machines exhibit speed variations caused by tape slippage and transport irregularities (a basic fact of analog life known as *wow* and *flutter*). If you were to synchronize a disk-based recorder to a time-encoded analog source containing excessive wow and flutter, the digital system would be called on to constantly speed up and slow down to precisely match speed fluctuations in the tape transport. The best solution, therefore, would be to record the original program material using a time-based reference that is extremely stable. Such a reference can be found in the world of video.

Black Burst

Whenever a video signal is copied from one machine to another, it is essential that the scanned data (containing timing, video, and user information) be copied in perfect sync from one frame to the next. Failure to do this results in severe picture breakup or, at best, the vertical rolling of a black line over the visible picture area.

Copying video from one machine to another generally isn't a problem because the VCR or VTR doing the copying is able to obtain its sync source from the playback machine without a hitch. A video postproduction house, however, often uses any number of video decks, switchers, and edit controllers in the production of a single program. Mixing and switching between these sources can definitely result in nonsynchronous chaos. The end result is likely to be a very unhappy client.

This sync nightmare can easily be solved by using a single timing source known as a *black burst generator*. This generator produces an extremely stable timing reference (called *black burst*, or *house sync*) that has a clock frequency of exactly 15,734.2657 Hz. The function of this signal is to synchronize the video frames and time code addresses received or transmitted by every video-related device in a production facility so that the frame and address's leading edge occurs at exactly the same instant in time (see Figure 8.15).

Figure 8.15. *Example of a system whose overall timing elements are locked to a black burst reference signal.*

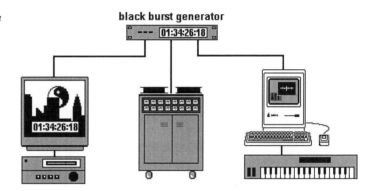

By resolving all video and audio devices to a single black burst reference, you are assured that relative frame transitions and speeds throughout the system will be consistent and stable. This holds true even for analog machines because their transports can be locked (in a slave fashion) to this reference, thereby smoothing out any inherent wow and flutter.

Synchronization Methods

Digital audio can obtain its clocking source from one of two places: an internal source or an external source. Whenever a digital device is recording audio as a stand-alone machine, the system's own internal quartz crystal oscillator serves as its clocking source. Whenever two machines are used to make a digital copy, however, the device doing the recording derives its clock pulse from the playback machine's timing circuitry (see Figure 8.16). In the case of the AES/EBU and S/PDIF digital transmission formats, this clock is embedded in the bitstream itself and doesn't require additional connections to be made.

Figure 8.16. *Whenever a signal is digitally copied, the device that is recording will derive its timing reference from the playback machine's clock pulse.*

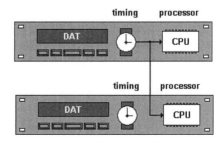

Theoretically, whenever a digital device is asked to play back audio that is synchronously locked to an external timing reference, its sample rate must be varied to match variations in that reference. In actual practice, however, digital audio systems can synchronize to various audio or visual media in a number of ways, depending on the actual application and the degree of timing accuracy that's required. These synchronization methods are: wild (no sync), SMPTE trigger, continuous sync, and locked trigger sync.

Wild

When two signals are reproduced without any form of synchronization, the signals are said to be *wild* or *on-the-fly,* meaning that they aren't directly referenced to any time-encoded source (see Figure 8.17). Under these conditions, the various transports can be manually triggered and any digital device's clock reference will be derived solely from its internal timing crystal. For example, a CD player can be used to insert a non-time-dependent music track into a video, or it could be used to insert a wild sound effect cue. Just press play at the right time and BLAM!

Figure 8.17. A wild signal is manually triggered without any reference to a time-encoded source.

SMPTE Trigger Sync

SMPTE trigger sync often is used as a straightforward way to achieve sync in almost all forms of media production. Trigger sync generally works by placing any number of sound or event cues into a playlist. Such a playlist can be assembled from within a MIDI sequencing program (in order to trigger soundcue events from a sampler), from within a digital audio editing program (triggering hard disk soundfile cues), or even from a video or other edit system for triggering external devices (such as a professional CD player). You should realize that once the event has been triggered, the digital device will play the event in a wild fashion; however, the timing will be relatively stable as it is referenced to the digital device's own internal clock (see Figure 8.18). Over longer periods of time, this form of relative sync will not be stable enough when absolute sync is required (as may occur with on-screen dialog in a video segment of over a few minutes in length). Under such conditions, an absolute synchronization process is necessary.

Figure 8.18. *Once a wild trigger event has occurred, the digital device plays back the event using its own internal clock as a reference.*

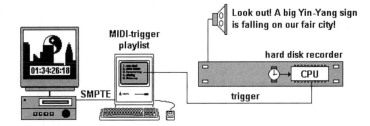

Continuous Sync

If longer segments are to be synced to a time-encoded source, or if a source's timing element is unstable (as would happen if you use a time-encoded tape machine that exhibits excessive wow and flutter), a method known as *continuous SMPTE* sync should be used in order to keep the two systems in close sync.

When using continuous sync (see Figure 8.19), once an audio file or segment has begun to play back audio, it begins to read and resolve its timing to incoming time code being supplied from the master device. In this way, the slaved digital audio system quickly chases to the proper SMPTE location point; thereafter, it remains in tight sync by varying its sample rate to precisely match the time code's speed. As a result, any speed variations in the master code are precisely matched in digital playback. Keep in mind that an unstable master timing reference can cause the audio quality to be degraded (due to excessive sample rate fluctuations or jitter). You may be able to remedy this by locking both systems to a stable master timing source, such as black burst, or by turning off the continuous sync function, thereby playing back audio in a standard SMPTE trigger fashion.

Figure 8.19. *Example of a system working in the continuous sync mode.*

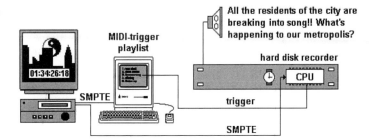

Locked Trigger Sync

Synchronization using the *locked trigger* method is much like triggering from a SMPTE-based playlist, in that a digital audio segment begins playback at a specified time code address. After playback has begun, however, the system begins to derive its timing clock from an external digital audio master source instead of from its own internal clock. This form of synchronization isn't available on all systems and can be used only when referencing the sync timing of one digital audio device (such as a hard-disk recorder) to a master timing reference (such as word clock, SMPTE, or black burst).

Proprietary Synchronization Methods for Modular Digital Multitrack Recorders

Modular digital multitrack recorders, such as the Tascam DA-88 and Alesis ADAT, encode a proprietary form of sync data directly into the datastream along with audio information. In order to enable these systems to lock to more traditional forms of sync (such as SMPTE, MTC, or black burst), a special interface (see Figure 8.20) must be used that can directly translate between the sync formats and resolve the digital recorders to an external timing source (or vice versa).

Figure 8.20. *JLCooper's dataSYNC MIDI synchronizer for the Alesis ADAT. (Courtesy of JLCooper Electronics)*

Pull-Down Sample Rates

Pull-down sample rates occur whenever a film soundtrack is transferred onto a videotape that is being recorded using the standard NTSC color format. As a result of this transfer, a discrepancy in frame rate of six frames per second will occur. This difference is due to the fact that film runs at 24 frames/second while the video runs at 29.97 frames/second. This difference can easily be compensated for by using a process known as *3:2 pull-down,* which duplicates every fourth frame of film onto the videotape copy.

Likewise, the audio tracks need to be "pulled down" in speed so as to match the picture's length and timing. If a digital track was originally recorded at 44.1 kHz, for example, the pull-down rate would end up being 44.056 kHz, whereas program material recorded at 48 kHz would pull down to 47.952 kHz.

CHAPTER 9

Amplifiers

In the world of audio, the amplifier has many applications. It can be designed to amplify, equalize, combine, distribute, or isolate a signal. It can also change the impedance of a signal. At the heart of any amplifier is a regulating device that can employ either a vacuum tube or semiconductor-transistor design.

Amplification

In order to best understand how the theoretical process of amplification works, let's draw an analogy using the term that originally was used for the vacuum tube: *valve* (still a common term in England and other Commonwealth countries). Assume that you have a high-pressure water pipe. Connected to this pipe is a valve that, when turned, can control the water pressure with very little effort (see Figure 9.1). Using this valve, you can control a large amount of water pressure with a much smaller amount of expended energy. In practice, both the vacuum tube and the transistor work in a similar manner. A vacuum tube (see Figure 9.2), for example, operates by having a DC power source supply a potential current between its plate and a heated cathode element. Located between these two elements is a wire mesh grid that acts as a control valve for allowing electrons to pass from the plate to the cathode. Small changes in signal level at the tube's grid, therefore, can result in much larger, corresponding changes between the plate and the cathode (see Figure 9.3).

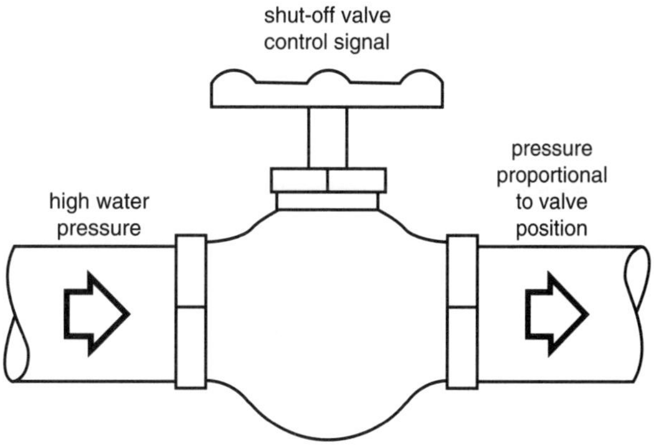

Figure 9.1. A current through a vacuum tube or transistor is controlled in a manner similar to the way water pressure is controlled by the valve tap on a water pipe.

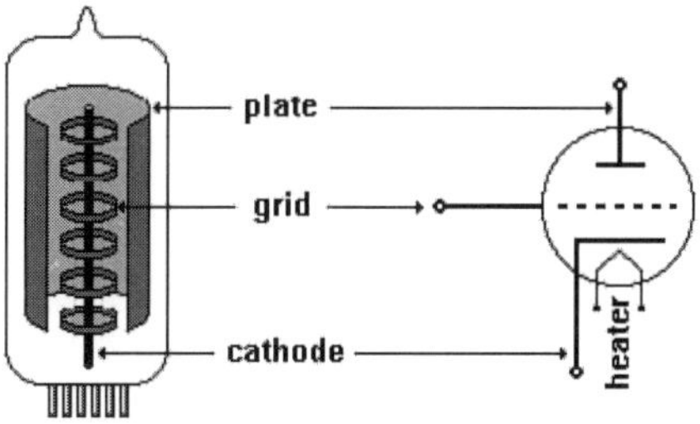

Figure 9.2. An example of a triode vacuum tube.

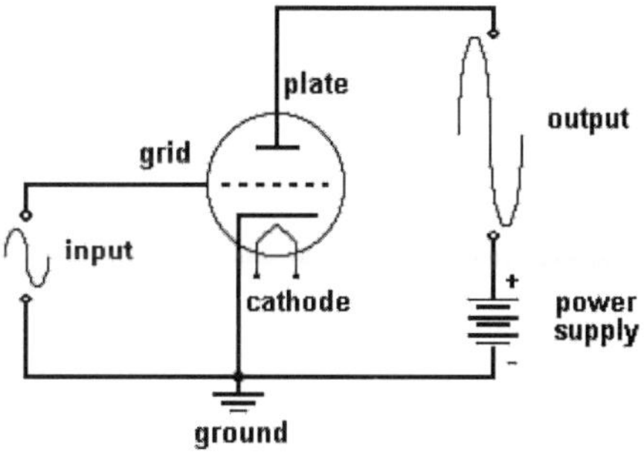

Figure 9.3. A schematic showing how small changes in current at the tube's grid can produce much larger, corresponding amplitude changes between its plate and cathode.

Although the transistor (a term derived from *trans-resistor*, meaning to change resistance) operates using a different electrical principle, the valve analogy used earlier is still relevant. Figure 9.4 shows a basic amplifier schematic with a DC power source set up across the transistor's collector and emitter points. Now think again about the valve. By presenting a control voltage to the transistor's base, it's possible to create a much larger and corresponding change in the resistance between the collector and emitter, resulting in an overall amplification at the device's output.

As a device, the transistor is not inherently linear. That is, a signal applied to the base doesn't always produce a corresponding output change. The linear operating region of a transistor lies between the cutoff region and the saturation point of the device (see Figure 9.5). It is in this region that a change in base current produces a corresponding change in the collector current and voltage.

When operating closer to the cutoff and saturation regions, the base current lines are not linear and the output will not be a true facsimile of the input. In order to limit the signal to this operating region, a DC bias signal is applied to the base of the transistor for much the same reason as a high-frequency bias signal is applied to a recording head—to lift the signal into the linear region. After bias has been applied and sufficient amplifier design characteristics have been met, an amplifier is limited in dynamic range by only two factors: *noise* (which is a result of thermal electron movement within the transistor and its associated circuitry) and *saturation*.

Figure 9.4. *A schematic showing how small changes in current at the transistor's base can produce much larger, corresponding amplitude changes between its collector and emitter.*

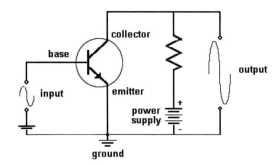

Amplifier saturation is the result of the input signal being at such a large level that the DC supply output voltage isn't sufficient to produce the required output without severe waveform distortion. This produces a process known as *clipping* (see Figure 9.6). For example, if an amplifier has a supply voltage of +24 volts (V), and is operating with a gain ratio of 30:1, an input signal of 1/2 V will produce an output of 15 V. Should the input be raised to 1 V, the required output level would be increased to 30 V. Because the maximum output voltage is limited to 24 V, wave excursions of greater levels will be chopped off, or clipped, at the upper and lower excursions of the waveform so that they remain at the maximum level until the signal falls below the maximum level. The result of amplifier clipping is the production of severe odd-order harmonics that are immediately audible with transistors and many integrated circuit designs.

The Operational Amplifier

The *operational amplifier,* or *op amp,* is a stable high-gain, high-bandwidth amplifier with a high input impedance and a low output impedance. These qualities enable use of the op amp as a basic building block for a wide variety of audio and video applications. Just add additional components to the basic circuit to fit the required design needs.

Figure 9.7 shows a typical op amp design used for amplification. In order to reduce the gain of the op amp to more stable, workable levels, a negative feedback loop often is used. *Negative feedback* is a technique that applies a portion of the output signal through a limiting resistor (that determines the gain) back into the negative or phase-inverting input terminal. Thus, part of the output is applied (fed back) out-of-phase to the input, reducing the overall output signal. Negative feedback has the additional effect of stabilizing the amplifier and reducing distortion.

Figure 9.5. *Operating region of a transistor.*

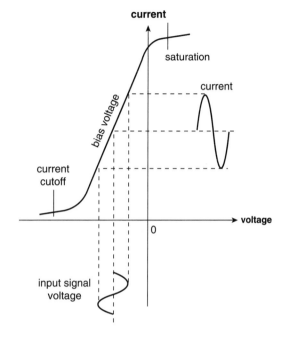

Figure 9.6. *A clipped waveform.*

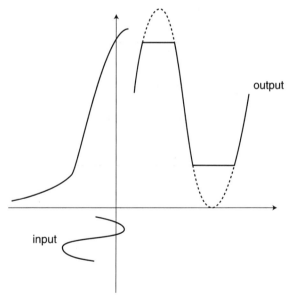

Figure 9.7. *Basic op amp configuration.*

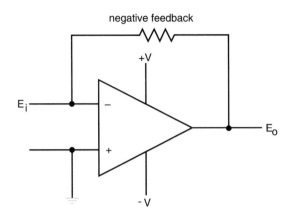

The Preamplifier

One of the mainstay amplifier types found at the input section of most professional mixer, console, and outboard devices is the *preamplifier* or *preamp*. This amp type generally can serve a number of purposes, such as boosting the low signal level of a microphone to a higher line level, providing variable gain control over a line level input signal, and providing a degree of signal isolation from extraneous input interference or improper grounding or signal voltage conditions.

Preamps often are an important component in audio engineering as they can set the "tone" of how a device or system will sound. Just as a microphone has its own sonic characteristics, not all preamp designs sound the same. A primary consideration is how the device sounds. Is it well designed from quality components? Does it use tubes or transistors? Is it quiet or noisy? The answer to this last question is extremely important when you are boosting a mic level signal to line level because a dynamic range in excess of 120 dB often is required for high-quality pickup conditions. Stated another way, a preamp with a 120-dB dynamic range will exhibit an 100-dB overall range when it is used in conjunction with a microphone with a self-noise of 20 dB.

Equalizers

An *equalizer* is a frequency-dependent amplifier. In most modern designs, equalization (EQ) is achieved through the use of resistive/capacitive networks located in the negative feedback loop of an op-amp, as illustrated in Figures 9.8 and 9.9. By changing the circuit design, any number of equalization curves can be achieved.

Figure 9.8. *Low-frequency equalizer circuit.*

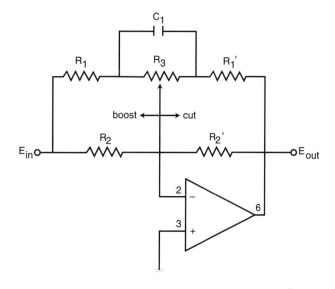

Figure 9.9. *High-frequency equalizer circuit.*

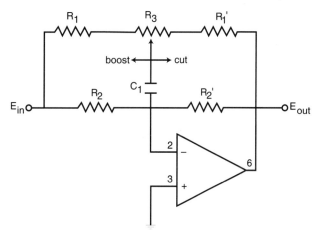

Summing Amplifiers

A *summing amplifier*, also known as an *active combining amplifier*, is designed to combine any number of discrete inputs while providing a high degree of isolation between these inputs (see Figure 9.10). The summing amplifier is an important component in console design because of the great amount of input signal routing that requires total isolation in order to separate each input from all the other inputs and still maintain signal-control flexibility.

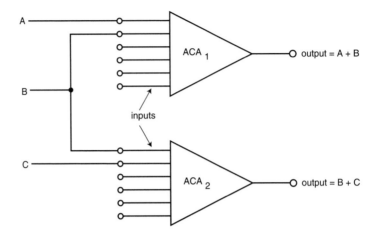

Figure 9.10. *Isolation between inputs of a summing amplifier.*

Distribution Amplifiers

The distribution of audio signals to many devices or signal paths often is necessary. Where increased power (such as in a headphone distribution path) is needed, a *distribution amplifier* is required. Under such circumstances, the amplifier may provide no gain (in which case it is termed a *unity gain amplifier*) but it amplifies the current delivered to one or more loads (see Figure 9.11).

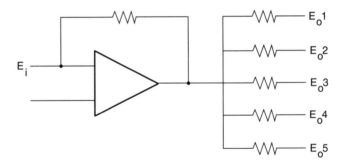

Figure 9.11. *Distribution amplifier.*

Isolation Amplifiers

An amplifier used to isolate signals combined at its input, as in a summing amplifier, may also provide electrical and ground isolation between the output of one device and the input of another. An example of such an amplifier would be an active direct box, which, in addition to reducing impedance, isolates spurious ground and voltage potentials of an electric instrument or amplifier from the console's mic inputs.

Impedance Amplifiers

Another amplifier application is to change the impedance of a signal. The impedance converter of a condenser microphone provides such an example, whereby an unusable impedance on the order of a billion ohms is reduced to a workable impedance of two hundred ohms.

Power Amplifiers

The function of a *power amp* (see Figures 9.12 through 9.14) is to boost the power of a signal to a level at which one or more loudspeakers can be driven at their rated volume levels. Power amplifier designs have their own special set of inherent problems. These include the fact that transistors don't like to work at the high temperatures often generated by amplifiers during continuous operation at high studio and concert levels. High operating temperatures also can result in changes in the response and distortion performance figures. Protective measures, such as fuse and thermal protection, must be taken to ensure reliability at high operating levels. Many of the newer amplifier models are able to provide protection for a wide variety of circuit conditions, such as a shorted load, a mismatched load, or an open (no-load) circuit. Most modern professional amplifiers are designed to work with speaker impedance loads that range from 4 Ω to 16 Ω (with most speaker models being designed to present a nominal 8 Ω load).

Figure 9.12. QSC's USA-850 professional power amplifier. (Courtesy of QSC Audio Products, Inc.)

Figure 9.13. Crown Macro-Tech 2400 stereo power amp. (Courtesy of Crown International)

Figure 9.14. *Hafler P1500 Trans-nova studio reference balanced power amplifier. (Courtesy of Hafler Professional, a division of Rockford Corp.)*

One important precaution you should take when you match amplifier and speaker combinations is to be sure that the amp is capable of delivering sufficient power to properly drive the speaker system. If the speaker's sensitivity (dB SPL/watt/meter) is too low for the amplifier's full output power to drive it to the desired SPL, there could be a tendency to "overdrive" the amp to levels where the peak signal will clip. This results in increased distortion. In addition, at the brief moment the signal is clipped, the signal present at the speakers contains a high-level DC component that could potentially damage the speaker's voice coil drivers.

Voltage-Controlled Amplifiers

Up to this point, the discussion has centered on amplifiers whose output levels are directly proportional to the signal level presented at the input. One exception to this is the *voltage-controlled amplifier* (VCA). In the case of a VCA, the program audio level is a function of a DC voltage (generally ranging from 0 to 5 V) that is applied to the control input of the device (see Figure 9.15). As the control voltage is increased (in relation to the position of the fader), the analog signal is proportionately attenuated. Thus, an external voltage is used to change the audio signal level. Console automation and automated analog signal processors make extensive use of VCA technology. Voltage-controlled equalization, which changes the equalization of an EQ amplifier as a function of a DC control voltage, is also used for the automation of equalization systems.

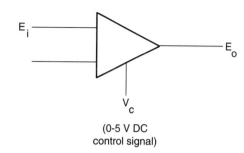

Figure 9.15. *The voltage-controlled amplifier.*

E_i

E_o

V_c

(0-5 V DC
control signal)

Digitally Controlled Amplifiers

Another system that relies on an external control input to vary amplification and attenuation levels is the *digitally controlled amplifier* (DCA). Although more expensive and less commonly used than its analog VCA counterpart, the DCA can be used in applications in which an analog signal source or sources must be directly controlled from a digital source (such as a computer, digital signal processor, or controller panel). DCAs can be used in such design applications as digitally controlled mixing consoles, gain-control circuits, computer-controlled power amps, and function genera-tors.

An example of a quality DCA is the Analog Devices LOGDAC AD7111 (see Figure 9.16). This chip is capable of deriving a level from an 8-bit digital word (input controller source) and attenuating the analog input signal over an 88.5-dB range in increment steps of 0.375 dB.

Figure 9.16. *Functional block diagram of the Analog Devices AD7111 digitally-controlled amplifier.*

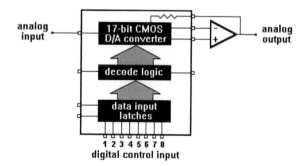

CHAPTER 10

The Audio Production Console

The purpose of the audio production console (see Figure 10.1) is to give the operator full control over volume, tone, blending, and spatial positioning of any or all signals that are applied to its inputs by microphones, electronic instruments, effects devices, and tape recorders. The audio production console should also provide a means of quickly and reliably routing these signals to any appropriate device in the studio or control room so that they can be recorded and properly monitored. A console and engineer may be likened to a palette and artist in that the console provides a creative control surface at which the engineer can experiment with and blend all the possible variables of sound.

Before the introduction of multitrack tape machines, all the sounds that were to be part of a recording were mixed together—at one time—during a live performance. If the recorded blend—or *mix*, as it is called—wasn't satisfactory, or if one musician made a mistake, the selection had to be performed again and again until the desired balance and performance were obtained. With the availability of multitrack recording, the production phase of a modern recording has radically changed into one that generally involves three stages: recording, overdubbing, and mixdown.

Figure 10.1. *A Solid State Logic SL-4000 G Plus console at the Town House, London. (Courtesy of Solid State Logic)*

Recording

The recording phase involves the physical process of recording live or electronic instruments onto tape. Logistically, this process can be carried out in a number of ways, including the following:

◆ All the musical instruments to be used in the song can be recorded onto tape in one live pass.

◆ Electronic instruments, which were previously arranged and sequenced to form the basic foundational tracks of a song, can be recorded onto the various tracks of a multitrack recorder in such a way that other live instruments, vocal tracks, and so on, can be added at a later time.

◆ Live musicians can be used to record the bed tracks of a song (either in a strictly live setting or to a set of previously sequenced MIDI tracks). Other instruments, vocal tracks, and so on can later be added to this.

The last two of these procedures are by far the most commonly encountered methods for recording popular music. The resulting foundation tracks, to which other tracks can be "laid down" at

another time, are called *basic*, *rhythm*, or *bed* tracks. These important tracks consist of instruments that provide the rhythm foundations of a song and usually are drums, bass and rhythm guitars, and keyboards. An optional guide vocal track can also be included to help the musicians and vocalist capture the proper tempo and feel of the song. The number and type of instruments chosen, the track layouts, and other important decisions generally are determined by the producer. In order to achieve the best possible performance and final product, the producer should also consider the personal wishes of the artist and any technical advice from the engineer.

When recording popular music, each instrument generally is recorded onto a separate track (or stereo tracks) of the master tape (see Figure 10.2). This is accomplished by plugging each mic into an input strip on the console (either directly into the mixer itself or into an appropriate input on a microphone panel located in the studio), setting the gain throughout the input strip's signal path to its proper level, and then assigning each signal to an output on the console, which is finally routed to a desired track on the multitrack tape recorder. The beauty behind this process is that although monitoring is important during this phase, final volume, tonal, and placement changes can be made at a later time during the mixdown stage.

Figure 10.2. When recording popular music, each instrument is generally recorded onto a separate track (or stereo tracks) on a multitrack tape recorder.

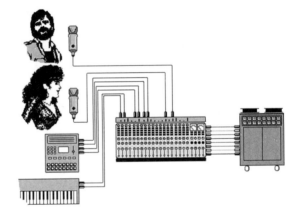

Should the need arise, several instruments can be recorded onto a single track, or a stereo pair of tracks, by assigning each signal's input strip to the same console output bus (a process known as *grouping*). These combined signals can then be balanced in level, equalization, and so on, either by monitoring the console's main output signal for the selected tracks, or by monitoring the signal returning from the tape machine itself (see Figure 10.3).

Unlike in the first example where each instrument was recorded onto its own track, much more care should be taken in determining the volume, tonality, and placement whenever a number of signals are grouped to a single track or track pair. As a general rule, it's much more difficult to make changes to recorded tracks that contain more than one signal source because changes to one instrument almost always directly affect the others.

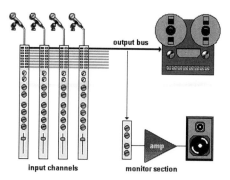

Figure 10.3. *Several instruments can be recorded onto a single track, or a stereo pair of tracks, by assigning each signal's input strip to the same console output bus.*

Each signal recorded onto multitrack tape should be recorded at the highest level possible without overloading the tape, and also without regard to the balance levels of independent instruments on other tracks. Recording at the highest level helps achieve the best signal-to-noise ratio possible for each track on the tape so that the final product isn't impaired by tape hiss or preamp noise.

The recording stage is vitally important to the outcome of the final recording. The rhythm tracks are often the driving backbone of a song, and recording them improperly gets the project off to a bad start. Getting the sounds on tape right (both musically and technically), without having to rely excessively on the "we'll fix it in the mix" approach, puts you on the path towards your goal of obtaining the best possible product.

Monitoring

Because the artist, producer, and engineer must be able to hear the instruments as they are recorded and played back, and because the levels have been recorded to tape irrespective of their overall musical balance, a separate mix must be made to monitor what is being laid down on tape.

As you learn later in this chapter, you can monitor a multitrack performance in several ways. No particular method is right or wrong; rather, you can choose a method to match your personal production style. No matter which style you choose, the overall technical result generally will be as follows:

- During the recording process, each signal being fed to a track on the multitrack recorder is also fed to the monitor mix section (see Figure 10.4) so that the various instrument groups can be blended (with regard to level, panning, and desired effects) and then fed to the control room's main monitor speakers.

- The same or (in most cases) an entirely separate monitor mix can be created that can be heard over headphones by the musicians in the studio. In fact, two or more separate cue mixes may be available, depending on the musicians' listening needs.

- The tape machine's output is used to feed the multitrack monitor section during recording, sync, and playback so that monitored balances and panning will not change.

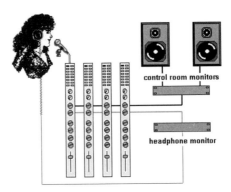

Figure 10.4. *During recording, each signal can be fed to the monitor mix section, at which point the various instruments are mixed and fed to the control room's main monitor speakers.*

Overdubbing

Instruments not present during the original performance can be added to the existing multitrack tape during a process known as *overdubbing*. At this stage, musicians listen to the previously recorded tracks over headphones and play along with these tracks. New performances can be recorded, for example, if one or more musicians made minor mistakes during an otherwise good performance, or if other instruments need to be added to the basic tracks in order to finish the project. The new performances are recorded in synchronization with the original performances and are recorded on unrecorded tracks or on tracks containing information that is no longer needed. When you overdub tracks onto an analog multitrack recorder, it's important to remember to place the tracks that are to be played back during the overdub process into the sync mode (a process whereby the record head acts as a playback head for the necessary channels in order to maintain the proper time relationship). Most modern multitrack recorders can be placed into a master sync mode, which will automatically switch between the input and sync monitor modes, thereby eliminating the above need for manual switching. For more information on sync playback, refer to Chapter 5, "The Analog Audio Tape Recorder."

Mixdown

After all the desired musical parts have been performed and recorded to the satisfaction of the artist and the producer, the *mixdown*, or mix, stage can begin. At this point, the inputs to the console are fed by the playback outputs of the multitrack recorder. Often, this is accomplished by switching the console to the mixdown mode or by changing the microphone/line switches on the appropriate console inputs to the line or "tape" position. The master tape is then played repeatedly while adjustments in level, equalization, effects, and panning are made for each track. Throughout this artistic process, the individually recorded signals are blended into a composite surround, stereo, or mono signal that is fed from the console outputs to the master mixdown recorder. When a number of mixes have been made and a single version has been approved, this recording is called the *final mix* and can then be assembled—along with other songs or productions in the project—into the final product.

The Professional Analog Console

Most analog audio production consoles used in professional recording studios, like the ones shown in Figures 10.5 through 10.7, are designed with similar controls and capabilities. They differ mainly in appearance, location of controls, on-board dynamic processing, signal-routing capacities, and how they incorporate automation and control-setting recall functions (if at all).

Figure 10.5. Mackie 32-8 console on stand with MB-32 meter bridge, 24-E expander console and sidecar. (Courtesy of Mackie Designs, Inc.)

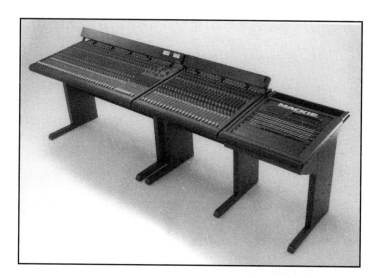

Figure 10.6. AMEK 9098 recording console. (Courtesy of AMEK U.S. Operations, Inc.)

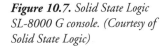

Figure 10.7. *Solid State Logic SL-8000 G console. (Courtesy of Solid State Logic)*

Before you delve into the details of how a console works, you need to understand one of the most important concepts in all of audio technology: the *signal chain* (or *signal path*). As is true with any audio system, the recording console can be broken down into separate signal paths. By identifying and examining the individual components that work together to form this chain, you can understand any mixer or console, no matter how large or complex.

The trick to understanding these systems is realizing that each component has an input and an output that has an associated audio source and destination (see Figure 10.8). In such a chain, the output of each source device must be connected to the input of the following device until the end is reached. Whenever a link in this source-to-destination path is broken, no signal can pass.

Although this may seem to be a very simple concept, it's one that may save you when paths, devices, and spaghetti cables get out of hand. It's as basic as knit one, purl two: An audio signal follows from one control or processing device to the next until the desired effect or destination is reached.

The signal flow for each input of a modern console follows vertically down a plug-in input strip known as an *I/O module* (see Figures 10.9a and 10.9b). *I/O* stands for input/output and is so named because all the associated electronics for a single track-channel combination are centrally located

on a single circuit board that's physically connected to each input strip. As the I/O module electronics are self-contained, they can be fitted into a modular mainframe in a number configurations, so as to better match the present and future production needs of a particular studio. The plug-in nature of I/O modules also make them interchangeable and easily removable for service.

Figure 10.8. *A simplified drawing of an I/O strip's signal chain.*

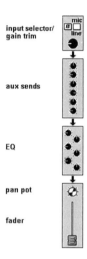

input selector/
gain trim

aux sends

EQ

pan pot

fader

The following sections fully describe the various I/O stages of a professional audio production console. Although consoles tend to vary in layout, this introduction should offer some good insight into I/O design.

Channel Input

The *channel input* section (see Figures 10.10a and 10.10b) serves to optimize the signal level at the input of an I/O module before it is further processed and routed. Either mic or line inputs can be selected—each having continuously variable level controls, called *pads* or *trims*. The range of the mic gain typically varies from +20 to +70 dB with the addition of an attenuation pad, while the line trim can often vary over a 20 to 30dB range.

Gain trims are a necessary component, as the output level of a microphone typically is very low (–45 to –55 dB) and requires a specially designed amplifier to raise and match the various mic output levels in order for the signal to be further processed without a degraded signal-to-noise ratio. When a mic or line signal increases to a point at which the preamp's output rises above +28 dBm, severe clipping will occur. To avoid this condition, the input gain must be reduced—either by simply turning the gain trim down or by inserting an attenuation pad into the circuit.

The preamp's output often is fed to a phase-reversal switch, which is used to effect a 180° change in-phase at the input signal to compensate for an out-of-phase microphone or cable. High- and low-pass filters also may follow the preamp, allowing the operator to filter out such extraneous signals as tape hiss or subsonic floor rumble.

Figure 10.9. Solid State Logic
SL-4000 G Plus console.
(Courtesy of Solid State Logic)

a. I/O strip.

b. I/O strip.

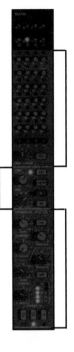

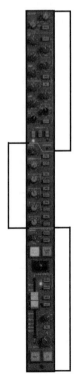

a

b

Channel Assignment Matrix—Capable of distributing the signal to any or all tracks on the multitrack recorder. LED indication of assignment is provided. Group panning provides stereo panning between odd/even channel pairs and between the left and right channels of any selected stereo mix bus.

Channel Input Controls—Include a line trim pot and a mic preamp pot with a wide gain range. Phase reversal and individual phantom power supply are supplied.

Dynamics—A full complement of compression, limiting, and expansion (including gating) is provided. This section lets the user dynamically process each signal easily and without having to come up with lots of outboard devices.

Equalizer—Used to compensate for variations or discrepancies in frequencies present in the audio signal. Equalizers often provide up to four continuously variable overlapping frequency-control bands with a variable bandwidth (Q) and a boost or cut value (dB). High- and low-frequency filters are also often provided to reduce unwanted signals (such as rumble, proximity effect, or tape hiss).

Auxiliary Section—The auxiliary (aux) sends often provide the overall effects (and possibly monitor cue) sends of a console. These sections are used to create separate and controlled submixes of any (or all) of the input signals to a mono or stereo output for routing to an effects or monitor destination.

Main Output Fader—Each input strip contains an associated output fader (which determines the level) and a pan pot (which determines the signal's left/right/front/rear placement in the stereo or surround field). Generally, this section includes Solo/Cut and signal grouping features.

Figure 10.10. *Channel input section.*

a. *AMEK 9098. (Courtesy of AMEK U.S. Operations, Inc.)*

b. *SSL SL-8000 G Plus. (Courtesy of Solid State Logic)*

a

b

Equalization

The next section in the strip, the *equalization section*, is fed directly from the input section. The equalizer is used to compensate for variations or discrepancies in frequencies present in the audio signal. On larger consoles, equalizers often provide up to four continuously variable overlapping frequency-control bands with a variable bandwidth (Q) and a boost or cut value of 18 dB. The EQ in/out button enables the engineer to silently place the selected EQ into the audio circuit.

As an example, the hi-mid band of the Mackie 8-Bus console (see Figure 10.11a) has a variable Q that ranges from 3 to 12 octaves. The frequency control ranges are as follows: low cut/boost—80 Hz (fixed), low-mid—45 Hz to 3 kHz, high-mid—500 Hz to 18 kHz, and high cut/boost—12 kHz (fixed). Other console equalizer examples can be found in Figures 10.11b and 10.11c.

Figure 10.11. *Equalization section.*

a. *8-Bus series. (Courtesy of Mackie Designs, Inc.)*

b. *AMEK 9098. (Courtesy of AMEK U.S. Operations, Inc.)*

c. *SSL SL-8000 G Plus. (Courtesy of Solid State Logic)*

a   *b*   *c*

Insert Point

Many mixer and console designs provide a break in the signal chain after the equalizer. At this point, a *direct send/return* or *insert access point* (see Figure 10.12) can be used to send the line level audio signal to an external processing device. The processed signal can then be directly and easily returned back into the input strip's signal path. Dynamic processors, equalization, and effects processing are typically inserted into the chain in a way that only affects the signal passing through the selected I/O channel. Console-wide signal processing (such as reverb and, in most cases, effects) usually aren't routed into the path in this manner; instead, they are often controlled through an auxiliary effects send section.

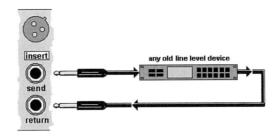

Figure 10.12. *A direct send/ return signal path.*

Physically, the send and return jacks may be accessible at the rear panel of the console as two separate jacks, as a single tip-ring-sleeve jack (in the form of a stereo jack providing the send, return, and common signal, respectively), or on the console's patch bay.

Auxiliary Send Section

The *auxiliary* (*aux*) *sends* often provide the overall effects (and possibly monitor cue) sends of a console. These sections (eight or more sends may be provided on a single console) are used to create separate and controlled submixes of any (or all) of the input signals to a mono or stereo output (see Figures 10.13 and 10.14). Unlike the direct insert, an auxiliary send can combine any number of signals and can then route this mix to an external device or amp.

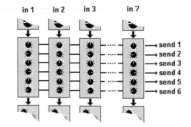

Figure 10.13. *Auxiliary sends are used to create separate and controlled submixes of any (or all) of the inputs to a mono or stereo output.*

Figure 10.14. *Auxiliary send section.*

a. 8-Bus series. (Courtesy of Mackie Designs, Inc.)

b. AMEK 9098. (Courtesy of AMEK U.S. Operations, Inc.)

c. SSL SL-8000 G Plus. (Courtesy of Solid State Logic)

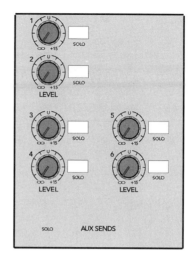

a

b   *c*

An auxiliary send can serve many purposes. For example, a number of mono sends can be used to drive reverb devices, signal processors, or other devices, while another mono send can be used to drive a speaker placed in that bathroom down the hall that has the great acoustics. A stereo pair of sends could be used to provide a headphone mix for a couple of musicians who decided they had to add another part onto some spare tracks. In fact, these sends can be used to accomplish almost any set of mixing tasks. Sends 3 and 4 could be used to send a live satellite feed to a Moscow TV station, which then beams the signal to NY, while send 1 feeds a miked 3" speaker hidden under a pillow. It's all up to you, your needs, and your creativity.

Dynamics Section

Many top-of-the-line consoles have designed a *dynamics* section into each I/O module on the console (see Figure 10.15). This section lets the user dynamically process each signal more easily and without having to scrounge up tons of outboard devices. Often the full complement of compression, limiting, and expansion (including gating) is provided. A complete explanation of dynamic control can be found in Chapter 11, "Signal Processors."

Figure 10.15. Dynamics section of an SSL SL-8000 G Plus. (Courtesy of Solid State Logic)

Monitor Section

Assuming that all the signals have been recorded onto tape at the highest levels possible without regard the relative musical balance on other tracks, a means for creating a separate monitor mix in the control room becomes necessary. Consequently, a separate stereo monitor section may be designed into the console (see Figure 10.16) to provide control over level, pan, and effects so that each track—either being sent to the multitrack recorder or returning from it—can be mixed to re-create a musical balance.

Figure 10.16. A monitor section can be used to obtained a balanced console mix during a recording session.

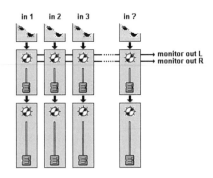

Methods for monitoring tracks during a recording often vary from console-to-console, as well as between individuals. Again, no method is more right or wrong than another. It simply depends on what type of equipment you are working with and your own working style. The following sections briefly describe a few of the most common monitoring approaches.

In-Line Monitoring

Many newer console designs incorporate an I/O small-fader section (see Figure 10.17), which can directly feed either the recorded signal being fed to the multitrack or the monitor mixer (depending on its selected operating mode).

Figure 10.17. Small-fader section.

In the standard monitor mix mode (see Figure 10.18a), the small fader is used to adjust the monitor level of the associated tape track. In the "flipped" mode (see Figure 10.18b), the small fader is used to control the signal level being sent to tape, while the larger, main fader can be used to control monitor mix levels. This function enables multitrack levels—which aren't often changed during a session—to be located out of the way, while the more frequently used monitor levels are located at the more accessible master fader position. This small fader feature is often found on American consoles and is becoming more common in newer console designs.

Figure 10.18. Small-fader monitor modes.

a. Standard monitor mode.

b. "Flipped" monitor mode.

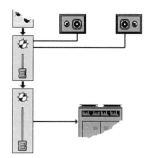

a

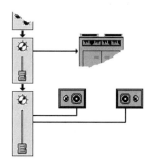

b

Separate Monitor Section

Certain English consoles (particularly those of older design) incorporate an entirely separate monitor section into the right side of the console (see Figure 10.19). This section is, in essence, a section of input for mixing output bus and tape return monitor signals. This type of design has the advantage of allowing a large number of extra inputs to be used for effects returns, electronic instrument inputs, and so on, during the mixdown phase. During the recording phase, the console surface is split over a large area and requires a bit of concentration to keep from confusing the "to tape" and monitor sections during a complex session.

Figure 10.19. Certain English consoles have a separate monitor section on their right-hand sides.

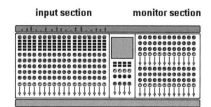

input section   monitor section

Direct Insertion Monitoring

Whenever a console is used that doesn't have either of the preceding monitoring facilities, a simple and effective third option is still available to the user. This approach uses each input strip's direct send and return insert point, in which the direct send of a particular channel (usually following the EQ section) is routed to the associated track input on the multitrack recorder (see Figure 10.20). The return signal is then routed back from the same-numbered tape track output into the direct return path, where it follows down the strip's path to the effects send, pan, and main-fader sections.

Using this monitoring approach, the input signal is fed to the tape track (with levels adjustable by way of the mic/line gain trim). The return path from the same-numbered tape track is then fed through the input strip's effects and output path so that the signal can be mixed (along with full effects) without regard to those signal levels going to tape. Placing the tape track into sync or playback will not affect the mix because the tape outputs are already routed to the mix in their proper balance. This approach can also be used effectively on larger consoles having other mix facilities. The only potential drawback is the large number of patch cords required to patch the inputs and outputs to and from the multitrack recorder.

Figure 10.20. *By directly inserting the tape send and return into the signal path, a recorded track can be monitored easily.*

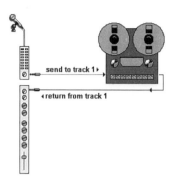

send to track 1 ▶

◀ return from track 1

Channel Assignment

Following the main "to-tape" fader, the signal will be routed to the strip's *channel assignment* matrix (see Figures 10.21a, 10.21b, and 10.21c), which is capable of distributing the signal to any or all tracks on the multitrack recorder. If a vocal mic were plugged into input #14, for example, the engineer might assign the signal to track 14 by pressing that track button on the matrix. If an overdub on track 15 were needed, all the engineer would have to do is unpress #14 and assign the signal to #15.

Figure 10.21. *Channel assignment section.*

a. *8-Bus series. (Courtesy of Mackie Designs, Inc.).*

b. *AMEK 9098. (Courtesy of AMEK U.S. Operations, Inc.)*

c. *SSL SL-8000 G Plus. (Courtesy of Solid State Logic)*

a

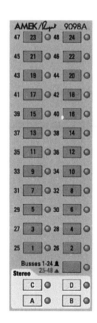

b

c

Many newer and smaller consoles offer only one button for even and odd paired tracks, which can then be selected individually using an associated pan pot. Therefore, by pressing the button marked 5/6 and panning to the left, the signal is only sent to output bus #5. This simple approach also allows for a simplified build up of submixes that enables multiple inputs to be sent to a pair of tracks on the multitrack recorder.

Output Fader

Each input strip contains an associated output fader—which determines the level—and a pan pot—which determines the signal's left/right placement in the stereo field. Generally, this section (shown in Figures 10.22a and 10.22b) includes a Solo/Cut feature with the following functions:

◆ *Solo.* Operates as a monitor function. When pressed, all other monitor channels are muted, allowing the listener to monitor only the selected channel without affecting the multitrack or main stereo outputs during the recording and mixdown process. During mixdown, this button can be programmed to cut all other channel signals (which may, for example, contain extraneous noises or unwanted musical passages).

◆ *Cut or mute.* Turns off all I/O signals of the selected channel. A master cut enables multiple channels to be muted by one master switch.

Figure 10.22. *Output fader section.*

a. AMEK 9098. (Courtesy of AMEK U.S. Operations, Inc.)

b. SSL SL-8000 G Plus. (Courtesy of Solid State Logic)

a

b

Output Bus

In addition to the concept of the signal chain, you should understand one other important signal path concept that is integral to all console designs: the bus (sometimes spelled *buss*).

Think of a *bus* as a single electrical conduit that runs the horizontal length of a console or mixer (see Figure 10.23). This single signal path often is a heavy copper wire or a single wire on a ribbon connector cable that runs the entire width of the console. It can be thought of as an electrical junction point that enables a signal to be injected onto the bus where it can be mixed in with other signals that are present. Signals can be routed off the bus to a particular destination, such as a console output, tape track output, and auxiliary effects send. Much like a city transit bus, this signal path follows a specified route and allows audio signals to get on or off at any point along its path.

Figure 10.23. *Example of an effects send bus.*

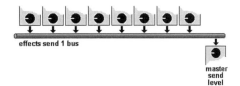

As you follow vertically down the input strip, you can see a number of places where audio is routed off the strip and onto a horizontal output bus path. Aux sends, monitor sends, channel assignments, and main outputs are all examples of buses. You can see that the same aux send control is horizontally duplicated across the console's surface. In the same way, these input controls are physically tied to a horizontal auxiliary send bus in which different amounts of signal from the various inputs are fed onto the bus and to its destination: an auxiliary send output. An example of a stereo bus is the mixer's main stereo output. After the signal is determined by each strip's fader, the signal is balanced from left to right (using pan pots) onto the left and right output buses. The main output is a combination of all the different input signals and can be sent to the master recorder.

Patch Bay

The components of the console interconnect at the *patch bay* (also called a *patch panel*). A typical patch bay is shown in Figure 10.24.

A patch bay is a panel that (under the best of conditions) contains a jack that corresponds to the input and output of every discrete component or group of wires in the control room. One jack is said to be *normalled* to another if the signal between the two jacks is connected directly from one to the other when there is no plug in either jack (see Figure 10.25). When a plug is inserted into one or both of the jacks, the normal between the jacks is said to have been broken (see Figures 10.26a and 10.26b).

Figure 10.24. The patch bay.

a. Single patch rack. (Courtesy of dbx Professional Products)

b. Example of a labled patch bay layout.

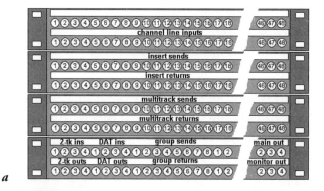

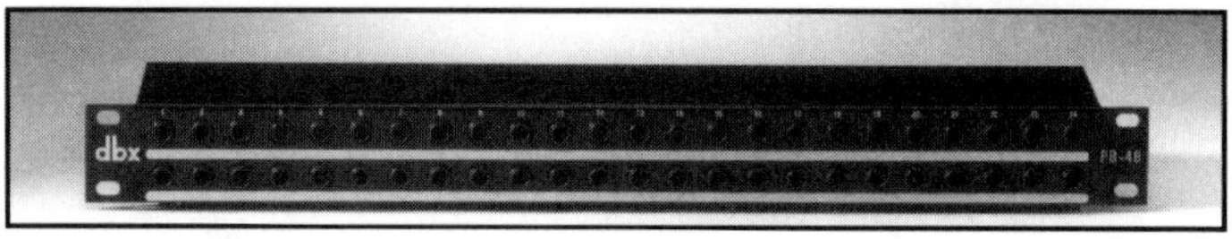

Figure 10.25. Normalled jacks; the preamp output is normalled to the EQ input.

Figure 10.26. A plug inserted into the preamp output jack or input jack breaks a normalled connection.

a. Access to the preamp's output.

b. Access to the equalizer's input.

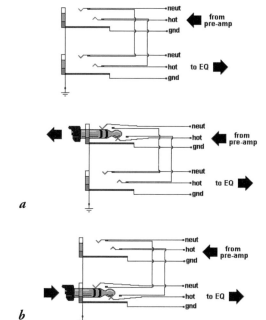

The purpose of breaking normalled connections is to enable the engineer to use patch cords to connect different or additional pieces of equipment between two normally connected components. For example, a limiter might be temporarily patched between a mic preamp output and an equalizer input. The same limiter could later be patched between a tape machine output and a console line input. Other uses of the patch bay are to bypass defective components or to change a signal path in order to achieve a certain effect. Patch bay jacks usually are balanced tip-ring-sleeve types that have two conductors plus ground. They are used with balanced circuits (see Figure 10.27a) and come in two standard sizes: military telephone and tiny telephone. Unbalanced jacks and plugs have tip-and-sleeve connectors only (see Figure 10.27b).

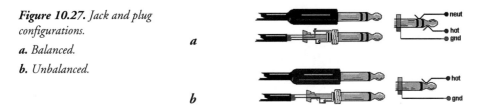

Figure 10.27. *Jack and plug configurations.*
a. *Balanced.*
b. *Unbalanced.*

Metering

Each input, line output, and possibly each effects send may be measured by a meter that displays the signal level. If the signal level being sent to tape is too low, tape noise will be a problem when the recording is played back. If the level is too high, the tape—and possibly the console or tape recorder amplifiers—may distort the signal. Proper recording level is achieved when the highest reading on the meter is near the zero level, although levels slightly above or below this will not cause difficulties (see Figures 10.28a through 10.28c).

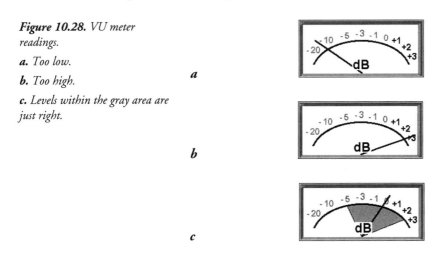

Figure 10.28. *VU meter readings.*
a. *Too low.*
b. *Too high.*
c. *Levels within the gray area are just right.*

Multitrack console metering bridges are available in two basic types: an LED or light-bar meter display and a VU meter bridge (see Figure 10.29). VU bridges often incorporate as many VU meters as there are tracks, and they often indicate both the bus output level (tape input level) in the record mode and the tape return level in the playback or overdub mode. On larger consoles, LED or light-bar displays often have the advantage of being switchable between VU (average) and peak level readings.

Figure 10.29. A set of LED, light-bar, and VU meter displays.

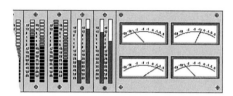

Digital Console Technology

With the advent of digital audio, signal routing and processing, as well as console design and technology, are undergoing a design revolution. Many of the newer recording console and mixer designs incorporate digital technology to route audio signal paths (track assignment, auxiliary send/return, channel on/off, EQ in/out, and so on) with the aid of a computer processor and interface. The use of digital technology makes it possible for many of the costly and potentially faulty discrete switches required for track selection and other functions to be replaced by a single, assignable digital switch or a series of highly reliable digital switching networks. The big bonus, however, is that because these routing and, in many cases, signal level functions are digitally encoded, it becomes a simple matter to store and save both snapshot routing and dynamic level setting information in computer memory for complete and virtually instantaneous recall at any time.

The digital console currently is available in two forms: the digitally controlled analog console and the fully digital console. Both design types are gaining in popularity; they are available in many configurations and with price tags running the full gamut of affordability. Although both these system types interface with the signal path in ways that differ from their analog counterpart, it's important to keep in mind that the signal chain is still conceptually the same: the audio signal follows from one control or processing device to the next until the desired effect or destination is reached.

Digitally Controlled Analog Consoles

In a digitally controlled analog console (see Figure 10.30), the signal path is distributed and processed in analog form; however, control over all console parameters is carried out in the digital domain. In most cases, this means that the console's control surface (containing all the knobs, faders, assignment buttons, and so on) will output its control parameters in the digital domain. This digital information is then transmitted to an interface that has direct control over the system's analog I/O routing, switching, and dynamic level functions.

Figure 10.30. *Euphonix CS2000 digitally controlled analog console. (Courtesy of Euphonix)*

Because control over signal parameters is carried out in the digital domain and the system is much less constrained by the analog signal path's layout requirements, the control interface and its layout can vary from the analog console's physical traditional form.

One way in which the physical controls of a digitally controlled analog console can be laid out is through a series of centralized parameter input sections, whereby each section represents a single processing or routing component in the signal path. By pressing an associated button on the channel input strip, the desired set of parameter controls becomes actively selected for that particular channel. For example, Figure 10.31 shows that by selecting a function button on input strip #5, the entire set of control settings becomes active and the current settings for that particular channel are displayed. Changes in these settings can then be heard in real time until the desired routing, levels, or processing have been achieved. Pressing another parameter button on another channel then activates that channel for further processing.

Figure 10.31. *A central set of parameter controls can be assigned to a specific input strip or function.*

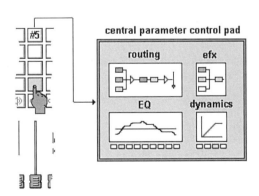

Another method for controlling the console parameters is to physically place a readout display at each control point on the input strip—at locations that might resemble those placed on an analog console (see Figures 10.32a and 10.32b). Using such a control/display system, you could simply grab the desired parameter and move it to its new position—just as though it were an analog console. Generally, this system is more expensive than the former as more readout indicators, control knobs, and digital control interface systems are required. The obvious advantage, however, is having instant access (both physically and visually) to all these parameters at once.

Figure 10.32. Physical controls over digital parameters offer instant readout and access to an audio function.

a. Numeric readout display.

b. Rotary LED display.

a　　　　　　　　*b*

Fully Digital Consoles

In the fully digital console, shown in Figures 10.33 through 10.35, the analog input signals are converted directly into digital data (or are directly inserted into the console's chain as digital data) and are distributed and processed by the system entirely in the digital domain. At the console's output, the signal may then be decoded back into analog or remain in its digital form for recording to a multitrack or mixdown recorder. Because all audio, routing, and processing data is digital, full system automation and snapshot recall can be recorded directly into computer memory or, in the case of more economical systems, can be stored as MIDI controller messages in a separate MIDI sequencer.

Figures 10.33. Yamaha ProMix 01 digital mixer. (Courtesy of Yamaha Corporation of America)

Figures 10.34. AMS Neve Logic
2 digital recording/production
console. (Courtesy of AMS Neve)

Figures 10.35. SSL Axiom
digital production system.
(Courtesy of Solid State Logic)

As with the digitally controlled console systems, the control surface may vary from the traditional analog layout. The parameter controls may be spread out or centrally located. In fact, new methods are evolving to help humans interface with these digital systems.

Console Automation

In multitrack recording, the final musical product generally isn't realized until the mixdown stage. Each console channel that's being fed from the multitrack tape (as well as from other sources) will often have its own particular volume, reverb, left-to-right pan, EQ and other settings. Multiply each of these settings by 24 or more tracks and add the auxiliary effects send and return volumes, compression/limiting, and other signal-processing functions, and it quickly becomes obvious that the mixdown process just might be more than one engineer can efficiently handle. As a result, mixdowns must be rehearsed repeatedly so that the engineer can learn which controls must be operated, how much they should be varied, and at what point any changes should come. It's not uncommon to spend 12 hours mixing a complicated piece of music before an acceptable mix can be obtained. Often mixes must be rejected because the engineer simply forgot to make one simple, but important, control change. The producer and engineer know how and when the control settings should be changed, but the memory and physical dexterity required to execute them can sometimes exceed human capabilities.

One solution to this problem is to mix with the aid of console automation. Such a system can remember and re-create some or all settings or changes made by the engineer. This allows for continual improvements until the desired final mix quality is achieved. In addition to providing an extra set of "hands," an automated mix can be a virtual lifesaver should you need to go back into the studio at a later date to make changes to the mix, or if a different mix is needed for another medium (such as mono radio or the group's latest MTV video).

Scanning

In practice, a dynamic automation system can range from being able to sense only the position of a volume fader and level-related switching functions to being a fully automated system that can store and recall all the dynamic functions on a production console. This amazing little feat is accomplished by scanning across each of a console's automated control functions in a sequential manner, determining the level setting or switching state that each control is currently at, and then converting this level (that often exists as a representative DC voltage level) into an analogous digital word. After being converted, the digital data can then be stored into a memory system for storage and future processing and/or playback. The time required to scan all the control inputs in the console one time is called the *scan rate* or the *updating rate* and usually is measured in milliseconds. As used here, *updating* refers to the repeated sampling by the encoder of control voltages—one time during each cycle—to create new data for storage. Updating also refers to the repeated modification of the reconstructed control voltages in response to each new cycle of data fed to the decoder.

The updating rate must be fast enough to provide smooth and accurate encoding of the engineer's actions. For example, if the engineer fades out a level control at a rate of 5 dB per second and the encoder scans the corresponding control voltage only once each second, the programmer will sense and encode 5-dB steps rather than a continuous fade (see Figure 10.36a). If the scan rate is increased to 50 ms, the programmer would process steps of 0.25 dB, which more closely approximate the continuous fade (see Figure 10.36b). The decoder output can be filtered to remove these small steps from the reconstructed DC voltage so that level changes are smooth. The higher the scan rate, the less filtering is needed because very small steps are perceived as a continuous change. The scan rate is limited by the storage capacity of the memory unit as well as by the time needed for the processor to encode or decode the analogous voltage signal.

Figure 10.36. A 5-dB-per-second fade of 1-second duration.

a. If updating occurs only once each second, the fade would be encoded as a single step change of 5 dB.

b. Should updating occur every 50 ms, the fade would be encoded over 20 steps of 0.25 dB each.

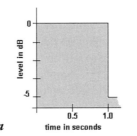

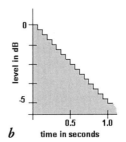

The programmer processes two different types of functions: switching and dynamic. Switching functions have only two states and can be used to turn signals on or off (see Figure 10.37a) where continuous control is not required. A dynamic function control can be used to vary a function between a minimum and a maximum value, such as program fader level (see Figure 10.37b).

Figure 10.37. Switching and dynamic automation functions.

a. A switching function.

b. A dynamic function.

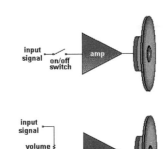

Voltage-Controlled Automation

Console automation is often made possible through the use of voltage-controlled amplifiers (VCAs), voltage-controlled equalizers (VCEs), data encoder/decoders, CPU, and storage memory (see Figures 10.38, 10.39a, and 10.39b). The VCA is used to control the level of audio signal as a function of an external DC control voltage, whereas a VCE changes the frequency response of an amplifier as a function of the DC voltages applied to its control inputs. In a console fully equipped for automated mixdown, VCAs, VCEs (optionally available on certain consoles), and digital routing facilities perform many of the fader, EQ, and switching functions. The physical controls, on the other hand, are simply used to apply varying DC control voltages to their respective voltage controllers.

Figures 10.38. *A simplified diagram of a VCA-based automation system.*

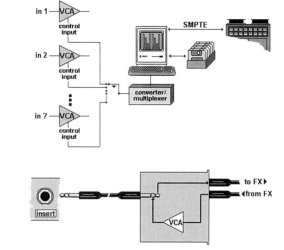

Figures 10.39. *Otto-34 and Ultramix Pro VCA-based automation. (Courtesy of Mackie Design, Inc.)*

a. *Otto-34 plugs into the direct insert jack of any console and provides VCA-based automation in addition to a 19" patch bay rack that gives direct user-access to these insert points.*

b. *Ultramix Pro automation software for the Mac.*

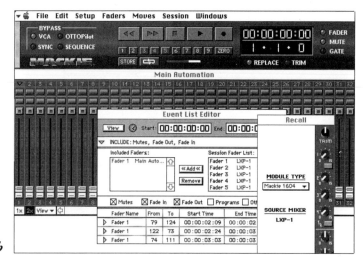

VCAs permit the free grouping of signals. Because there is no leakage of the audio signal into the VCA control input, or vice versa, a single control voltage source can be used to control any number of VCAs while maintaining complete separation of the audio signals being controlled. The advantage to this grouping is that after the balance of several instruments is set relative to each other, their overall level can be controlled by a single fader rather than having to change each individual channel volume. When using a nonautomated console, this can be done in two ways: either the signals are mixed to one channel on the console and the mix is then fed through a fader to control the overall level (see Figure 10.40a), or the signals are kept separate and controlled through a ganged, multichannel fader (see Figure 10.40b). In the latter method, the number of tracks controllable by one fader depends on the number of channels available in the multichannel fader.

In Figure 10.40c, a single fader is used to control the voltage levels of each control fader which, in turn, feed the individual VCA control inputs. In Figure 10.40d, the output of each VCA is proportional to the sum of its channel and the group control signals. Attenuation is 0 dB when the sum of the control voltages is 0 V. As the sum increases, so does the attenuation.

Figure 10.40. *Varying the levels of several signals with one control.*

a. Mixing analog signals together using a submaster.

b. Retaining analog separation by using a ganged multichannel fader.

c. A grouping fader supplies a single master DC source to all VCA control inputs.

d. A single DC source acts as a master controller voltage for a number of grouped VCA control faders.

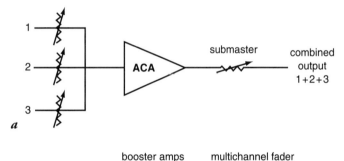

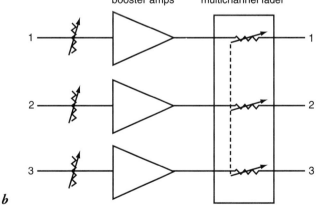

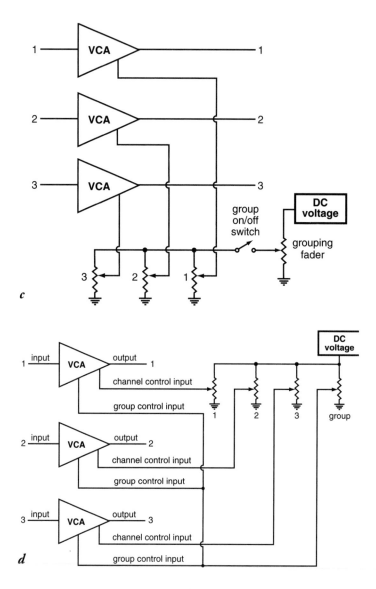

When using a VCA-based control voltage, the levels of a number of instruments placed into a single audio channel can be controlled by a single-channel submix or grouping or subgroup fader. This is done by controlling the DC voltage feeding the grouping fader; the VCAs can all be switched on or off simultaneously with one switch. For example, if a horn section recorded on several different tracks is to begin playing in the middle of a song, but a few of the horns come in a bit early, the engineer could effectively mute the VCAs until the moment the horns begin to play and then could turn them all on with one automated master switch. By the same token, if one track

had no noise and began to play at the proper time, a noise gate could be connected to it and the noise-gate control voltage could be used to turn on all the other tracks when the noiseless track began to play. Control inputs may be grouped together by assigning a fader to a particular group number, which in turn provides the DC voltage source for all VCAs that are connected to that particular group.

Modes of Operation

Voltage-controlled console automation operates in three basic modes: write, read, and update.

Write Mode

In the write mode, the programmer scans the DC control input level of each of the voltage-controlled devices of the console in a certain sequence many times per second, encoding this control voltage information into a digital form that can be stored for later use. Most modern systems store this data to a computer so that it can be stored to and recalled from disk. Because each control voltage represents the position of a console control, the data track contains a record of how the console controls were set during each scan cycle. In order for the settings to be reconstructed, the data must be synchronized with the program material. For this reason, most automation system's time-based references are locked via SMPTE time code to the multitrack tape machine.

Read Mode

In the read mode, the console controls no longer supply DC voltages to their respective control inputs. Instead, the programmer reads the encoded control information from memory and feeds it through its decoder section, which converts it back into DC voltages and sequentially applies these voltages to the appropriate voltage-controlled device inputs. As the master tape and data track are played, the DC voltages control the console level settings, producing a mix identical to the one that originally produced the data track. If, after listening to the tracks mixed automatically in the read mode, the engineer and producer decide certain functions need to be changed, the write mode can be initiated on only these functions while automatic control of the others is retained by leaving them in the read mode. The DC voltages created by the decoder and by manual operation of the console are then encoded and stored as a merged file that can be updated until the desired mix is achieved.

If only part of the mix needs to be reprogrammed for a certain function, the write mode can be entered on that function at the point at which the changes are to be made. In order to provide a smooth transition from automatic to manual control with no jumps in level, the manual control must be set so that the DC voltage it creates equals that which is re-created by the decoder. This is accomplished by manually adjusting the control with the aid of nulling indicator lights (or fader

level indicator bars shown on a computer or video monitor screen) connected to read the difference between the automatically and manually generated DC voltages for each function (see Figures 10.41a and 10.41b). After a level match is achieved, the engineer can switch a function from read to write without the levels changing. After the write mode is entered, the controls can be moved from the level match position to the new settings desired.

Figure 10.41. VCA relative voltage level indicators.

a. Null indicator lights indicate the relative fader-to-control voltage level.

b. On-screen bar graphs provide an easy to read visual indication of the relative fader-to-control voltage levels. (Courtesy of AMEK U.S. Operations, Inc.)

a

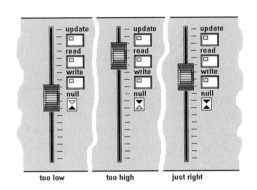

b

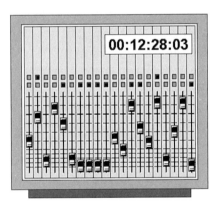

Update Mode

The update mode eliminates the need to match the manual settings to the automatic one when punching into an existing data track. In this mode, settings are changed by adding to or subtracting from the automatic control voltage rather than completely rewriting them. The advantage to this process is that if some involved level changes were made correctly except for the loudness and softness of the track, the overall volume can be changed without having to again perform all the other level changes. In this mode, the new mix changes are automatically merged with the previously desired level changes and are then stored into memory. A level match between the new and old control voltages is achieved by setting the manual control to a position indicated by the manufacturer of the automated console to cause no increase or decrease when the update mode is entered. Once in the update mode, the fader can be moved as desired and, on leaving this mode,

the level is returned to its pre-update level setting. At this point, either the read mode can be initiated or the update mode can be retained while leaving the fader in its neutral (no movement) position.

Automated Servo-Driven Fader

Another means of achieving console automation is using the *servo-driven fader automated system* (often referred to as *moving faders*).

Unlike a VCA-based system in which the DC voltage levels control automation levels, the fader actually is a true resistive attenuator that is driven automatically by a servo motor interface. Thus, during the playback of an automated mix, the faders move on their own in accordance with the requirements of the mix.

Servo-driven faders don't have separate provisions for read, write, or update modes because most are instinctive automation systems. Often, all that is required to update the levels to a new position is to take hold of the touch-sensitive fader and move it to the desired location. The processing computer then automatically stores this information as updated material. The multiple faders can often be assigned to a number of groups that can be controlled from a single master or by grabbing any fader in the group. Switching functions are also addressable and can be stored in memory. Automated data can be stored to disk as a virtual mix (allowing for an unlimited amount of dynamic updates) or as multiple snapshots of fader and/or mute functions that can be instantly recalled.

MIDI-Based Automation

A more recent form of console and mixer automation involves the use of MIDI for the storage and control over various console automation elements. Immediate advantages to automating console functions via MIDI include the following:

◆ MIDI automation data can be stored to a conventional MIDI sequencer.

◆ Most modern effects devices can be MIDI controlled, allowing effects to respond to program-change messages during a mix.

◆ MIDI data is an established, cost-effective standard in worldwide use for professional and project studio production.

Certain console manufacturers make use of MIDI as a cost-effective means for implementing basic or intermediate forms of automation. For example, certain mixer and console designs allow for snapshots of console mute functions to be saved in internal memory as a specific program change number. By recalling these program changes from a MIDI sequencer, basic mute automation can be accomplished. Other console designs use MIDI controllers to provide dynamic control over VCA or moving fader movements in real time with the aid of an add-on internal or external automation package. The various topics relating to MIDI-based mixing are fully discussed in Chapter 7, "MIDI and Electronic Musical Instrument Technology."

Signal Processors

Signal processors increasingly have become a central component of modern audio and music production. It is the function of a signal processor to change, augment, or otherwise modify an audio signal. These audio signals can be modified in either of the following two ways:

- *In the analog domain*—Audio signal levels are directly processed without being converted into a digital form.
- *Using digital signal processing (DSP)*—Analog signals that have been converted into a digital, binary form can be mathematically recalculated according to a program algorithm so as to alter the nature of the sound in a specific way.

Both signal processing methods are frequently used in all phases of audio production and exert an ever-increasing amount of control over amplitude level processing (dynamic range) and the spectral content of a sound (equalization), as well as the augmentation or re-creation of room ambience, delay, time/pitch alteration, and other types of special effects. These special effects range from the sublimely subtle to the outrageously "in yo' face" wacky and include all the possible combinations in between.

As we step into the next millennium, and as digital signal processing has begun to step out of its initial infancy, we're beginning to see huge advances in the way integrated personal computer systems and dedicated processors can directly process signals with ever-increasing amounts of power. These same advances also have had an impact on the way that complex effects processing can be automated in the audio production environment.

In this chapter, you take an in-depth look at the many types of traditional and not-so-traditional effects processing tools that are currently available. The chapter begins with a good, hard look at the broad range of analog-based signal processors that are the cornerstones of today's studio. These signal processors include devices such as equalizers, dynamic range controllers, acoustic and electro-acoustic reverb systems, and even some tape-based tricks that may come in handy. After this initial discussion, the chapter continues on to study the ever-important field of digital signal processors.

Equalization

The most common form of signal processing is *equalization* (also known as *EQ*). The audio equalizer, shown in Figure 11.1, is a device or circuit that enables a recording or mix engineer to control the relative amplitude of various frequencies within the audible bandwidth. Put another way, the audio equalizer lets you exercise tonal control over the harmonic or timbral content of a recorded sound. EQ may need to be applied to a single recorded channel, group of channels, or to an entire program signal for any number of reasons, including the following:

◆ To correct specific problems in a recording or in a room (generally to restore a sound to its natural tone) in order to overcome deficiencies in the frequency response of a microphone or in the sound of an instrument

◆ To alter a sound purely for musical or creative reasons

◆ To allow contrasting sounds from several microphones or tape tracks to blend better in a mix

◆ To increase the separation between mics or recorded audio tracks by seeking to reduce those frequencies that cause leakage

Figure 11.1. NTI's EQ³ Sound Enhancement System. (Courtesy of Night Technologies International)

Equalization refers to the alteration in frequency response of an amplifier so that the relative amplitude levels of certain frequencies are more or less pronounced than others. EQ is specified as plus or minus a certain number of decibels at a certain frequency. You could say, for example, that a signal has been boosted by "+4 dB at 5 kHz."

Only one frequency was specified in the preceding example; in reality, however, a range of frequencies either above, below, or centered around the specified frequency also are affected. The amount of boost or cut at frequencies other than the one named is determined by whether the curve is peaking or shelving, by the bandwidth of the curve, and by the amount of boost or cut at the named frequency. A +4 dB boost at 5 kHz, therefore, also may add a degree of boost at 4 kHz and 6 kHz (see Figure 11.2).

Figure 11.2. *A 5-kHz peak boost curve.*

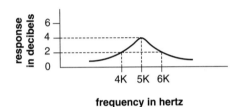

Older equalizer systems and newer redesigns often base their design around filters that employ passive components (such as inductors, capacitors, and resistors) and incorporate amplifiers only to make up for internal losses in level, called *insertion loss*. Figure 11.3a shows typical signal levels in a passive equalizer set for a flat response, while Figure 11.3b shows the signal level structure of an equalizer that has a boost in its low-frequency response.

Figure 11.3. *Typical signal levels in a passive equalizer.*

a. *EQ is set for flat response.*

b. *EQ filter is set for 6-dB boost at 100 Hz.*

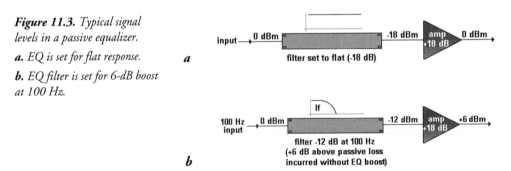

Most equalization circuits today, however, are of the active filter type that changes its characteristics by altering the feedback loop of an op amp (see Figure 11.4). Available in many configurations, active equalizers are favored over their passive counterparts for their low cost, small size, light weight, loading indifference, and superior isolation (generally having a high-input and low-output impedance), wide gain range, and line-driving capabilities.

Figure 11.4. *An example of an active equalization circuit.*

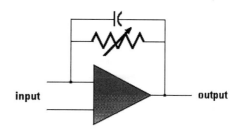

Peaking Filters

The most common equalization curve is the *peaking* curve. As its name implies, a peak-shaped bell curve is created that can either be boosted or cut at a selected center frequency. Figure 11.5 shows the curves for a peak equalizer set to boost or cut at 1000 Hz. The *quality factor* (Q) of a peaking equalizer refers to the width of the bell-shaped curve (see Figure 11.6). A curve with a high Q will have a narrow bandwidth with few frequencies outside the selected bandwidth being affected, whereas a low-Q curve is very broad-band and affects many frequencies.

Figure 11.5. *Peaking equalization curves.*

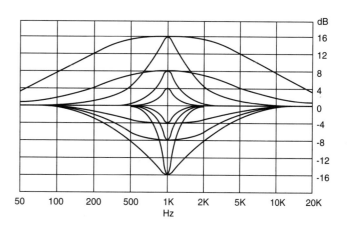

Figure 11.6. *The number of hertz between the points that are 3 dB down from the center frequency is the bandwidth of a peaking filter.*

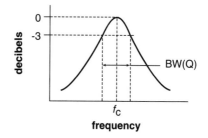

Bandwidth is the range of frequencies between the –3-dB (half-power) points on the curve. The Q of a filter is an inverse measure of the bandwidth. To calculate Q, divide the center frequency by the bandwidth. For example, a filter centered at 1 kHz that is 1/3-octave wide has its –3-dB frequency points located at 891 Hz and 1123 Hz, respectively, yielding a bandwidth of 232 Hz (1123 – 891). This EQ curve's Q, therefore, is 1 kHz divided by 232 Hz, or 4.31.

Shelving Filters

Another type of equalizer is the *shelving filter*. Shelving refers to a rise or drop in frequency response at a selected frequency, which tapers off to a preset level and continues at this level to the end of the audio spectrum. Shelving can be inserted at either the high or low end of the audio range and is the curve commonly found on a home stereo's bass and treble controls (see Figure 11.7).

Figure 11.7. *High/low, boost/cut curves for a shelving equalizer.*

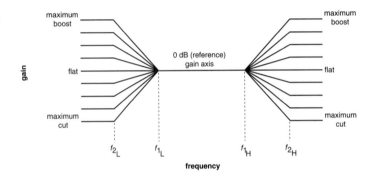

High-Pass and Low-Pass Filters

Equalizer types also include *high-pass* and *low-pass filters*. As their names imply, certain frequencies are passed at full level, while others are attenuated. Frequencies that are attenuated by less than 3 dB are said to be *inside the passband*; those attenuated by more than 3 dB are *in the stopband*. The frequency at which the signal is attenuated by exactly 3 dB is called the *turnover* or *cutoff* frequency and is used to name the filter. Ideally, attenuation would become infinite immediately outside the passband; in practice, however, this isn't attainable. In the simplest case, attenuation increases at a rate of 6 dB per octave. This rate is called the *slope* of the filter. Other common slopes are 12 and 18 dB per octave. Figure 11.8, for example, shows a 700-Hz high-pass filter response curve with a slope of 6 dB per octave, while Figure 11.9 shows a 700-Hz low-pass filter response curve with a slope of 12 dB per octave.

High- and low-pass filters differ from shelving EQ in that their attenuation doesn't level off outside the passband. Instead, the attenuation continues to increase. A high-pass filter in combination with a low-pass filter can be used to create a bandpass filter, with the bandwidth being controlled by the filter's turnover frequency and the Q controlled by the filter's slope (see Figure 11.10).

Figure 11.8. A 700-Hz high-pass filter with a slope of 6 dB per octave.

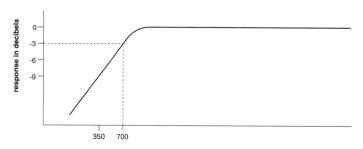

Figure 11.9. A 700-Hz low-pass filter with a slope of 12 dB per octave.

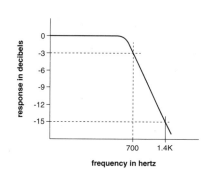

Figure 11.10. A bandpass filter is created by combining high- and low-pass filters with different cut-off frequencies.

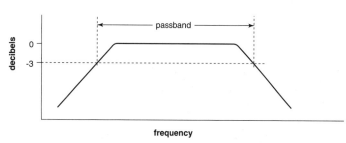

Equalizer Types

Four basic types of equalizers are in use today that employ one or more of the previously described filter types: the selectable frequency equalizer, the parametric equalizer, the graphic equalizer, and the notch filter.

The *selectable frequency equalizer*, as its name implies, has a set number of frequencies from which to choose. The equalizer usually allows a boost or cut at the selected frequency with a predetermined Q and, typically, has an independent low-mid-high range. This form of equalization is most often found on older console designs and on newer low-cost production consoles.

With the *parametric equalizer* (see Figure 11.11), the center frequency in each band is continuously adjustable, rather than selectable in discrete steps. Control over the center frequency (Q) is either selectable or continuously variable, although certain manufacturers have no provisions for a variable Q. The amount of boost or cut is also continuously variable. Generally, each set of frequency bands overlaps into the next band section, providing smooth transitions between frequency bands or allowing multiple curves of a high Q to be placed in a narrow frequency range. Due to its increased flexibility and performance, the parametric equalizer has become the standard design for input strips in most consoles.

Figure 11.11. Drawmer 1961 vacuum tube equalizer. (Courtesy of Drawmer)

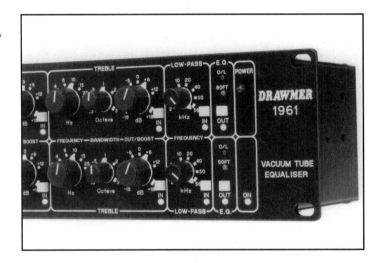

The graphic equalizer (see Figure 11.12) provides boost/cut level control over a series of center frequencies that are equally spaced according to music intervals. An "octave band" graphic equalizer may, for example, have 11 equalization controls spaced at the octave intervals of 20, 40, 80, 160, 320, 640 Hz, and 1.25, 2.5, 5, 10, and 20 kHz, with 1/3-octave equalizers having up to 36 controls. The controls for the various equalization bands generally are linear sliders, arranged vertically, side-by-side. The physical positions of these controls provide a "graphic" representation of the overall frequency response curve. This type of equalizer is often used in applications requiring the fine tuning of a system to compensate for the acoustics of a room, such as an auditorium or studio control room.

In addition to its use in modifying sound, an equalizer can be used to remove hum and other undesirable discrete-frequency noises. A notch filter is used for this purpose. The filter can be tuned to attenuate a particular frequency and has a very narrow bandwidth so it has little effect on the

rest of the program (see Figure 11.13). Notch filters are used more in film-location sound than in studio recording because the problems encountered in location work aren't usually present in a well-designed studio.

Figure 11.12. *The Rane GE60 stereo graphic equalizer. (Courtesy of Rane Corporation)*

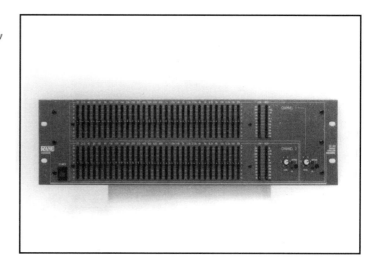

Figure 11.13. *Notch filter response curves. (Courtesy of Orban Associates, Inc.)*

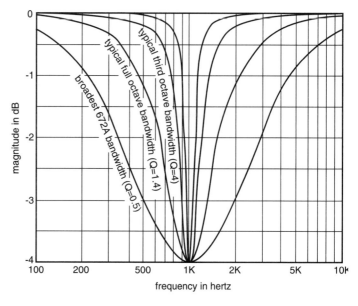

Applying Equalization

Although most equalization is done by ear, it's helpful to have an idea about which frequencies affect an instrument in order to achieve a particular effect. On the whole, the audio spectrum can be divided into four frequency bands: low (20-200 Hz), low-middle (200-1000 Hz), high-middle (1000-5000 Hz) and high (5000-20,000 Hz).

When the frequencies within the 20- to 200-Hz range are modified, the fundamental and the lower harmonic range of most bass information is affected. These sounds often are felt as well as heard, so boosting in this range can add a greater sense of power or punch to music. Reductions in this range will weaken or thin out the lower frequency response.

The fundamental notes of most instruments are within the 200- to 1000-Hz range. Changes in this range often result in a dramatic variation in overall signal energy, with an increase adding to the overall impact of a program. Due to the ear's sensitivity in this range, a minor change in level often results in a major audible effect. The frequencies around 200 Hz can give the bass a greater feeling of warmth without a loss of definition, while those frequencies in the 500- to 1000-Hz range may make an instrument sound horn-like. Too much boost in this range often causes listening fatigue.

Higher-pitched instruments are most often affected in the 1000- to 5000-Hz region. Boosting these frequencies often results in an added sense of clarity, definition, and brightness. Too much boost in the range of 1000 to 2000 Hz may have a "tinny" effect on the overall sound, while the upper mid-frequency range (2000-4000 Hz) affects the intelligibility of speech. Boosting in this range may make music seem closer to the listener, but too much of a boost tends to cause listening fatigue.

The high-frequency region (5000-20,000 Hz) is composed almost entirely of the harmonic components produced by most instruments. For example, boosting frequencies in this range often add sparkle and brilliance to a string and woodwind instrument. Boosting too much may produce sibilance on vocals and make the upper range of certain percussion instruments sound harsh and brittle. Boosting at around 5000 Hz has the effect of making music sound louder. A boost of 6 dB at 5000 Hz, for example, can make the overall program level sound as though it has been increased by 3 dB; conversely, attenuation can make music seem more distant.

Table 11.1 provides a further analysis of the ways in which frequencies and equalization can interact with various instruments. (For more information, refer to the section "Microphone Placement Techniques" in Chapter 4, "Microphones: Design and Application".)

Table 11.1. Instrumental frequency ranges of interest.

Instrument	Frequencies of Interest
Kick drum	Bottom depth at 60-80 Hz, slap attack at 2.5 kHz
Snare drum	Fatness at 240 Hz, crispness at 5 kHz
Hi-hat/cymbals	Clank or gong sound at 200 Hz, shimmer at 7.5 kHz to 12 kHz
Rack toms	Fullness at 240 Hz, attack at 5 kHz
Floor toms	Fullness at 80-120 Hz, attack at 5 kHz
Bass guitar	Bottom at 60-80 Hz, attack/pluck at 700-1000 Hz, string noise/pop at 2.5 kHz
Electric guitar	Fullness at 240 Hz, bite at 2.5 kHz
Acoustic guitar	Bottom at 80-120 Hz, body at 240 Hz, clarity at 2.5-5 kHz
Electric organ	Bottom at 80-120 Hz, body at 240 Hz, presence at 2.5 kHz
Acoustic piano	Bottom at 80-120 Hz, presence at 2.5-5 kHz, crisp attack at 10 kHz, "honky tonk" sound (sharp Q) at 2.5 kHz
Horns	Fullness at 120-240 Hz, shrill at 5-7.5 kHz
Strings	Fullness at 240 Hz, scratchiness at 7.5-10 kHz
Conga/bongo	Resonance at 200-240 Hz, presence/slap at 5 kHz
Vocals	Fullness at 120 Hz, boominess at 200-240 Hz, presence at 5 kHz, sibilance at 7.5-10 kHz

One way to zero in on a particular frequency using an equalizer (especially a parametric equalizer) is to set the amount of boost to near maximum and change the center frequency until the desired instrument range to be EQed is found. The amount of boost can then be decreased until the desired effect is obtained. Attenuation of a frequency range can be achieved in a similar manner.

If boosting one range of an instrument creates the need to boost the other ranges, the overall effect is simply that of raising the overall level. This is more easily done by simply increasing the input fader's level. If a simple increase in gain doesn't sound satisfactory, it may be that one range of frequencies is too dominant and requires attenuation.

As far as recording with EQ goes, there are a number of different opinions on this subject. Some use EQ liberally to make up for placement and mic deficiencies, whereas others use it sparingly, if at all. One example in which EQ might be used sparingly is if the engineer doing the recording knows that another engineer will be mixing a particular song or project. In this situation, the mixing engineer might have a very different idea of how the instruments should sound. If large amounts of EQ were recorded to tape during the session, the mixdown engineer may have to work very hard to counteract the original EQ settings. On the other hand, if everything was recorded flat, the producer and artists may have difficulty passing judgment on a performance or hearing the proper balance during overdubbing. Such a situation may call for equalization in the monitor mix.

When several mics are to be combined onto one channel of the tape, they can be EQed individually only during the recording phase. Thus, recording the signals flat will, as a rule, prevent later optimization of each signal picked up by each mic during mixdown. In addition, although recording with EQ doesn't change the perceived noise level, changes in EQ during playback can potentially add residual tape and/or amplifier noise to the track. Boosting highs during playback, for example, would make the hiss of an analog tape track more pronounced than if the highs were boosted before recording. If the same engineer is to record and mixdown the tape, then recording with EQ generally isn't a problem. In any event, unless a special effect is desired, EQ should be used moderately, and microphone selection should be used to obtain a good instrument sound. If an instrument is poorly recorded in an initial recording session, it can rarely be fully corrected later during mixdown.

In summary, an equalizer is a powerful tool. Used properly, it can greatly enhance or restore the musical and sonic balance of a signal. Experimentation is the key to equalizer use, and no book can replace the trial-and-error process. Keep in mind that an equalizer can't be regarded as a panacea for improper microphone technique. Rather, EQ should be used simply as a tool for correcting the minor problems of room acoustics or to modify—to your tastes—a system or pickup signal's frequency response.

Dynamic Range Control

Along with the other unpredictable things in life that get out of hand from time-to-time, the level of a signal can vary widely at a moment's notice. For example, if a vocalist lets out an impassioned scream following a soft passage, you can almost guarantee that the mic's signal will jump from its optimum recording level into severe distortion. Conversely, if you set an instrument's mic to accommodate the loudest level, its signal may be buried during the rest of the mix. For these and other reasons, it becomes obvious that some form of control over the signal's dynamic range may be necessary under certain conditions.

Metering

Amplifiers and magnetic tape are limited in the range of signal level they can pass without distortion, so audio engineers need a way to determine whether the signals they are working with will be stored or transmitted without distortion. The most convenient way to make this determination is to use a visual device, such as a meter. If preventing distortion on the tape was the only concern, peak-indicating meters could be used to display the maximum amplitude fluctuations of a waveform. The human perception of loudness, however, doesn't have much relationship to the peak level of signals. The meter may read higher at a certain point in the program, but the signal might not sound any louder (see Figure 11.14). If a meter is used to set or maintain a certain perceived volume level, peak indication isn't much use.

Figure 11.14. *A peak meter reads higher at point A than at point B, even though the loudness level is the same.*

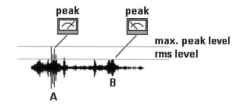

Because the perception of the ear to loudness is proportional to the rms (average) value of a signal, a meter was designed that could read this level so that volume and meter indication would coincide (see Figure 11.15). The scale chosen for the meter was calibrated in volume units—hence the name *VU meter* (see Figure 11.16). Zero VU is considered the standard operating level. Although VU meters do the job of indicating volume level, they ignore the short-term peaks that can overload tape. These peaks can be from 8 to 14 dB higher than the indicated rms value. This means that the electronics must be designed so that unacceptable distortion doesn't occur until at least 14 dB above 0 VU. Typical VU meter specifications are provided in Table 11.2.

Figure 11.15. *A VU meter reads the rms level and ignores instantaneous peaks that don't contribute to loudness.*

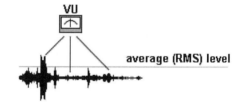

Figure 11.16. *A VU meter's upper scale is calibrated in volume units for use in recording. The lower scale is percentage modulation for use in broadcasting.*

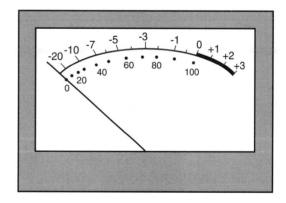

Table 11.2. VU meter specifications.

Characteristic	Specification
Sensitivity	Reads 0 VU when connected across a +4-dBm signal (1.228 V in a 600-ohm circuit)
Frequency response	±0.2 dB from 35 Hz to 10 kHz; ±0.5 dB from 25 Hz to 16 kHz
Overload capability	Can withstand ten times 0-VU level (+24 dBm) for 0.5 second and five times 0 VU (approximately +18 dBm) continuously

Volume units indicated on the meter are equal to dB for sine waves; for other waves or complex signals, however, the VU meter actually reads between the rms and peak values of the signal. For these waves, volume units are larger than dB.

The difference between the maximum level that can be handled without incurring distortion and the average operating level of the system is called *headroom*. Some studio-quality amplifiers are capable of signal outputs as high as 26 dB above 0 VU and thus have 26 dB of headroom. Magnetic tape, however, has limited headroom. Its dynamic range is such that providing the headroom necessary to prevent distortion of the peaks would make tape noise too audible during the rest of the program. The 3% distortion level for magnetic tape recorded on a tape machine is typically only 8 dB above 0 VU, whereas console amplifiers have distortion of less than 0.4% at this level. The proper record level for most program material is 0 VU, although higher levels are possible providing that short term peaks—which would cause distortion—aren't excessively high.

Rather than utilizing a meter pointer, some meters now use light-emitting diodes (LEDs) or liquid crystal displays (LCDs) to provide level indication through the illumination of lights that correspond to different signal levels. These readout indicators follow peaks better than any meter can and give a virtually instantaneous display of the signal level. Such indicators often simultaneously display both rms and peak level values.

Dynamic Range Processors

The overall dynamic range of music is potentially on the order of 120 dB, whereas the dynamic range of the digital medium is approximately 90 dB (see Figure 11.17). The range of analog magnetic tape is on the order of 60 dB, excluding the use of noise-reduction systems that can add another 15 to 30 dB, which still falls short of music's full 120-dB range. The overall dynamic range of a compact disc also is often 80 to 90 dB. With such a wide dynamic range, however, unless the CD is played in a noise-free environment, either the quiet passages would get lost in the ambient noise of the listening area (35-dB to 45-dB SPL for the average home) or the loud passages would be too loud to bear. Similarly, if a program of wide dynamic range were to be reproduced through a medium with a narrow dynamic range, such as AM radio (20 to 30 dB) or FM radio (40 to 50 dB), a great deal of information would get lost in background noise. To prevent these problems, the dynamic range can be reduced to a level appropriate both for the medium through which it

is to be reproduced and for comfortable listening in the average home. This gain reduction can be accomplished either by manually riding the fader's gain or by using a device that automatically changes the dynamic range of a signal, known as a *dynamic range processor*.

Figure 11.17. *Dynamic ranges of several audio media.*

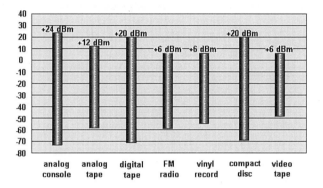

Compression

A *compressor* is, in effect, an automatic fader. When the input signal exceeds a predetermined level (see Figure 11.18), called the *threshold*, the gain is reduced by the compressor and the signal is attenuated. By attenuating the louder signal levels, you are, in fact, reducing the program's overall dynamic range. Because the range between the loudest and the softest signals is "compressed" by increasing the signal's overall gain, the average (rms) level will be greater. Thus the signal will be perceived as being louder than it otherwise would be.

Figure 11.18. *A compressor reduces levels that exceed a selected threshold by a specified amount.*

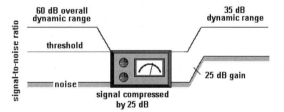

The increase of input signal (in dB) needed to cause a 1-dB increase in the compressor's output signal is called the *compression ratio* or the *slope* of the compression curve. For a ratio of 4:1, therefore, an 8-dB increase of input produces a 2-dB increase in output. Because the recorded signals vary in loudness and may be above the threshold at one moment and below it the next, the speed that controls the rate by which the gain is to be reduced (once the signal exceeds the threshold) and then restored (after the input signal falls below the threshold) must be determined.

These speeds may be varied by altering the attack and release times, respectively. The attack time is defined as the time it takes for the gain to decrease by a certain amount—usually to 63% of its final value. The release time is defined as the time required for the gain to return to 63% of its original value.

As stated earlier, the perception of the ear to the loudness of a signal is proportional to its rms value; large short-duration peaks, therefore, don't noticeably increase the loudness of a signal. What is desired is to allow the signal volume to rise and fall over the course of the program but to lesser extents than the volume would if it were uncontrolled. If waveform peaks were permitted to trigger gain reduction, the volume would actually decrease rather than increase. This would noticeably change the dynamics of the program and in an unacceptable way. To avoid the triggering of compression by these peaks, the attack time is set so that the waveform must exceed the threshold level for a long enough duration to increase the average level.

If the release time was set too short for the program material, that is, if full gain was restored each time the signal fell below the threshold, audible side effects that sound somewhat like thumps, pumping, and breathing noises would be heard due to the rapid rise of background signals as the gain was increased. Also, if peaks in rapid succession were fed into the device, the program gain would be restored after each one and the level of the program would rise after each peak. Because the level-sensing mechanism is sensitive to both positive and negative waveform excursions, extremely short release times could cause gain reduction twice each cycle and introduce harmonic distortion into the signal.

To eliminate these side effects, longer release times are used so that repeated waveform excursions above the threshold will cause gain reduction only once. The gain will remain reduced through all these excursions and will return gradually to normal. This makes the increase in background noise less obvious. If the release time is too long, however, a loud section of the program may cause gain reduction that persists through a soft section, making the soft section inaudible. The release time is defined as the time needed for the gain to return to a certain percentage of its no-gain reduction value (usually 63%).

Compressors usually have built-in metering to allow monitoring of the amount of gain reduction that's taking place. The meter usually sits at 0 VU when the input signal is below the threshold and falls to the left to indicate the number of decibels of gain reduction when the input signal exceeds the threshold (see Figure 11.19). Some compressors use meters or LED readouts that display gain reduction directly (see Figure 11.20) and move either upscale or downscale (depending on the system's design) to show gain reduction in dB.

Figure 11.19. A compressor with a dual-function (VU/GR) gain-reduction meter.

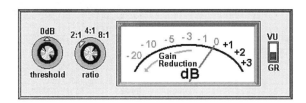

Figure 11.20. *The gain-reduction indicators on some compressors are designed to read gain reduction directly.*

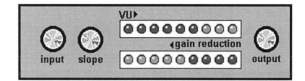

Limiting

If the compression ratio is made large enough, the compressor becomes a *limiter* (see Figures 11.21 and 11.22). A limiter is used to keep signal peaks from exceeding a certain level in order to prevent the overloading of amplifier signals, recorded signals on tape or disc, broadcast transmission signals, and so on (see Figure 11.23). An extreme case of a limiter is a *clipper*, which chops off the top of any waveform exceeding the threshold level. A clipper could be said to have an infinite compression ratio. Most limiters have ratios of 10:1 or 20:1, although they are available with ratios up to 100:1. Because such a large increase in the input signal is required to produce an increase in the output of the limiter, the likelihood of overloading the equipment following the limiter is greatly reduced.

Figure 11.21. *The Symetrix 501 peak-rms compressor/limiter. (Courtesy of Symetrix, Inc.)*

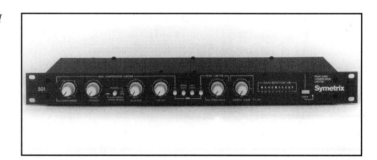

Figure 11.22. *Manley Electro-Optical limiter. (Courtesy of Manley Labs, Inc.)*

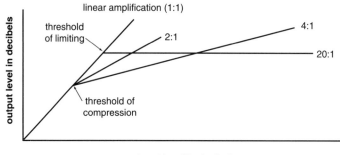

Figure 11.23. *The output of a compressor is linear below the threshold point and follows the slope of the selected compression or limiting curve above the threshold.*

(graph labels: output level in decibels; input level in decibels; linear amplification (1:1); threshold of limiting; threshold of compression; 2:1; 4:1; 20:1)

Limiting usually is used in recording to prevent short-term peaks from reaching their full amplitude. Short-term peaks would add little level to the program in proportion to the distortion they would cause (if they saturated the tape) or the noise they would allow to enter the system (if the signal was recorded at a level low enough so that the peaks wouldn't distort). Extremely short attack and release times are used so that the ear can't hear the overall average gain reduction. Limiting is used only to remove occasional peaks because gain reduction on many successive peaks would be noticeable. If the program contains many peaks, the threshold should be raised and the gain reduced manually so that only occasional extreme peaks are limited.

Expansion

Expansion is the process by which the dynamic range of a signal is increased. Depending on the system's design, the expander operates by decreasing the gain of a signal as its level falls or by increasing the gain as the level rises. When using an expander (see Figure 11.24) that operates using the former and more commonly encountered system, as the signal level falls below the expansion threshold, the gain is proportionately decreased (according to the expansion ratio) in such a way that the low-level signals are reduced (see Figure 11.25). These devices can also be used as noise-reduction systems by adjusting them so that the noise to be removed is below the threshold level, while the desired signal is above the threshold.

Figure 11.24. *The Aphex Model 622 Logic Assisted Expander/Gate. (Courtesy of Aphex Systems, Inc.)*

Figure 11.25. An expander
increases a signal's dynamic range
by reducing levels that fall (by a
specified amount) below a selected
threshold.

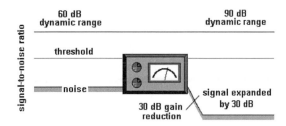

The Noise Gate

One other type of expansion device is the *noise gate,* shown in Figure 11.26. This device acts as an infinite expander and allows a signal above the selected threshold to be passed through to the output at unity gain and without dynamic processing. Once an input signal falls below this threshold level, the device effectively shuts down the signal by applying full attenuation to the output. In this way, the desired signal is allowed to pass while background sounds or unwanted noise during pauses in the music are not. This device also is effective in the reduction of leakage.

Figure 11.26. The Drawmer
DS404 Quad Gate. (Courtesy of
QMI)

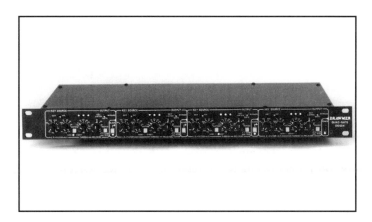

Sometimes, a special key input (see Figures 11.27a and 11.27b) is included as a *sidechain* input to a noise gate. A sidechain input is an external control source that is capable of affecting a device's main audio path in a specific manner. This enables an external analog signal source, such as a miked instrument, synthesizer, or oscillator, to control or trigger the output of the gate's main audio path.

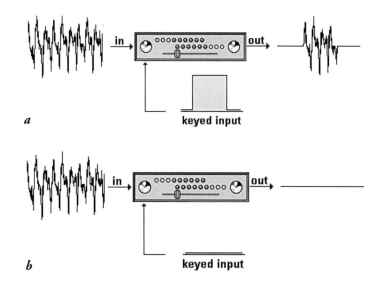

Figure 11.27. *Diagram of a basic keyed-input noise gate.*
a. *With input present at key.*
b. *Without input present at key.*

Applying Dynamic Range Control

The selection of proper attack and release times and degrees of compression, expansion, and limiting depends on the program material. In multitrack recording, dynamic range modifications usually deal only with single instruments or groups of instruments. In radio, television, and disc cutting, entire songs are compressed and the need to apply the proper parameters is more critical.

Limiting usually is used only for recording speech or instruments with *transients*—momentary high peak levels—so that the signal can be recorded at a high level without overloading the tape.

Compression may be used for any of the following reasons:

◆ To minimize the changes in volume that occur when an instrumentalist or vocalist has too great a dynamic range for the music or momentarily changes his/her distance from a mic.

◆ To balance out the different volume ranges of an instrument. The notes produced by an electric bass, for example, often vary widely in volume from string to string. Compression can be used to produce a smoother bass line by matching the volumes of the various strings. As another example, some instruments (such as horns) are louder in some registers than in others due to the amount of effort required to produce the notes. Compression can equalize these volume levels.

◆ To enable a signal to be made significantly louder in the mix while increasing the overall signal-level reading on the meter only slightly. This is accomplished by increasing the ratio of average-to-peak levels.

◆ To reduce sibilance by inserting a filter into the compression circuit that causes the circuit to trigger when an excess of high-frequency signal is present. A compressor used in this manner is called a *de-esser*.

Some producers strive to cut their recordings as "hot" as possible; that is, they want the recorded levels to be as far above the normal operating level as possible without blatant distortion. The logic behind the thinking here is that in this competitive business louder recordings, when broadcast, stand out from the soft recordings in the top-40 playlist. In fact, a recording 1 or 2 dB louder than another appears to have more bass and more highs because of the effect of the Fletcher-Munson curve. To achieve these hot levels without distortion, dual-channel compressors and limiters often are used during the mastering phase to remove peaks and to raise the average level of the program so that the disc is louder than it otherwise would be. A combination unit is shown in Figure 11.28.

Figure 11.28. Valley Audio's 433 "Dynamite" stereo dynamic processor. (Courtesy of Valley Audio Products, Inc.)

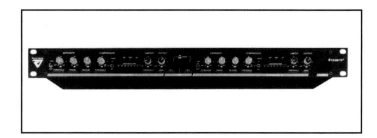

Compressing a mono mix is done in much the same way as compressing a single instrument. The adjustment of the threshold, attack, release, and ratio controls, however, is more critical in preventing "pumping" by prominent instruments in the mix. Compressing a stereo mix gives rise to an additional problem. If two individual compressors are used, a peak in one channel reduces the gain on that channel and causes sounds centered between the two speakers to jump toward the channel not undergoing compression. To avoid this center shifting, most compressors have a provision for connecting them in stereo with a second compressor of the same make and model. This interconnection procedure of ganging the two channels together combines the outputs of the signal-level sensing circuits of the two units. In that way, a signal causing gain reduction in one channel causes equal gain reduction in the other channel, thereby preventing the center information from shifting in the mix.

Digital Signal Processing

In modern day audio production, an ever-increasing amount of signal processing is being carried out in the digital domain through the use of digital signal processing (DSP). A major advantage to working with DSP is that software programming can be used to configure a digital processor in order to achieve a variety of effects, such as reverb, echo, and delay, as well as numerous other processing functions, such as equalization, pitch shifting, and gain changing.

The task of processing a signal in the digital domain is accomplished by combining logic circuits in a building block fashion. These logic circuits follow basic binary computational rules according to a specialized program algorithm. When combined in a building block fashion, these logic circuits can be used to alter the numeric values of sampled audio in a predictable way.

After a program has been configured from either internal RAM or the system's software, complete control over a program's setup parameters can be altered and measured as discrete numbers or as percentages of a full value. Because these values are both discrete and digital, the settings can be precisely duplicated or, better yet, saved to RAM or to hard disk as a file that can be easily called up at any time.

The following sections, "Real-Time and Non-Real-Time DSP," "DSP Basics," and "The Real World of DSP Design," are excerpts from the book *Hard Disk Recording for Musicians* by David Miles Huber (© 1995, Amsco Publications, NYC) and have been reprinted with permission.

Real-Time and Non-Real-Time DSP

The number-crunching process involved in performing DSP calculations can be performed in one of two ways: real time or non-real time.

Real-time calculations are capable of processing a signal in the here and now. In other words, real-time systems can alter, mix, or otherwise process sampled audio as it's being recorded or reproduced. In such a signal processing environment, the signal alteration is often done while reproducing audio from disk without affecting the original soundfile data. This non-destructive editing also allows any inadvertent changes to be easily undone at a later time.

Non-real-time signal processing, on the other hand, is often used by hard-disk systems and digital audio workstations that don't have or don't need a dedicated real-time processor block to perform some or all processing tasks. After a non-real-time data or signal processing function is invoked, the system's processor or coprocessor dedicates itself to performing the task in non-real time (which means that you have to wait while the system performs the necessary calculations). When these calculations have finished, the final results usually are written to hard disk as a separate file. If this processed section is used to replace an existing nonprocessed segment (such as a faded or crossfaded area), it generally is attached or "tagged" to marker points in the original, unprocessed soundfile areas. Then on playback, the processed section is butt-joined with the original areas to reproduce the program output in a seamless manner.

DSP Basics

The scope and capabilities of digital signal processing are limited only by speed, number-crunching power, and human imagination. Yet the process itself is made up of only three basic building blocks (see Figure 11.29):

- ◆ Addition
- ◆ Multiplication
- ◆ Delay

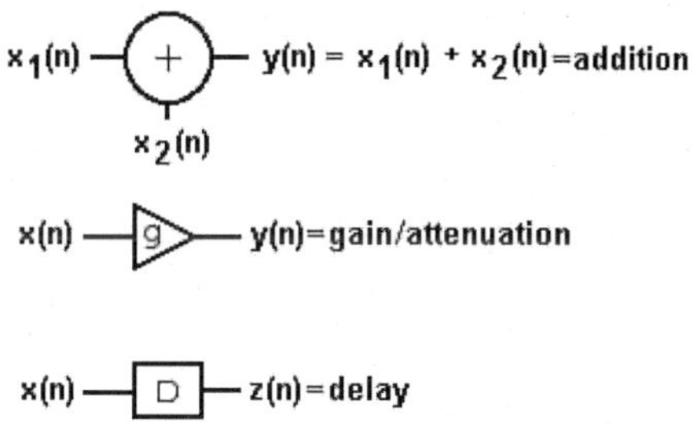

Figure 11.29. *The DSP process is made up of only three basic building blocks: addition, multiplication (gain/attenuation), and delay.*

$x_1(n) \rightarrow (+) \rightarrow y(n) = x_1(n) + x_2(n) =$ addition
$x_2(n)$

$x(n) \rightarrow |g\rangle \rightarrow y(n) =$ gain/attenuation

$x(n) \rightarrow [D] \rightarrow z(n) =$ delay

One of the best ways to understand these building blocks of digital logic is through an application. Just for fun, let's try building a simple 4-in x 1-out digital mixing application that might be used to combine the output channels of a digital workstation.

Addition

As you might expect, a digital adder sums together the various bits at the input of the circuit in order to create a single combined result. With this straightforward building block, you can get started on your mixer by combining the four input signals into one output channel (see Figure 11.30).

Figure 11.30. *Inputs to your digital mixer can be summed together into a single data stream.*

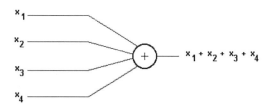

x_1
x_2
x_3
x_4
$(+) \rightarrow x_1 + x_2 + x_3 + x_4$

Multiplication

The multiplication of sample values by a numeric coefficient enables the gain (level) of digitized audio to be changed either up or down. Whenever a sample is multiplied by a factor of 1, the result is unity gain or no change in level. Multiplication by a factor of less than 1 yields a reduction in gain (attenuation). Likewise, the multiplication by a number greater than 1 results in an increase in gain.

Now that you know this, you can add gain controls to your mixer. This gives you some real control (see Figure 11.31). Gain control can be accomplished by adding variable digital faders that can directly supply the processor with the appropriate gain coefficients with which to multiply the input signal.

Figure 11.31. Gain can be added to determine the mix ratio of the combined signals.

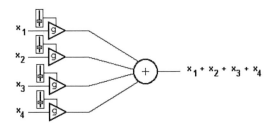

Before moving on, take a look at how multiplication can be applied to the everyday production world by calculating how the gain of a recorded soundfile can be changed over time. Examples of this applied gain change are fade-ins, fade-outs, crossfades, and changes in level.

To better understand this process, look at how a non-real-time fade-out might be created using a hard-disk-based system. Suppose that you have a song that wasn't faded during a mixdown session but now definitely needs to be faded out over its final chorus section. Your first task is to define the part of the song that needs to be faded (see Figure 11.32a), call up the fade function, and then perform the fade.

After the fade has been completed, it's the signal processor's job to continually multiply the affected samples by a diminishing coefficient. The result is to reduce the gain over the length of the defined fade region. After the new sample has been calculated, the results are written automatically to disk as a separate file (see Figure 11.32b). On playback, the fade is tagged onto the original soundfile, ending at the appropriate point (see Figure 11.32c).

Figure 11.32. Example of a 1-second fade.

a. Original soundfile.

b. Defined area to be faded is calculated and written to disk as a separate file.

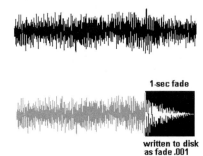

Figure 11.32. *(continued)*
c. *Faded file is tagged to the original file and is reproduced with no audible break or adverse effect.*

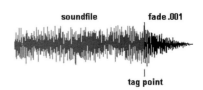

soundfile   fade .001

c   tag point

Delay

The final DSP building block deals with time—that is, the use of delay over time in order to perform a specific function or effect. In the world of DSP, delay is used in a wide variety of applications. This discussion, however, focuses on two types of delay:

♦ Effects-related delay
♦ Delay at the sample level

Most modern musicians and those associated with audio production are familiar with the way different delay ranges can accomplish a wide range of effects. They also are familiar with the use of digital delay for creating such sonic effects as doubling and echo delay. These effects (discussed later in this chapter) are created from discrete delays that are 35 milliseconds or more in length (1 millisecond, or ms, equals 1 thousandth of a second).

A digital delay is accomplished by storing sampled audio directly into RAM. After a defined length of time (milliseconds or seconds), sampled audio can be read out from memory at a later (and thus delayed) point in time (see Figure 11.33).

Figure 11.33. *Delay uses memory to store data, where it can be read out at a later time.*

Blam! ➡ [chip] ▯ delay ➡ Blam!

As the delay time is reduced below the 10-ms range, however, a new effect begins to take hold. The effect of mixing a variable short-term delay with the original undelayed signal creates a series of peaks and dips in the signal's frequency response. This effect (known by the generic effect name of *flanging*) is the result of selective equalization (see Figure 11.34). If you have a digital delay hanging around, you can check out this homemade flange effect by combining the delayed and undelayed signal and then listening for yourself.

Figure 11.34. *The effect of mixing a short-term delay with the original undelayed signal creates a* comb *(multiple notch) filter response that has the generic effect name of* flanging.

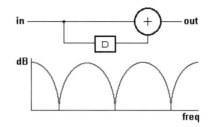

By further reducing the delay times downward into the microsecond range (1 microsecond, or ms equals 1 millionth of a second), you can begin to introduce delays that affect the digitized signal at the sample level. In doing so, control over the filter characteristics can be improved to the point that selective equalization is accomplished. Figure 11.35 shows two basic equalization circuits that provide low- and high-frequency shelving characteristics, respectively, in the digital domain.

Figure 11.35. *Simple EQ circuits and possible response curves.*

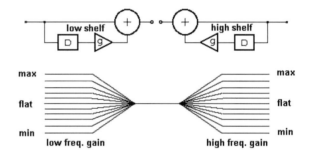

The amount of equalization to be applied (either boost or cut) depends on the multipliers that control the amount of gain to be fed from the delay modules. By adding more stages of delay and multiplication to this basic processing concept, complex stages of equalization can be assembled to create digital equalizers that are more complex and parametric in nature.

It should be pointed out that delays of such short durations aren't created using RAM delay-type circuits (which are used in creating longer delays). Rather, logic circuits known as *shift registers* are used. These sample-level delay circuits are better suited to the task because they are simpler in design and are more cost-effective.

Getting back to the digital mixer example, you can now finish your project by adding some form of equalization to the final stage. For example, Figure 11.36 shows your project as having a very simple high-pass filter. This filter cuts out any low-end rumble that gets into your system courtesy of the air conditioner down the hall or the local transit authority.

Figure 11.36. *Sample-level delays can be added to your mixer to provide equalization.*

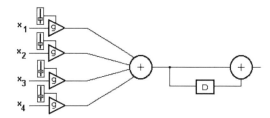

Echo and Reverberation

Now that you have looked at single delay effects, you can add to your bag of effects tricks by successively repeating delays to create echoes. Repeated echoes are created by feeding a portion of a delayed signal's output back into itself (see Figure 11.37). By adding a multiplier stage into this loop, it is possible to vary the amount of gain to be fed back and thus control both the level and the number of repeated echoes.

Figure 11.37. *By adding a simple feedback loop to a delay circuit, it's possible to create an echo effect.*

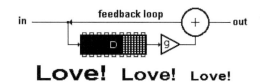

Although reverb could be placed in a separate category, reverb is actually nothing more than a series of closely spaced echoes. In nature, acoustic reverb can be broken down into three subcomponents:

◆ Direct signal
◆ Early reflections
◆ Reverberation

The direct signal is heard when the original sound wave travels directly from the source to the listener. Early reflections are the first few reflections reflected back to the listener from large, primary boundaries in a given space. Generally, these reflections are the ones that give the listener subconscious cues as to the perception of size and space.

The last set of reflections makes up the signal's reverberation characteristic. These signals are broken down into the many random reflections that travel from boundary to boundary within the confines of a room. These reflections are so closely spaced in time that the brain is unable to discern each individual reflection, so they are perceived as a single decaying signal.

By designing a system that uses a number of delay lines carefully controlled in both time and amplitude, it's possible to create an almost infinite number of reverb characteristics. To illustrate this point, let's return to our basic building block approach to DSP and build a crude reverb processor (see Figure 11.38).

Figure 11.38. A crude reverb processor design.

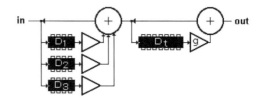

As you know, the direct signal is the first signal to arrive at the listener. Thus, it can be represented as a single data line that is added to each stage—flowing directly from the input to the output. Included in this first DSP stage are a number of individually tunable delay lines. By tuning these various modules to different times between a range of 10 ms to over a second (perceived bathroom size to Grand Canyon size), early reflections can be simulated. Following this section, one or more delay lines with echo feedback loops (which are designed to repeat echoes at a very fast rate) can be placed into your system. The echoes will decay over a predetermined time.

Such a reverb unit might sound crude because its simplicity would severely limit the type and quality of sounds that it might offer. By adding more stages and placing the gain controls under microprocessor command (using various user-defined algorithms), a large library of high-quality sounds can be created.

The Real World of DSP Design

As you have seen from the preceding discussion, the basics of DSP are fairly straightforward. In application, however, these building blocks can be programmed into a processing block using some rather complex combinations and elaborate algorithms to arrive at a final result.

In addition to the complexities that can exist, there are restrictions and overhead requirements that must be used to safeguard the process from cranking out erroneous, degraded, or even disastrous results. For example, whenever a number of digital samples are mixed together using an adding

circuit, the added results could easily equal a large number that is beyond the system's maximum signal limit. Without proper design safeguards, a condition known as *bit wraparound* could occur, causing the signal to output a loud "pop!"

Whenever signal processors are called on to add samples together or to multiply numeric values by lengthy coefficients, it's possible for errors to accumulate or for the final results to be greater than 16-bits in wordlength. To reduce these errors to acceptable levels or to prevent the "chopping off" of potentially important least significant bit values, high-quality processors often can calculate wordlength values with a resolution of up to 24 or 32 bits.

Fortunately, modern design has removed most of these obvious gremlins. The deeper you dig into DSP, however, the more you find that the challenge in designing a quality system isn't in getting it to work but in eliminating the pesky glitches that often are the side effects to performing complex sonic functions.

The remainder of this chapter examines many of the common signal processing devices and applications used in audio production, including the following:

♦ delay

♦ reverberation

♦ pitch- and time-shift related effects

♦ psychoacoustic enhancement

Delay

One very common application of effects using DSP is the altering of time by introducing various forms of delay into the signal path. Creating a delay circuit is a relatively simple task to accomplish in the digital domain. Although dedicated delay devices—often referred to as *digital delay lines,* or DDLs—are readily available on the market, fact is, almost every digital signal processor that can perform multiple functions can create this effect.

You might recall from Chapter 6, "Digital Audio Technology," that delays in the milliseconds and seconds range rely on the storage of sampled audio into RAM. After a defined length of time, the data is read out and mixed in with the original, undelayed signal (see Figures 11.39a and 11.39b). The maximum delay time that can be delivered by such a device is limited only by the sample frequency and the size of the memory block.

When program-material delays of 35 to 40 ms and greater are used, the listener perceives them as discrete delays. When mixed with the original signal, this can add depth and richness to an instrument or range of instruments. Use care, however, when adding delay to an entire musical program because the delay tends to muddy its intelligibility.

Figure 11.39. *The DDL introduces one or more discrete repeats of the input signal at user-defined intervals.*

a. *Single delay signal.*

b. *Signals fed back into memory can create numerous, repeated delays.*

a

b

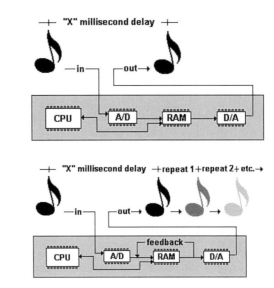

Reducing these delay times to the 15 ms to 35 ms range will create delays that are spaced too closely together to be perceived by the listener as discrete delays. Instead, such short-delayed signals create a "doubling" effect (see Figure 11.40). When such short delays are mixed with an instrument or group of instruments, the brain is fooled into thinking that more instruments are playing than actually are. The effect, at the least, is a subjective increase in the sound's density. This effect (known as *doubling* or *automatic double tracking*) can be used on background vocals, horns, string sections, and other ensembles to make the ensemble sound double its actual size. This effect also can be used on foreground tracks, such as vocals or instrument solos, to create a larger, richer, and fuller sound.

Figure 11.40. *In certain instances, doubling can fool the brain into thinking that more instruments are playing than actually are.*

Whenever delays that are varied below the 15-ms range are mixed with the original undelayed signal, a *combing* effect is created. Combing is the result of equalized peaks and dips that occur in the signal's frequency response (see Figure 11.41). Either by manually or automatically varying the time of one or more of these short-term delays, a constantly shifting phase effect known as *phasing* or *flanging* can be created. Depending on the application, this effect ranges from being relatively subtle (phasing) to moderate-to-wild shifts in time and pitch (flanging).

Figure 11.41. Peaks and dips in a signal's frequency response (as shown in gray areas) result from the combination of several short-term delay modules that shift over time to create the effects of phasing or flanging.

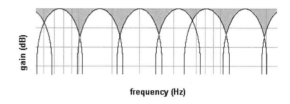

By combining two identical (and often slightly delayed) signals that are slightly detuned in pitch from one another, you can create another effect known as *chorusing*. Chorusing is an effects tool that often is used by guitarists and other musicians to add depth, richness, and harmonic structure to their instruments.

Reverberation

In professional audio production, natural acoustic reverberation is an extremely important tool for the enhancement of music and sound production. A properly designed acoustical environment adds a degree of quality and natural depth to a recorded sound that often affects the performance as well as its overall sonic character. In those situations where added room ambience is required, however, high-quality digital reverberation can be used to fill out and add a new sense of dimensional space to the production.

As you learned in Chapter 3, "Studio Acoustics and Design," reverberation is the pattern of closely spaced and random multiple echoes that are reflected from one boundary to another within a determined space (see Figure 11.42). This effect results in perceptible cues as to size, density, and nature of a space and adds to the perceived warmth and depth of recorded sound.

Figure 11.42. Signal level versus time for reverberation.

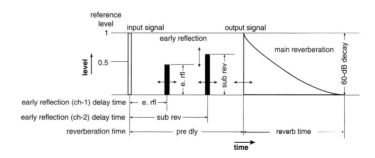

Digital Reverb Devices

In recent decades, the digital reverb unit (see Figures 11.43 through 11.45) has become the most prominent device used in audio production because of its high quality, small size, flexibility, and cost-effectiveness. In the section on DSP in Chapter 6, you learned that digital reverb is accomplished through the regeneration of an input signal by a series of closely spaced digital delays. Through the performance of digital algorithms, these delays follow a defined set of random patterns that result in a reverb of a predictable character.

Figure 11.43. *The Yamaha REV 100 digital effects processor. (Courtesy of Yamaha Corporation of America)*

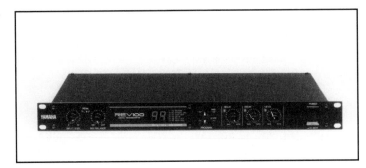

Figure 11.44. *The Sony DPS R7. (Courtesy of Sony Corporation)*

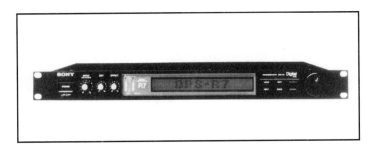

Figure 11.45. *The Lexicon 480L digital effects system with LARC. (Courtesy of Lexicon, Inc.)*

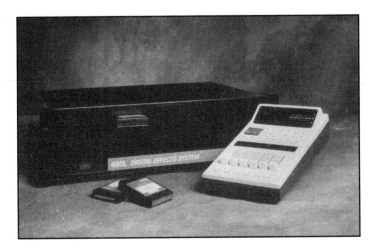

This microprocessor-based device can exercise extensive control over such reverberation characteristics as level, decay time, pre-echo delay, low-EQ, high-EQ, and variable EQ crossover points. Most digital reverb processors are designed so that the user can recall a number of factory-programmed settings and then tailor these settings in real time to fit his or her present needs. The user can then save these customized settings to the device's memory as a preset for recall at a later time.

Pitch- and Time-Shift-Related Effects

Certain signal processors also can alter the speed and pitch of an audio program. This complicated task often requires that the system be capable of carrying out complex calculations (especially if the processing is performed in real time by a dedicated effects box).

Generally, the process of pitch and time shifting has a window of effectiveness. That is, it can correct or alter program material only within a limited pitch-shift range without introducing certain amounts of distortion into the signal. This range varies from system to system and with different types of program material (such as voice, music, and complex waveforms). In the final analysis, your ears must judge whether the possible drawbacks of induced distortion outweigh the benefit of the effect itself.

Currently, the options for pitch and time shifting include the following: changing pitch without changing duration, changing duration without changing pitch, and changing both the duration and the pitch.

Pitch Shifting

Pitch shifting can be used to vary the pitch of a program either upward or downward to transpose the relative pitch of an audio source without affecting the program duration. This process can take place either in real time or in non-real time. One way this process works is by writing sampled audio data to a temporary storage medium (known as a *circular buffer*) where it can be resampled at either a higher or a lower rate (see Figure 11.46). In order to revert the resampled signal back to its original rate, the next task is to alter the signal so that it matches the sample rate as it existed at the device's input. The resulting change in pitch will be a ratio that exists as the difference between the internally resampled rate and the outgoing sample rate. Thus data that was resampled to a higher rate will be lowered in pitch, and vice versa. Two popular pitch-changing devices are shown in Figures 11.47 and 11.48.

Figure 11.46. Pitch shifting occurs by resampling data that is temporarily stored in a circular buffer at either a higher or a lower rate.

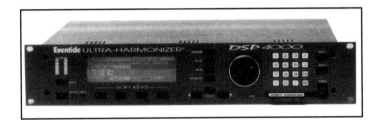

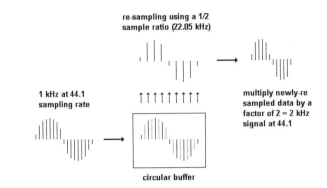

Figure 11.47. Eventide DSP 4000 Ultra-Harmonizer. (Courtesy of Eventide Inc.)

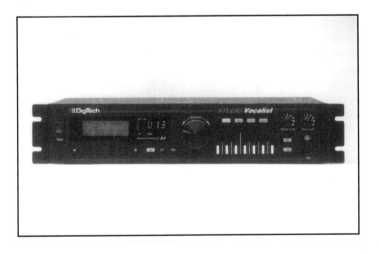

Figure 11.48. Digitech's Studio Vocalist. (Courtesy of Digitech)

Time Compression/Expansion

By combining variable sample rates and pitch shifting, it's possible to alter the aspects of a program's duration (varying the length of a program by raising or lowering its playback sample rate), as well as alter its relative pitch (shifting it either up or down). In this way, two possible time/pitch combinations can occur:

◆ A program's length can be altered with a corresponding change in pitch due to sample rate change.

◆ A program's duration can be altered while pitch is shifted either up or down in order to maintain the same relative pitch as that of the original program material.

Time-compression/expansion facilities, available on newer digital audio workstations, samplers, and stand-alone processors, are becoming a standard time-processing option for use in the audio-for-video, film, and broadcast industries. This feature provides control over the running time of film video and audio soundtracks while maintaining the original, natural pitch of voice, music, and effects.

To illustrate how such time stretch-and-squeeze techniques can be applied, suppose that you want to stretch a 27-second public service radio spot to exactly 30 seconds. Time compression/expansion techniques can reduce the sample rate by the precise ratio needed to achieve a 30-second length. Then, by applying pitch-shifting techniques, the signal can be raised back to its original pitch value.

Psychoacoustic Enhancement

Certain signal processors rely on psychoacoustic cues in order to function. In other words, these devices operate by affecting the ways in which sound is perceived by the brain. The earliest and most common of these devices are those that enhance the overall presence of a signal or entire recording by synthesizing upper-range frequency harmonics and then mixing them back in with the originally recorded signal. Although the addition of these synthesized harmonics does not significantly affect the program's overall volume, the perceived effect is a marked increase in the recording's upper range that makes it sound brighter and more present. The most well-known of these processor types is the Aural Exciter from Aphex Systems, Inc., shown in Figure 11.49.

Other psychoacoustic processors use equalization and time manipulation (usually in the form of small delay changes that introduce phase shift) to enhance the spatial aspects of sound (those relating to placement within the stereo field).

An example of a psychoacoustic equalization processor is the Vitalizer from Sound Performance Laboratory, shown in Figure 11.50 The Vitalizer uses multiple filter loops that correlate frequency, amplitude, and phase response in such a way that it demasks individual instruments by separating sounds in adjacent frequency bands and applying subtle shifts in delay. In addition to sophisticated processing functions in the sub-bass and mid-high frequency regions, the Vitalizer can be used to

add fourth-order harmonics to the signal to heighten and define the upper-frequency range without added distortion. Psychoacoustic surround processing occurs by detecting off-center signals and feeding them into the opposite channel in reverse-phase, thereby reportedly creating a sound that broadens the subjective stereo field width.

Figure 11.49. *Aural Exciter Type C with Big Bottom. (Courtesy of Aphex Systems, Inc.)*

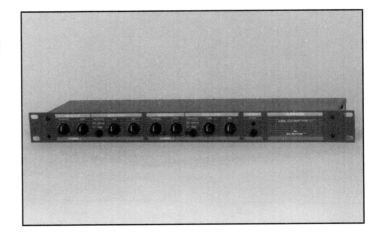

Figure 11.50. *The Vitalizer psychoacoustic equalization processor. (Courtesy of Beyerdynamic)*

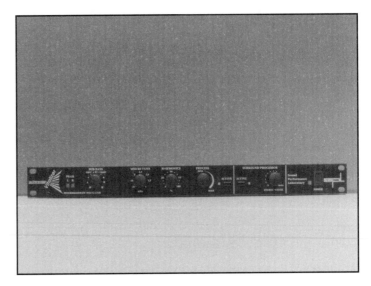

Other digital psychoacoustic processors (see Figures 11.51 and 11.52) deal exclusively with the subject of spatialization (the placement of an audio signal within a three-dimensional acoustic field), even though the recording is being played back over stereo speakers. Often these devices have multiple inputs that can be manipulated in a 360-degree field through a continuous alpha-dial or joystick. By varying these positioning parameters (in either real time or by encoding automated placement information in the form of MIDI data into a sequencer), each processing path can create

a phase and amplitude path that can fool the brain into perceiving that the performance is in a surround soundfield. This, of course, is the theoretical objective; in practice, the effect is carried off with degrees of success that vary from system to system.

Figure 11.51. Spatializer three-dimensional digital processor. (Courtesy of Desper Products, Inc.)

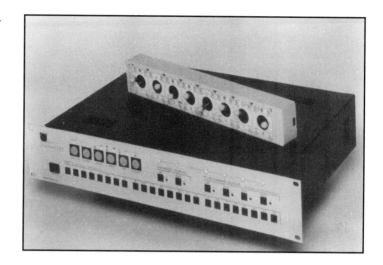

Figure 11.52. Roland SRV-330 dimensional space reverb. (Courtesy of Roland Corporation, US)

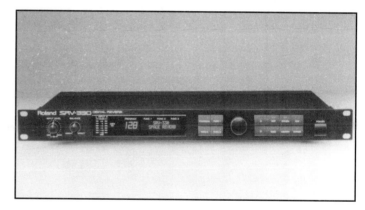

Multiple-Effects Devices

Because many digital signal processors can be easily programmed to perform various functions, an ever-increasing number of signal processors are multiple-effects devices (see Figures 11.53 through 11.55). Multi-effects, in this case, can have two basic meanings:

◆ A single-effects device that offers a range of processing functions; however, only one can be called up at a time.

◆ A single-effects device that offers a range of processing functions of which a limited number can be called up and performed at one time.

Figure 11.53. *Boss SE-70 multi-effects processor. (Courtesy of Boss)*

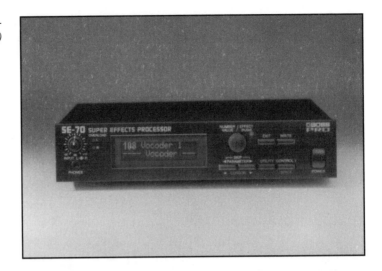

Figure 11.54. *Lexicon PCM-80 multi-effects processor. (Courtesy of Lexicon, Inc.)*

Figure 11.55. *T.C. Electronic M5000 multi-effects processor. (Courtesy of T.C. Electronic of Denmark)*

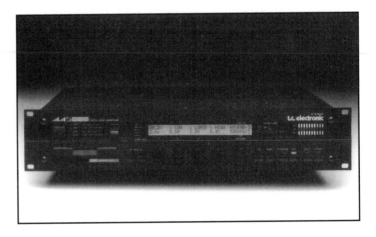

Both device types are common and are invaluable tools for the project or recording studio in which versatility in processing is often a big plus. The following list offers some of the possible effects-processing functions provided to the user by certain multi-effects devices:

Reverb	Delay, chorus, phasing, and flanging
Equalization	Compression, limiting, expansion and gating
Pitch shift	Time change
Sample rate conversion	Spectral and spatial enhancement
Sampling or one-shot sampling	Overdrive distortion
Wah pedal	Rotary speaker and auto-panning
Tremolo and vibrato	Effects morphing

Dynamic Effects Editing Using MIDI

One of the most common methods for automating effects devices during a mix—from a MIDI sequence or on-stage during a live performance—is by using the MIDI program-change commands. In the same way that a characteristic sound patch can be stored in an instrument's memory location register for later recall, most MIDI-equipped effects devices offer a number of registers in which effects type and parameter "patch" data can be stored. Through the transmission of a MIDI program-change command, it's often possible for a desired effects patch to be automatically recalled by a device that is set to receive such a program-change command over a particular MIDI channel (see Figure 11.56).

Figure 11.56. Effects settings can be automated throughout a system with the use of MIDI program-change commands.

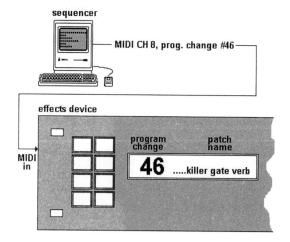

Program-change commands (and, occasionally, continuous controller messages) enable complex signal processing functions to be easily modified or switched during the normal playback of a MIDI sequence. Often, a sequencer allows the simple insertion of a program-change number—that corresponds to the desired effects patch—into a separate sequenced track or onto a track that contains related performance data.

In addition to real-time program changes, effects parameters can often be changed in real time through the use of real-time, system-exclusive messages (that is, program type, reverb time, EQ, or chorus depth). Control over these messages often is accomplished in real time through the use of an external MIDI controller or hardware- and software-based data fader controllers (see Figure 11.57).

Figure 11.57. *Dynamic control over effects parameters by way of an external MIDI command controller.*

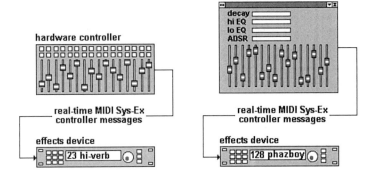

Computer-based editing software known as *patch editors* (see Figures 11.58 and 11.59) also offer dynamic editing of effects parameters by allowing the user to edit and fine tune effects parameters through the use of on-screen, mouse-driven visual graphics and scales or, alternatively, through the display of numeric values that directly represent the device's control parameter settings. Control over these parameters is accomplished in real time through the use of device-specific system-exclusive (Sys-Ex) messages. After the desired effect or multi-effect has been assembled and fine tuned, these parameter settings can be saved as one of the device's preset registers for recall at a later time, either from the device's front panel or through program-change messages.

In the recent past, program editors often have been designed to operate with a specific device or limited range of devices because these editors were required to communicate system-exclusive data that pertained solely to the device being edited. With the large number of MIDI-controlled effects devices appearing on the market, however, it is increasingly more common to find universal editors that are capable of directly editing a wide range of devices in real time. As such, these universal programs have to effectively manage a wide range of devices from various manufacturers. Consequently, they often contain a more generic visual interface for controlling a wide range of effects devices and musical instruments than their dedicated counterparts.

Figure 11.58. *Midi Quest for Windows V4.0 Universal editor/ librarian. (Courtesy of Sound Quest, Inc.)*

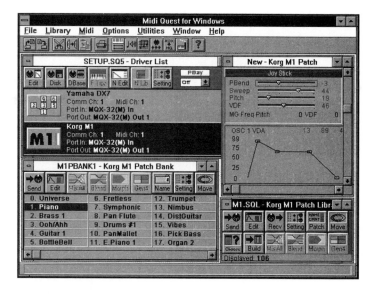

Figure 11.59. *Galaxy Universal editor/librarian for Macintosh. (Courtesy of Opcode Systems, Inc.)*

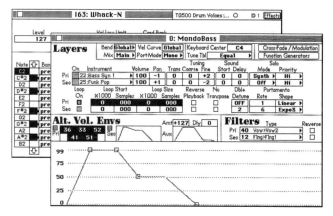

After a device's bank of preset locations has been filled, a program editor often allows these system-exclusive messages to be transmitted to the host computer by way of a Sys-Ex MIDI data dump. In this way, multiple preset banks can be stored and recalled, which allows a much larger number of effects "patches" to be stored in a computer-based library. Such editing programs, often referred to as *patch librarians*, also permit the user to organize effects patches into groups of banks according to effects type or in any other order that the user wants.

As with most electronic instrument patches, effects patches can be acquired from a number of different sources. Among these are patch books (containing written patch data for manual entry), patch data cards (ROM cards or cartridges containing patch data from manufacturers or

second-party developers), data disks (computer files containing patch data from manufacturers or second-party developers), and computer bulletin board files (containing patch data that can be downloaded via a computer modem).

In the final analysis, MIDI automation of effects processing can be a super-charged production tool that can boost your musical and production effectiveness with a minimum of fuss and money.

CHAPTER

12

◆

Noise Reduction

Because of the increased dynamic range and the demand for better quality sound brought about by digital audio and the compact disc, it often is necessary to pay close attention to the background noise level that's produced by analog magnetic tape, amplifier self noise, and the like. The overall dynamic range of human hearing encompasses a full 130 dB; however, this range can't be adequately recorded onto analog tape (see Figure 12.1). In order to record a dynamic range in excess of 60 dB, some form of dynamic range compensation is required. The limitations imposed on conventional analog ATR and VTR audio tracks are dictated by tape noise, which is perceptible when the overall signal level is too low, or by tape saturation, which is caused by distortion when recording at excessively high levels. Should an optimum recording level produce an unacceptable amount of noise, the engineer is then faced with two options: record at a higher level (with the possibility of increased distortion) or change the signal's overall dynamic range.

Analog tape noise may not be a strong limiting factor when you are dealing with one or two tracks of an audio production, but the combined noise and other distortions brought about by combining 8, 16, 24 or more tracks can range from bothersome to downright unacceptable. The following types of noise may need to be eliminated:

◆ Tape and amplifier noise
◆ Crosstalk between tracks
◆ Print-through
◆ Modulation noise

Figure 12.1. *The dynamic range of several audio devices.*

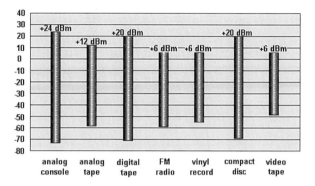

These analog-based noises have been reduced or eliminated by implementing the following actions:

- ◆ Improving the dynamic range of professional recording tape and/or increasing tape speed in order to record at higher flux levels
- ◆ Designing the tape heads and electronics with reduced crosstalk between recording channels
- ◆ Using thicker tape base for the reduction of print-through

With the availability of tapes combining low noise and high output (which have generally increased the signal-to-noise ratio by 3 dB), noise levels have been reduced even further.

Modulation noise is a high-frequency component that causes fuzziness as well as the generation of sideband frequencies (see Figure 12.2). This is due, in part, to irregularities in the coating of magnetic recording tape. Modulation noise only occurs when a signal is present and increases with rising signal levels. This type of noise has a major impact in shaping what could be called the *analog sound*.

Figure 12.2. *Modulation noise of a sine wave.*

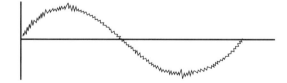

The Compansion Process

A common noise reduction process in use today is the *compander system*. The most notable of these systems are manufactured by Dolby Labs and dbx.

The compansion encode-decode process gets its name from the fact that the incoming signal is compressed before it is recorded onto tape; upon reproduction, the signal is reciprocally expanded back to its original dynamic range (with a resultant reduction in background tape noise).

To better understand this process, take a look at Figure 12.3. This figure shows a signal that is companded across its entire audio bandwidth. Before being sent to the input of the recorder, the overall dynamic range of the input signal is compressed into a restricted but tightly controlled dynamic range. This is done so that the newly compressed dynamic range can be recorded onto tape at a level significantly higher than the residual tape noise. During playback, the signal is then downwardly expanded back to its original range, while tape noise is likewise expanded down to levels that can range from less offending to inaudible.

Figure 12.3. An example of a *full-bandwidth compansion noise-reduction process.*

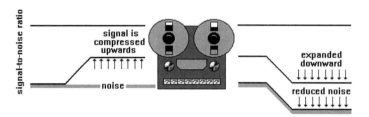

Although similar in theory to the preceding example, other system types operate by breaking the audio signal into a number of separate bandwidths. This type of system makes it possible for only those signal bandwidths that require compansion (that is, frequency ranges in which loud signal passages are lacking that would otherwise mask the presence of existing tape noise) to be affected, while leaving largely unaffected those that do not require compansion (that is, ranges in which sufficiently loud signal passages are present).

In addition to the reduction of noise associated with analog tape, companders can also help reduce extraneous noise and hum from long-distance transmission or broadcast lines.

The Dolby Noise-Reduction System

Dolby is available in four system types for audio production: Dolby SR, Dolby A, and the consumer Dolby B and C.

Dolby spectral recording (SR) is an encode-decode process that readily lends itself to any analog audio recording or transmission application. Its use practically eliminates the influence of noise and nonlinearity on reproduced sound, providing an improvement of up to 24 dB over a system not equipped with noise reduction.

The Dolby SR signal shaping processor is a sidechain that runs parallel to the device's main audio path. The output of this sidechain is either added or subtracted from the main signal, depending on whether the circuit is enabled as an encoder or decoder.

At the lowest signal levels, or in the absence of a signal, Dolby SR applies a fixed-gain/frequency characteristic that reduces noise and other low-level disturbances. This results in an optimized form of equalization that doesn't change as long as the signal level stays below a certain threshold. Only when the level of any part of the signal spectrum increases significantly does the circuit change its own processing characteristics. When this happens, Dolby changes the gain only at frequencies where change is needed and only by the amount required.

Although most of its electronics are used for spectral analysis, Dolby SR's principal operating system consists of five groups of fixed- and sliding-band filters with gentle slopes arranged by level and frequency. Those with fixed-bandwidth are electronically controlled to vary their gain; those with fixed-gain can be adjusted to cover different frequency ranges. The filters are cross-linked by a technique known as *action substitution*, which enables both types of filters to be selected in the exact proportion needed.

By selecting and combining filters, the SR control circuit can create an infinite number of filter shapes through which the signal can pass in the encoding process. During decoding, filter shapes are automatically created that are the exact opposite of those used during encoding. This results in an accurate, linear response in level, phase, and frequency.

The operation of the Dolby SR system depends on its being used both when a tape is recorded and when it is played back, with each tape track being assigned a separate channel. The same unit that processes the signal for recording is used on playback and usually is switched between modes automatically by the recorder's transport switching logic.

Dolby's original professional compansion system, Type A, provides 10 dB of noise reduction below 5 kHz, gradually increasing the gain of low-level signals rather than decreasing the gain of the high-level signals. The circuit uses a sidechain limiter that prevents the output signal from rising much above −40 dBm (see Figure 12.4). The output of this limiter is added algebraically to the uncompressed input signal. Because the output of this limiter rises only slightly above −40 dBm, the added effect of this signal—when combined with the input signal—depends on the input signal's level. When the input signal is low, the output of the limiter is large in comparison, and so the addition of these signals results in an overall level boost. When the input signal is high, the limiter's output is negligible when the two signals are added.

Figure 12.4. *Output of the sidechain limiter rises only slightly above −40 dBm.*

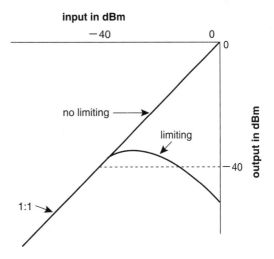

The same procedure is followed during the expansion process, except that the output of the limiter is precisely subtracted from the input signal, thus reciprocally reducing the lower gain levels to their original values.

Dolby A effectively reduces program noise by dividing the spectrum into four separate bands (see Figure 12.5). Each band of frequencies has its own dynamic-range processor, such that the presence of a high-level signal in one band does not interfere with detection in the other bands.

Figure 12.5. *The four filter bands of the Dolby A system.*

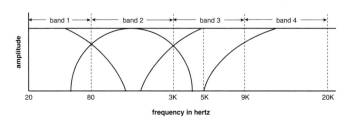

The outputs of the four filters and signal processors are combined in such a manner that low-level signals (below –40 dBm) are boosted by 10 dB from 20 Hz to 5 kHz, with the boost rising gradually between 5 kHz and 14 kHz to a maximum of 15 dB. As the level of a signal in one band rises, its noise reduction decreases, but the effective masking increases; so the noise level appears to be constant. The bands are not sharply defined, so when the noise reduction in Band 2 is disabled by the presence of loud signals between 80 Hz and 3 kHz, some noise reduction (in addition to masking) is provided by Band 1, and above 1.8 kHz, by Band 3. If Band 3 also has its noise reduction turned off by loud signals between 3 kHz and 9 kHz, Band 4 will contribute noise reduction from 5 kHz and up. Bands 1 and 4 rarely have their noise reduction shut down completely, except by very loud organ tones or by cymbal crashes. Although the actual amount of noise reduction throughout the audio spectrum changes from one moment to the next, the noise level perceived by the ear remains constant.

To ensure that the tape is played back at the same level at which it was recorded, a 400- or 700-Hz, 0-VU level tone is recorded at the beginning of each tape so that the level can be adjusted properly when the tape is played back. This technique is used to ensure that signals below the threshold during recording will be the same amount below the threshold on playback. A ±3-dB tolerance will exist before a difference in the signal playback becomes noticeable. If the signal is played back too loudly through the Dolby unit, too little expansion will take place because the limiter is blocked by the high-level signal and the signal will sound compressed and overly bright. If the signal is played back at too soft a level, the expander will expand too much, and the signal will sound dull and have too great a dynamic range. The signal record level doesn't matter as long as the playback level is the same.

Dolby B, commonly found in consumer cassette decks, is designed to reduce tape hiss. The Dolby B system acts only to compand the upper-frequency component and has no operating effect on

lower-frequency noises such as hum or rumble. The effect of type-B noise reduction is 3 dB at 600 Hz, rising to 10 dB at 5 kHz, at which point it levels off in a shelving fashion. Dolby C is a more recent consumer version of Dolby SR that offers up to 20-dB of overall noise reduction when used on cassette tapes.

The dbx Noise-Reduction System

The dbx system (Figure 12.6) is a full-bandwidth compansion system that provides from 20 to 30 dB of noise reduction. As with any compansion system, dbx channels are connected between the console and tape machine. The compressor employs a 2:1 ratio between levels of –90 dBm and +25 dBm, with its unity gain point occurring at +4 dBm (0VU). Unlike the Dolby system, all signals are compressed and expanded at a 2:1 ratio, regardless of signal level, and thus aren't level sensitive.

Figure 12.6. The dbx 911 noise reduction module. (Courtesy of dbx Professional Products)

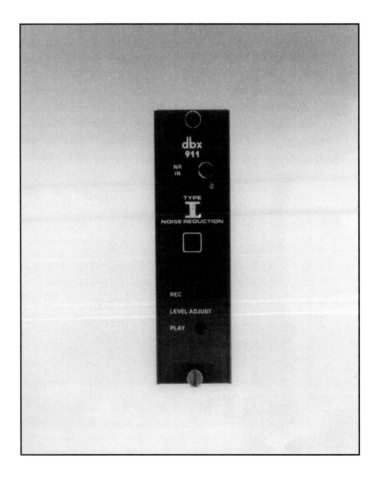

The dbx noise-reduction process is shown in Figure 12.7. Assume that there is a 60-dB signal-to-noise ratio in the tape recorder and a 60-dB dynamic-range program is recorded. During the softest passages of the program, the noise added by the program is just as loud as the program and therefore is quite audible. With the dbx system, the program passes through the dbx record section and is compressed 2:1 into a 30-dB dynamic range. This compressed signal is then recorded on tape where, in effect, it is mixed with the tape noise. The tape noise, however, is now 30 dB below the softest passage of the music. On playback, the expander reduces the signal level 30 dB on the softest passages and reduces the noise 30 dB below that—or 90 dB below 0 VU. Thus, the noise reduction is 30 dB.

Figure 12.7. Expansion of dynamic range from 60 dB to 90 dB when using dbx noise reduction.

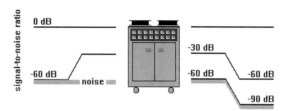

The difference in level between the noise due to the tape and the softest signal recorded on the tape determines the signal-to-noise ratio. A 2:1 ratio was chosen over a higher ratio because of the reduced effect that tape dropouts have on the expander. Because the expander reduces its output signal twice as much as the signal played back from the tape, a tape dropout of 2 dB causes a 4-dB dropout signal with the 2:1 ratio. A 3:1 ratio, for example, would cause a 6-dB dropout. A ratio of 1.5:1 would make the dropout problem less noticeable, but it would do so at the cost of noise reduction (only 20 dB would be possible). The 2:1 ratio was considered the best compromise between the amount of noise reduction and the sensitivity to dropouts.

The dbx system operates on the full frequency range from 20 Hz to 20 kHz. Dolby Laboratories found that having the compressor operate over the full frequency range caused a loss of noise reduction throughout the frequency spectrum in the presence of loud signals, even though they might have a restricted frequency range. The noise not masked by the signal became apparent on loud passages. The dbx system overcomes this problem to a certain extent by using a filter to pre-emphasize the high frequencies by 12 dB before the compressor and by using another filter to de-emphasize the high frequencies after the expander. So, even though the signal may be at 0 VU when no compression or expansion takes place, there is an effective 10 dB of noise reduction at the high frequencies. This 10 dB is only achieved when most of the signal energy is below 500 Hz. When the program material extends into the high frequencies, the effect of the filter is diminished and is replaced to a certain extent by the masking effect of the program. A special control circuit for the compressor and expander reduces the gain when the level of high-frequency signals becomes very high to avoid the tape saturation that would be caused by NAB high-frequency boost.

Single-Ended Noise-Reduction Process

The compansion systems require processing during both recording and playback to achieve noise reduction. When using these devices, noise can't be removed from a signal; it can only be prevented from entering. The noncomplementary or single-ended noise-reduction process acts basically as a frequency-dependent expander (or noise gate) that deletes noise from audio sources by utilizing a downward dynamic-range expander in conjunction with a program-controlled dynamic low-pass filter. Expansion and dynamic filtering can often be used either together or separately to provide the greatest possible amount of noise reduction.

Both analog and digital single-ended noise reduction systems are currently available on the market. Digital systems are most commonly found as a program algorithm in certain digital signal processors. Analog systems, on the other hand, are stand-alone devices that are designed around specialized voltage-controlled amplifier (VCA) and associated level detection circuitry (see Figure 12.8).

Figure 12.8. Behringer's Denoiser systems Model SNR 202 and 208. (Courtesy of Behringer Specialized Studio Equipment Ltd.)

Single-ended noise reduction systems work by breaking the audio spectrum into a number of frequency bands. Whenever the program's signal level within each band falls below a user-defined threshold, the signal will be attenuated. This downward expansion/filtering process accomplishes noise reduction by taking advantage of several basic psychoacoustical principles. The first principle is that music is capable of masking noise that exists at lower levels within the same bandwidth. The second principle is that reducing the bandwidth of an audio signal reduces the perceived noise because the greater the spectral distribution of noise, the greater the human ear's sensitivity to that noise. The dynamic filter examines the incoming signal for high-frequency content and, in the absence of high-frequency energy, the filter's bandwidth is decreased. When high-frequency energy returns, the filter opens back up as far as necessary to pass the entire signal.

Noise Gates

A noise gate (see Figure 12.9) can be a very effective noise-reduction device when used to reduce background noise on certain program material. A noise gate can be viewed as a high-ratio expander that acts as a unity gain amplifier in the presence of a signal above a certain threshold, passing the signal at full volume. When the audio signal at the input of the device falls below the preset threshold level, the signal is effectively turned off at the output. Thus, noise is eliminated from the signal when there is no other signal present to mask it. For critical program material, it may be necessary to fine-tune the attack and release controls to eliminate unwanted "pumping" or "breathing" of the noise floor below the desired signal.

Figure 12.9. Symetrix 564E Quad/Expander Gate. (Courtesy of Symetrix, Inc.)

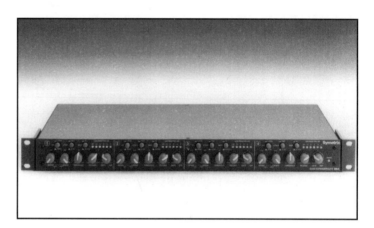

Digital Noise Reduction

Advanced forms of digital signal processing (DSP) have recently been designed that can reduce the noise content of previously recorded material. Such noise may include such analog side effects as tape hiss, needle ticks, pops, and even certain types of distortion that were inherent in the original recording process. Certain algorithms are even capable of smoothing out the grainy side effects of digital recordings that have been recorded with a low bit-rate resolution (such as an 8- or 12-bit sample.)

Digital noise processing can be used in a number of applications, ranging from the simple removal of analog tape hiss, hum, or an obtrusive background ambience, to the clean up and restoration of older program material for transfer to CD.

Although stand-alone digital noise processors do exist, at the time of this writing, the most popular systems available exist as software for the Macintosh computer. Two examples are the Dynamic Intelligent Noise Reduction (DINR) plug-in module from Digidesign and the No-Noise desktop program from Sonic Solutions.

Digidesign's Intelligent Noise Reduction system, shown in Figure 12.10, is a signal processing plug-in module that lets you remove hum, tape hiss, and other extraneous noises from your

recordings by analyzing the audio in question (a brief, isolated noise passage is ideal but not necessary). This analysis is then used to create a "sonic model," or noise template, that is digitally subtracted (by a user-selectable amount) from the original soundfile.

Figure 12.10. The Digidesign Intelligent Noise Reduction system. (Courtesy of Digidesign.)

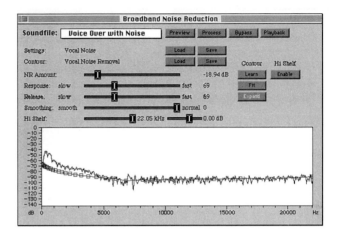

After the sound has been analyzed, DINR can process the signal in either real time or non-real time. Using a feature known as the UltraMaximixer, noise processing can be performed by the computer in real time without committing the results to disk. This feature is often used to play through a soundfile that's being processed while the results are being recorded directly to DAT or other digital medium. Alternatively, the soundfile data can be rewritten to disk in non-real time for retrieval later.

The No-Noise system from Sonic Solutions, shown in Figure 12.11, can be compared to an intelligent multiband expander. In this case, however, *multi* means that the audio spectrum is divided into more than 2,000 frequency bands. The system performs more than 53 million computations for each second of sound, and as with the Digidesign program, recent improvements in coprocessor speeds enable complex computations such as these to be made in either real time or non-real time.

The No-Noise process is divided into three functional stages: visual analysis, de-clicking and de-noising. These stages are described in the following sections.

Visual Analysis

In the visual analysis stage, the soundfile is searched both visually and audibly to determine the extent to which No-Noise must be applied to rid the program of noise and click pollution.

Figure 12.11. *The No-Noise desktop application .(Courtesy of Sonic Solutions.)*

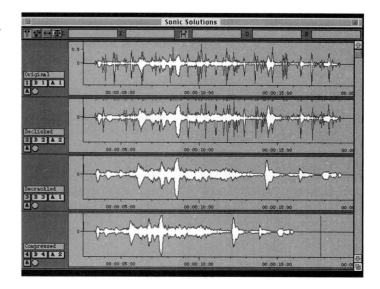

De-Clicking

After the soundfile has been analyzed, No-Noise can be used to automatically remove clicks and pops. The program doesn't edit these noises out of the signal. Instead, it reconstructs and repairs the problem portion by performing a frequency analysis, both before and after the click. The program can then sample enough of the surrounding material to make a plausible guess as to the original waveform content. It then pastes a resynthesized segment of audio over the nasty offender.

De-Noising

After transient noises have been eliminated, the next task is to lower the background noise floor so that tape hiss, surface noise, or recurrent background noise (hum, buzzing, air-conditioner noise, and the like) is reduced or eliminated.

The noise removal aspect of No-Noise involves a set of complex computations that make intelligent estimates as to when a background noise will or will not be audible at any point in time. Specifically, the process breaks down the audio spectrum into over 2,000 frequency bands in order to predict whether louder sounds (program) will mask a softer sound (noise).

The No-Noise noise removal process is performed by comparing the signal with a noise "fingerprint," known as a *signature,* taken from a sample of pure background noise that might exist at the beginning or end of the program or at a momentary pause in the program.

After the system is in action, it searches both forward and backward in time over a range of samples for each instant of music to determine what is noise and whether its removal would damage the music. If levels within one of the analyzed frequency bands fall below or equal to the level of the noise signature, the processor decides that the content is most likely noise and digitally expands this narrow bandrange downward, thus reducing the final noise content.

Keep in mind that the process of reducing noise often provides a challenge to a DSP-based noise removal system: each program type is different and requires its own approach to clean up. Processing variables, such as depth of reduction and audible side effects, are entirely up to the user.

With any digital noise removal system, it's possible to overprocess a recorded soundfile—often resulting in distorted sideband artifacts that sound like a flock of warblers gone crazy. The level at which this "digital warbling" becomes apparent depends on the type of program you are processing. In the end, it's best to process and then listen to a short segment before committing yourself to processing an entire soundfile.

13 CHAPTER

◆

Monitor Speakers

In the recording process, judgments of, and adjustments to, sound are primarily based on what is heard through the monitor system. In fact, within the professional audio and video industries, the word *monitor* refers to a device that acts as a subjective standard or reference.

Despite recent advances in design, speakers are still one of the weakest links in the audio chain. This weakness is due to potential nonlinearities in the frequency response of speakers. In addition, large peaks and dips in frequency response, when combined with the acoustic character of a room, can often occur at the listening position. These variations in response tend to render the sound of many monitor systems inaccurate. The only place a speaker can truly exhibit a flat response or a response that represents its intended design spec is in an *anechoic chamber*—a room that absorbs all of the speaker's output and reflects none of it back. In such a room, there can be no constructive or destructive reflections to interfere with the direct output of the speaker. We don't listen to music in anechoic chambers, however, so the design and sonic characteristics of the listening room must be taken into account when choosing a monitor speaker (see Figure 13.1).

Figure 13.1. An example of a
properly designed control room for
reduced reflections. (Courtesy of
Acoustical Physics Laboratories)

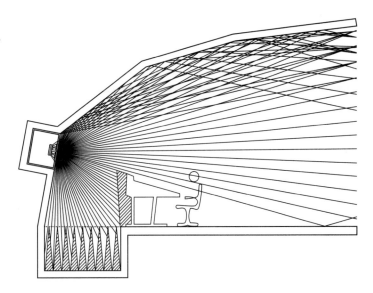

Speaker and Room Considerations

Unless the rooms are of identical dimensions and furnishings, a speaker will sound different (that is, it will have a different frequency-response curve) in every room in which it is placed. This means that speakers must be tested and auditioned in the room in which they are going to be used.

The problem of sound variation from room to room makes it difficult to interchange recording studio control rooms. Even if high standards of acoustic construction and tuning are followed, no two rooms will sound exactly alike. While a tape is being recorded in a specific control room, the producer and artists become accustomed to hearing the material sound a certain way. If the tape is then taken from this studio and mixed at another studio using different speakers—or even the same set of speakers, but with a different speaker/room response—there can be quite a discrepancy in the way the recorded instruments will sound.

To reduce or eliminate these variations, many studios equalize or *tune* their speakers to the room's acoustics so that the adjusted frequency response curve is reasonably flat and, therefore, reasonably compatible with most other control rooms. This tuning is accomplished by placing a 1/3-octave bandwidth graphic equalizer between the console's control-room monitor outputs and the power amplifier (see Figure 13.2). *Pink noise,* which has a flat energy spectrum curve throughout the audio range, is fed into the speaker system. Then the acoustic outputs are measured one at a time in 1/3-octave increments using an instrument known as a *spectrum analyzer* (see Figure 13.3). This device is used to visually display the speaker's frequency response as measured through a specially calibrated omnidirectional microphone. The spectrum analyzer can provide an accurate and instantaneous reading of the combined speaker/room frequency response at the microphone's specific location.

Figure 13.2. *Example of a 1/3-octave graphic equalizer connected within the monitor chain.*

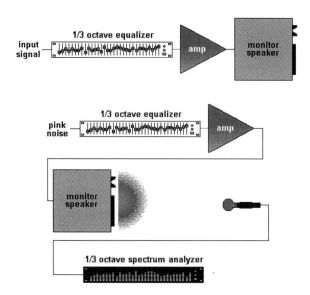

Figure 13.3. *A real-time spectrum analyzer can indicate how to adjust the graphic equalizer's response curve for an optimum speaker frequency response at the listening position.*

An omnidirectional mic is used because the ear is omnidirectional in nature and thus hears the reflected sound of the room as well as the direct sound from the speaker. Because the mic can only be at one spot at any one time, the resulting response curve will only be accurate at the listening position where the mic was placed. Although modern control room designs have improved the distribution of sound to be more even throughout the listening area, the response curve of a tuned system often still varies from one spot in a room to another. The response at the engineer's and producer's listening positions (generally, over the entire front length of the console) often is an overall compromise so that both will hear a similarly accurate sound balance.

Pink noise, rather than individual sine waves, is used for testing because it is of a random nature and does not stimulate standing waves in a room, as would a sustained tone. The presence of such standing waves would introduce inaccurate readings on the analyzer that would vary with the position of the microphone in the room.

Speaker Design

Just as one speaker sounds different from another in different acoustic environments, speakers of different designs vary widely in sound character. Enclosure size, number of components and their size in each enclosure, crossover frequencies, and design philosophy contribute to differences in sound quality. Professional speaker enclosures usually are one of two design types: air suspension and bass reflex.

An air-suspension speaker enclosure is an air-tight sealed system that separates the air in the enclosure's interior from the air outside the enclosure. This system offers a gradual bass rolloff and

good bass transient response (see Figure 13.4a). In the bass-reflex or vented-box design, a tuned bass port hole is designed into the face of the speaker enclosure, enabling the air mass inside the enclosure to mix freely with the air outside the enclosure (see Figure 13.4b) in such a way as to act as a Helmholtz resonator, which acoustically reinforces the speaker's output at the lower octaves.

Figure 13.4. *Speaker enclosure designs.*

a. *Air suspension.*

b. *Bass reflex.*

a

b

With so many variables to consider in speaker and room arrangements, there's no such thing as the "ideal" monitor speaker system. Choice of system is more a matter of subjective taste. Those monitors that are widely favored over a long period of time tend to become regarded as the industry standard; but this can easily change as preferences vary. Again, the best judge of what will work for you should be your own ears.

Crossover Networks

Because individual speaker elements (drivers) are more efficient in some frequency ranges than in others (that is, they produce more undistorted output for the same input level signal), different drivers are often used in combination to give the desired response. Large-diameter drivers, such as 15" units, produce low-frequency information more efficiently than high-frequency information; medium-sized speakers, such as 4" and 5" units, produce mid-range frequencies better than the highs or lows; and small speakers, such as 1/2" to 1 1/2" diaphragm size, produce highs better than any other range.

These speakers are connected by crossover networks, which prevent any signals outside a certain frequency range from being applied to the speaker. These networks usually have one input and either two or three outputs. Input signals above the crossover frequency are fed to the mid- or high-frequency driver, while signals below the crossover frequency are fed to the bass driver or drivers. The passive crossover network uses inductors and capacitors and is designed so that a signal at the crossover frequency is sent equally to the respective outputs (or according to the proportional needs of the system). This design provides a smooth transition from speaker to speaker. If a speaker

system has only one crossover frequency, it is called a *two-way system* because it divides the signal into two bands. If the signal has two crossover frequencies, it is called a three-way system.

The Westlake Audio BBSM-8 monitor speaker, for example, is a ported three-way system that uses two 8" woofers for the bass end, a 3.5" cone midrange driver for the mid-frequencies, and a 1" soft dome tweeter for the highs. The lower crossover frequency is 600 Hz, and the higher crossover is at 5kHz.

Certain speaker designs incorporate a crossover level control that determines how much energy is to be sent to the middle- and high-frequency drivers, allowing the user to partially compensate for differences in room acoustics. For example, an absorptive room often requires more high-frequency energy than a live room in order to produce the same audible effect.

Electronic crossover networks, called *active crossovers* (see Figure 13.5), differ from conventional crossover systems in that the console's line level monitor output signal is split into various frequency bands (depending on whether the speaker is a 2- or 3-way system). Each equalized line level signal is then fed to its own power amp, which in turn is used to drive the respective bass, mid, or high driver elements. Using this approach has the following advantages:

◆ The signals are at low levels in the active crossover, so active filters without inductors can be used. This removes a source of intermodulation distortion.

◆ Power losses due to the resistance of inductors in the passive crossover network are eliminated.

◆ Each frequency range has its own power amp, so the full power of the amplifier and the respective speaker's efficiency is available, regardless of the power output requirements that are placed on a single speaker component at any one point in time.

Figure 13.5. Model AC 22 active crossover. (Courtesy of Rane Professional Audio Products)

For example, consider a 100-watt amplifier feeding high- and low-frequency range drivers through a passive crossover network (see Figure 13.6a). If the low frequencies are using 100 watts (W) of power, and a high-frequency signal comes along that requires an additional 25 W of power from the amp, the amplifier can't supply it. Both the low- and high-frequency signals will become distorted. These monitoring requirements, however, could also be met without incurring distortion by using a 100-W amp to drive the low-frequency speaker and a separate 25-W amp to drive the high-frequency speaker, with the input signal being fed to the power amps through an active crossover network (see Figure 13.6b).

Systems using active crossovers and multiple-power amps are called *bi-* or *tri-amplified* systems, depending on the number of power amps used per channel.

Figure 13.6. *Passive and active bi-amplified crossover systems.*

a. *Passive crossover system.*

b. *Active crossover system.*

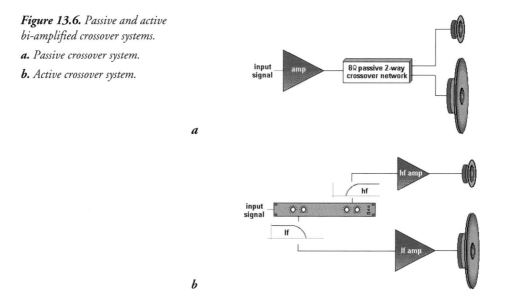

The crossover points for both a passive and an active crossover network are generally 3 dB down from the flat portion of the response curve. Frequency slopes outside the filter's passband are usually 6, 12, 18, or 24 dB per octave with 12 dB/octave being the most common (see Figure 13.7). Depending on the speaker's design, almost any frequency can be selected to be a crossover point; however a few of the most commonly selected frequencies are 500 Hz, 800 Hz, 1200 Hz, 5000 Hz, and 7000 Hz.

Figure 13.7. *Frequency response of a 3-way crossover network with crossover frequencies of 500 Hz and 5 kHz.*

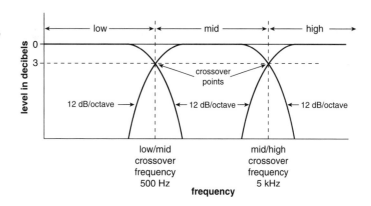

Speaker Polarity

With regard to the way that a pair of speakers is wired, their polarity may be said to either be in-phase or out-of-phase with respect to one another. Speaker polarity is said to be electrically in-phase (see Figure 13.8a) whenever the same signal applied to both speakers will cause their cones to move in the same direction (either positively or negatively). If they are wired out-of-phase (see Figure 13.8b), one speaker cone will move in one direction while the other moves in the opposite direction.

Figure 13.8. *Relative cone motion; speakers in-phase and out-of-phase.*

a. *In-phase.*

b. *Out-of-phase.*

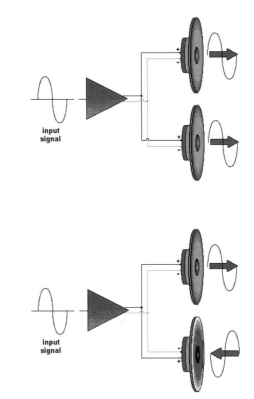

Speaker polarity can be tested easily by applying a mono signal to both speakers at the same level. If the signal's image appears to originate from directly between the speakers, they have been wired in-phase. If the image is hard to locate, appears to originate beyond the outer boundaries of both speakers, or shifts as the listener moves his or her head, however, they generally have been wired out-of-phase. This effect is especially noticeable with low frequencies. An out-of-phase speaker condition can be corrected easily by checking the speaker wire polarities (hot leads to the + or red post, and negative leads to the – or black posts on both the amp and speakers) and reversing the improperly polarized cable. Speaker cable generally is color coded, with white or red being positive

(+) and black being negative (–). If no color coding is present, heavy-duty power cable and other types of cabling suitable for speakers often have a notched ridge or set of ridges on one side. This ridge usually is connected to the negative posts on both the amp and the speakers.

Speaker wire should always be heavy duty. No. 18 wire is the proper size for lengths less than 25' to 50', and No. 14 is used for lengths between 50' and 100' (No. 14 is thicker than No. 18). Two reasons for increasing the thickness of the conductor as cable length increases are as follows:

◆ All cable has resistance, and resistance builds up as length increases. The more resistance in a cable, the more power is dissipated in the cable and is unavailable to drive the speaker.

◆ The higher the cable resistance, the lower the effective damping factor of the amplifier. The amplifier damping factor is related to how well the amplifier can control the motion of the speaker cone. The lower the damping factor, the less control the amp has over the speaker, possibly resulting in a loss of bass tightness, definition, and clarity. Thick conductors have lower resistance and thus minimize these problems.

Monitoring

In mixing, it's important that the engineer be seated exactly between the stereo speakers and that their volume be adjusted equally. If this isn't done, signals that have been panned to the center of the stereo sound field may appear to be off to one side or the other. Likewise, if the engineer is closer to one speaker than the other, that speaker will seem to be louder than the other and the engineer may be tempted either to pan the instruments toward the far speaker or boost that entire side of the mix to equalize the volumes. The resulting mix will sound centered when played in that specific control room, but when the mix is listened to in another environment, it will be off center. As a quick check against this, the engineer should always make sure that an audible volume difference between speakers is accompanied by a corresponding visual difference on the VU meters monitoring the signal sent to tape. Another guard against off-center levels is to check their balance by monitoring pink noise through both speakers at equal loudness and then placing a microphone in the center listening position. The resulting output levels from each speaker can then be read and matched using an SPL meter or VU meter from a spare console input.

Although the main output meters generally don't read the same at all times when mixing in stereo, the presence of a solo in one channel should read only a few dB higher than the other channel unless the other channel is being kept very low for a specific purpose. The maximum readings on the meters should be about the same (often in the –3- to +2-dB range on the VU). Center channel balance can be best checked by switching on the console's lineup oscillator, which allows the main output pair to be calibrated on the stereo output meters for identical left and right output levels.

Mixing

Several other problems remain to be considered with respect to monitoring. Even if the monitor speakers in use are absolutely flat in the control room, few of the people who buy recordings have flat speaker/room curves and, as a result, won't hear exactly the same mix that was heard in the control room. The buying public will often hear different frequency balances due to response variances between the almost innumerable types of speakers and listening rooms. Faced with this fact, the best we can do as professionals is to rely on our best judgment, our experience, and our ears to create a mix that will do the best possible justice to the project under a wide range of listening conditions.

To obtain an acceptable balance, you can choose between several accepted means of monitoring during either the recording or mixdown process (see Figure 13.9). In a modern production facility, a combination of two or more monitor types or systems generally is available. A console sometimes lets you select between speaker systems, with each set commonly having its own associated amplifier to provide a greater degree of level matching.

Figure 13.9. Accepted means of control room monitoring: A—Far-field monitoring; B—Near-field monitoring; C—Small-speaker monitoring; D—Headphones.

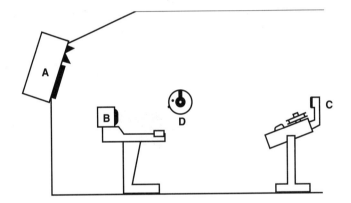

Far-Field Monitoring

Far-field monitoring involves listening over the studio's main monitor pair. These speakers (see Figures 13.10 and 13.11) generally are flush-mounted in the control room's front facing wall. In certain situations, however, they may be freestanding. This method of monitoring may be used within the recording phase as the speakers have been optimized for the room and are also generally difficult to blow out. This last reason can come in handy should a microphone drop or a vocalist decide to get your goat by screaming into a mic. Far-field monitoring may provide the most accurate studio sound when monitoring at moderate to high listening levels, but few homes are equipped with speakers that can deliver such relatively accurate sound at high SPLs. For this reason, many professionals have begun using monitor speakers that more realistically represent the type of listening environment that John and Jill Q. Public most likely have.

Figure 13.10. *Tannoy System 215 DTM II professional monitor speakers. (Courtesy of Tannoy-TGI North America, Inc.)*

Figure 13.11. *The Westlake family of professional monitor speakers. (Courtesy of Westlake Audio)*

Near-Field Monitoring

The term *near-field* refers to the placement of smaller "bookshelf-style" speakers on or slightly behind the metering bridge of a recording or production console. These speakers (see Figures 13.12 through 13.15) are placed at a closer working distance to the engineer and producer so that a greater portion of the direct sound mix is heard relative to the room's overall acoustic characteristics.

Figure 13.12. *Yamaha NS-10M studio monitor speakers. (Courtesy of Yamaha Corporation of America)*

Figure 13.13. *Genelec 1030A*
Active (self-powered) Monitor.
(Courtesy of QMI)

Over the course of the 1980s and 1990s, near-fields have become an accepted standard for monitoring in almost all the fields relating to audio production for the following three reasons:

◆ Quality near-field monitors more accurately represent the sound that would be reproduced by the average home speaker system.

◆ The placement of these speakers at a position closer to the listening position reduces unwanted room reflections and resonances. In the case of untuned rooms, this creates a more accurate monitoring environment.

◆ The cost of these moderate-sized speaker systems is significantly less than their larger studio reference counterparts (not to mention the reduced amplifier cost due to the need for less overall wattage).

Figure 13.14. Audix Nile V monitor speakers. (Courtesy of Audix Corp.)

As with any type of speaker system, near-fields differ widely in both construction and fundamental design philosophy. Care should be taken when choosing the speaker system that best fits your production needs and personal tastes.

Small Speakers

As radio and television airplay are both major forces in boosting recording sales, it's often a good idea to monitor your final mix through small, inexpensive speakers that mimic the nonlinearities and poor bass response of many consumer broadcast products. Such speakers can either be bought or easily made, and occasionally they are incorporated into console and two-track ATR designs.

Before listening to a mix over such small speakers, it's a good idea to take a break to allow both your ears and your brain to recover from prolonged listening at moderate to high-pressure levels over larger speakers.

Figure 13.15. *The KRK model 7000B professional monitor speaker. (Courtesy of KRK Monitoring Systems)*

Headphones

Headphones are also an important monitoring tool because they entirely discount studio acoustics and allow the engineer or producer to hear only what is recorded. Room acoustics aren't involved, so headphones offer excellent spatial positioning in that they let the artist, engineer, or producer place a sound source at critically specific positions within the stereo field.

Monitor Volume

It's important to keep in mind that the Fletcher-Munson curves will always have an effect on the frequency balance of a mix, in that your ears will perceive recorded sound differently at various monitoring levels. If you have set a balance while listening at loud levels, your ears will easily perceive the extreme high and low frequencies in the mix. When the mix is played back at lower levels (such as over the radio), your ears will be much less sensitive to these frequencies and the bass and extreme highs will be deficient (leaving the mix sounding distant and lifeless). Conversely, if you set a balance while listening at levels that are too low, the extreme frequencies will be unduly exaggerated using EQ; when they are played back at moderate to high listening levels, the bass and treble will be overemphasized.

Unlike the 1970s, when excruciatingly high SPLs tended to rule in many rock studios, more recent decades have seen the reduction of monitor levels to a more moderate 75 dB to 85 dB SPL. These levels offer a good compromise level for mixing because they more accurately represent listening levels that are encountered in the average home (meaning that the Fletcher-Munson curves will be more closely matched). Ear fatigue and potential ear damage due to prolonged exposure to high SPLs by industry professionals can also be avoided.

Compatibility

Another monitoring concern is mono-stereo, stereo-surround sound, and mono-surround sound compatibility.

It's important to remember that a percentage of your potential customers may first hear your mix over FM and AM radio in mono. Therefore, if a recording sounds good in stereo but poor in mono, it may not sell well because it has failed to take this medium into account. The same might go for a surround sound mix of a music video or feature release film in which proper attention wasn't paid to phase cancellation problems in mono and/or stereo (or vice versa). The moral of this story is this: To prevent potential problems, a mix should be carefully checked in all its release formats to ensure that no out-of-phase components are included that would cancel out instruments and potentially degrade the balance.

CHAPTER

Product
Manufacture

After a tape has been approved in a mixdown master-tape form, the next step is to transform this master tape into a form that can be mass-produced and sold. With the technology available today, that product may take the form of a compact disc (CD), cassette tape, vinyl record, or CD-ROM. Each of these products has a specific manufacturing process and method of quality control. For each method, careful attention is required throughout the entire manufacturing process.

A common misconception is that once you walk out the door of a recording, project, or production studio with your masters in your arms (or, nowadays, more likely in your hip pocket), the creative process of producing a product is over. All that you have left to do is hand the DAT or other medium over to a duplication facility and voilà! Often this scenario is far from the truth. Now that you have the program content in hand, you have to think through and implement the following additional stages if your product is to make it into the hands of the consumer:

- ◆ Mastering
- ◆ Product manufacture
- ◆ Marketing and sales

Mastering

Now that you've completed your carefully crafted project, the next stage is to begin preparations for turning the final master into a finished product that can be mass duplicated. At this point, your project is ready for *mastering*.

The mastering process is an art form that uses specialized, high-quality audio equipment and the talents of a mastering engineer. Working with this specialized equipment (see Figure 14.1), the engineer processes and transfers a final, sequenced master to a production medium that can be easily replicated by the appropriate duplication equipment.

Figure 14.1. Bob Ludwig's Gateway Mastering Studio. (Courtesy of Gateway Mastering Studio)

Now that so much care and trouble have gone into the creation of your final project master, suppose that you simply walked into XYZ mastering labs with a DAT in hand. Your intention is to have it mastered into a form preferred by an East Coast CD manufacturing plant. While listening to the master, one of two possible scenarios is likely to be played out:

1. The master sounds great on the engineer's reference monitors as well as on the general listening near-field speakers.
2. The overall EQ has some general problems, and a couple of cuts have some severe level and EQ problems because of a rush job in a late-night session.

Of course, you already had your suspicions that the latter was going to be the case. If it were a perfect master, the engineer would simply congratulate you on a great job and call home to say he'd be early for dinner. But since you know you have some work ahead of you, you can at least be glad that you chose a reputable mastering engineer who can equalize and process the project to

minimize or eliminate the problems. After a few hours of signal processing and inputting the reference codes for CD mastering, you can finally sit back and listen to a project you can be proud of.

The preceding scenario isn't all that uncommon. The talent of an experienced mastering engineer can be an invaluable tool towards ensuring that the job gets done correctly. If you don't have the luxury of being with the mastering engineer during this phase (be it mastering for CD, cassette, or record), it's always a good idea to have a *reference cut* (also known as a *test pressing*) made from the resulting master. The production of such a single "one-off" copy enables you to hear the final results and approve the product or give instructions as to what changes you want made (such as changes in EQ, level, or dynamic range). If time and money permit, it's also wise to have a test reference made of any corrected versions.

Making a test pressing is well worth the time and money; the alternative is to receive the full product shipment on your doorstep, only to find that it's not what you wanted—a far more expensive, frustrating, and time-consuming option. Is usually isn't wise to simply assume that a manufacturing or duplication facility's mastering room will dutifully hand you a product that lets your artistic endeavors shine through.

Product Manufacture

Before you begin a project, decide in what form you want to sell the product: CD, cassette, or LP. The financial outlay, artwork type, distribution, and sales approach all depend on the product's final form.

The CD, of course, is the most common form of saleable product. Even though manufacturing costs have dropped over recent years, CD manufacture is still the most expensive route; the returns (excluding artist royalties, of course), however, are also the highest. The cassette is a cost-effective medium for getting a product out. Depending on your audience, the trade-offs in quality versus pricing may not warrant the manufacture of both cassette and CD products. The vinyl record, although technologically out-of-vogue, is still the medium of choice for certain collectors, DJs, and consumers. The choice of media obviously depends on factors such as financing, distribution, and your target audience.

Choosing the Right Manufacturer

After a project's final medium has been chosen, the next manufacturing-related task is to seek out the right manufacturer or duplication facility. If the project is being underwritten and distributed by an independent or major record label, they will be fully aware of their production needs and most likely will have an established production and manufacturing network in place. If you are having the project manufactured yourself for self-distribution, you can expect to face the daunting task of searching out a manufacturer that fits your budget and quality needs.

A number of resources exist for finding such manufacturers. For starters, the Mix Master Directory (CBM Music and Entertainment Group, 6400 Hollis St., Suite 12, Emeryville, CA 94608; 1-800-233-9604) publishes an annual directory of industry-related products and services. This directory provides a comprehensive listing of manufacturers and is cross-referenced by product category. The Recording Industry SourceBook (CBM Music and Entertainment Group, 6400 Hollis St., Suite 12, Emeryville, CA 94608; 1-800-233-9604) also includes a full listing of these companies.

Companies often contract out for special services (such as art layout or printing) or they may supply you with a list of other companies that can provide these services. If necessary, however, a number of full-service manufacturing facilities exist that offer complete production packages to the public. These packages include mastering, art layout, manufacturing, and packaging for a single flat rate. Generally, these will be offered for limited quantity runs of 1000 to 5000; however, combination production packages aren't uncommon that offer runs of 500 CDs and 500 cassettes. Again, when using a full-service production team, it's usually a good idea to add the extra time and expense of requesting an advance proof of the final artwork and test pressing before committing to mass duplication.

Compact Disc Production and Manufacture

Unless you've been living on Mars, you probably are fully aware that the digital revolution has become strongly entrenched in the consumer audio market through the introduction of the compact disc, or CD. This silvery polycarbonate disc (see Figure 14.2) is available in a standard 4 3/4" format and is capable of playing times of about one hour. The mostly defunct 3" CD-3 format offers playing times up to about 15 minutes. Both formats have their information digitally encoded in the reflective underside of the disc in the form of microscopic *pits*. A pit is approximately 1/2 micrometer wide, and a standard disc can hold about two billion pits. These pits are placed on a disc in a spiraling fashion, similar to that of a record, except that 60 CD spirals can fit in the groove of a single, long-playing record. The spiral of a CD differs from that of a record in that it travels from the inside to the outside of the disc. The pits are impressed within the plastic substrate of a disc and covered with a thin coating of aluminum so that the laser light can be reflected. When the disc is placed in a compact disc player, a low-level infrared laser is reflected in the form of a stream of digital audio information back to a photo-sensitive pickup. This data is modulated on the disc so that each pit edge represents a binary 1, and the absence of a pit edge represents a binary 0 (see Figure 14.3). Upon reflection, the data is demodulated and converted into its originally recorded analog form.

Figure 14.2. The compact disc. (*Courtesy of Disc Makers*)

Figure 14.3. The transitions between a pit edge (binary 1) and the absence of a pit edge (binary 0).

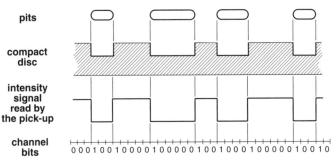

The CD Mastering Process

After the program material to be mastered has been processed (EQ, dynamic processing, and so on) and sequenced into its proper order, the next step in compact-disc production requires the use of a compact-disc mastering system (see Figure 14.4). The mastering system may be a fully dedicated CD mastering system, which is often used by CD manufacturing facilities to prepare a master for manufacture. Or the mastering system may take the form of a computer program integrated into certain Mac-based digital audio workstations. Mastering software enables the owner of a digital audio workstation to assemble and master a project into a completed form that can be directly accepted by either a mastering lab or a manufacturing facility.

Figure 14.4. *Diagram of a CD mastering system. (Courtesy of Sony Corporation)*

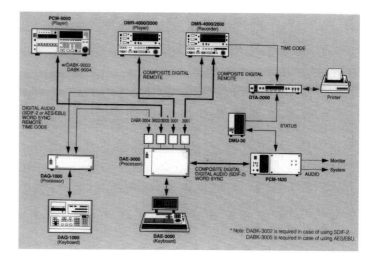

Figure 14.5. *Masterlist CD mastering software. (Courtesy of Digidesign)*

No.	Start Time	Name	Length	Stop Time	L Gain dB R	Xfade	Length
1	0:02:00:00	Teach Me A Song	0:04:37:16	0:06:37:16	0.0 0.0	I	0 ms
2	0:06:38:01	I Didn't Know What Time	0:03:22:09	0:10:00:10	0.0 0.0	I	0 ms
3	0:10:02:10	Lucky To Be Me	0:04:33:15	0:14:35:26	0.0 0.0	I	0 ms
4	0:14:37:11	Moon And Sand	0:05:36:28	0:20:14:09	0.0 0.0	⟋	5000 ms
5	0:20:14:09	Never Let Me Go	0:04:39:06	0:24:53:16	-1.0 -1.0	I	0 ms
6	0:24:54:16	I'm All Smiles	0:05:29:11	0:30:23:27	0.0 0.0	I	0 ms
7	0:30:25:27	I Chase The Sun	0:04:24:00	0:34:49:28	0.0 0.0	I	0 ms
8	0:34:51:13	Old Chair	0:03:06:15	0:37:57:28	0.0 0.0	I	0 ms
9	0:37:59:13	What We Thought Was Real	0:02:57:26	0:40:57:10	-1.0 -1.0	I	0 ms
10	0:40:58:10	So Wrong (Lullabye)	0:05:56:27	0:46:55:07	-1.0 -1.0	I	0 ms
11	0:46:58:07	It's Alright With Me	0:03:48:07	0:50:46:15	0.0 0.0	⟋	4000 ms
12	0:50:42:15	Detour Ahead	0:06:36:14	0:57:19:00	-1.0 -1.0	I	0 ms

If a digital master is supplied that already is based on the 44.1-kHz sampling frequency, the digital signal may be fed directly into the processor. Should the digital signal be based on another sampling clock rate, it is necessary to translate from that sampling rate to the 44.1-kHz rate by way of a sample rate converter. An alternative is to convert the digital signal to analog. Analog masters must be digitally encoded and fed into the processor at this point.

After the digital audio information has been fed into the system, the subcode channels—which are used for frame identification—can be programmed into the storage system. Subcodes are event-point codes that are stored at the head of each compact disc. These codes tell the CD player's microprocessor how many selections are on the disc and their locations. At present, eight subcode channels are available on the CD format, although only two (the P and Q subcodes) are used.

The encoder then splits the 16 bits of information into two 8-bit words and applies error correction in order to correct for lost or erroneous signals. The system then translates these 8-bit words into a 14-bit word format for ease of recording onto disc (a process known as *eight-to-fourteen modulation* or *EFM*). Next, the system begins constructing a methodical system for compact-disc operation known as a *data frame*. Each data frame begins with a frame-synchronization pattern

(27 bits) that tells the pickup beam where it is on the disc. Next is a word of subcode (17 bits), followed by 12 words of audio data (17 bits each), 8 parity words (17 bits each), 12 more words of audio, and then 8 more parity words.

CD Cutting

After all the index and modulation coding has been completed, the data can be stored on 3/4" U-matic videotape using a Sony PCM-1630 processor, or it can be stored directly to a CD recorder, such as the Sony PCM-9000 master disc recorder, shown in Figure 14.6. Either method produces a final, encoded master that will be accepted by most CD manufacturing facilities.

Figure 14.6. *Sony PCM-9000 master disc recorder. (Courtesy of Sony Corporation)*

If the final master data has been transferred using the PCM-1630/U-matic medium, the next step is to cut the original CD master. The heart of such a CD cutting system is an optical transport assembly that contains all the optics necessary to write the digital data on a reusable glass master disc prepared with a photosensitive material. At this point, the master is cut using a 15 to 50 milliwatt laser. After the glass master has been exposed to the record laser, it is placed in a developing machine that etches away the exposed areas to create a finished master. An alternative process, known as *nonphotoresist*, etches directly into the photosensitive substrate of the glass master without the need for the development process. Alternatively, if the data has been written directly to CD by way of a CD recorder, most manufacturing plants can accept the resulting CD as the initial master.

CD Pressing

After the glass or CD master disc has been cut, the compact disc manufacturing process begins under extremely clean room conditions. First, the disc is electroplated with a thin layer of electro-conductive metal. This processed master can then be played and checked by a special player for

miscoding, surface errors, and other potential problems. From this, a negative metal master is used to create a positive metal mother, which, in turn, is used to replicate a number of negative metal stampers (metal plates that contain a negative image of the CD's data surface). The resulting stampers make it possible for clear plastic discs (containing the positive encoded pits) to be mass duplicated, coated with a thin layer of foil for increased reflectivity, and encased in clear resin for stability and protection. When this has been accomplished, all that remains is the process of automatically screen printing the disc's top side and the final packaging. What follows is in the hands of the record company, the distributors, marketing, and you.

The CD-ROM

One of the most important developments in the computer, information, and entertainment fields has been the introduction of the CD-ROM. Unlike its older brother, the CD-Audio disc, CD-ROMs can contain up to 680 MB of any type of computer-based data, including graphics, digital audio, MIDI, text, or raw data. Also unlike the CD-Audio disc, the CD-ROM isn't tied to any specific data format, which means that it is up to the manufacturer or programmer to specify what is contained on the disc. Consequently, a CD-ROM can contain an entire encyclopedia's worth of text, audio, and graphics; digital audio sample files (containing thousands of sampler-specific samples with encoded loop data); software demos; books (like this one)—you name it.

With the recent price drops in PC-based CD-ROM drives and the huge proliferation of CD-ROM titles, this media form has become a driving force in all communications media, and its impact will continue to be felt well into the new millennium.

CD-ROM Authoring

The demands for multimedia product have spawned the computer-related industry of CD-ROM authoring. The term *authoring* refers to the creative, design, and programming aspects of putting together a CD project. At its most basic level, the authoring process can be accomplished in an integrated fashion by one or more people working with a single multimedia program package. Or these same individuals can work with a number of related programs that enable the various media to be assembled and prepared in such a way that a final CD-ROM master can be cut using a CD recorder. At a more advanced level, any number of individuals and companies (each having particular strengths and talents) can join forces to create a unified product.

Because a CD-ROM doesn't have to conform to any specified standards, each product is free to create its own visual, audio, and textual environments (unless the CD-ROM conforms to a company-specified standard, such as CD-I or CD-TV). This factor tends to lead to a wide range of data applications and approaches to educational and game-related topics.

Graphics

Computer-related graphic images are displayed using a horizontal and vertical matrix of dots known as *pixels*. The number of pixels that can be used to fill an entire screen determines the overall image resolution. For example, a screen containing a pixel ratio of 640 x 480 has a reduced resolution over a screen that can display 1024 x 768 pixels. The collection of pixels that creates an image is known as a *bitmap*.

Color resolution depends on the memory capabilities of a computer as well as on the number of colors written into the disc. For example, a 256-color system requires 8 bits to encode the color information, whereas a 16,777,216-color system requires 24 bits to encode the colors. Full-motion graphics—running at frame rates up to 24 fr/sec—require millions of bits of information per frame, so additional video RAM and coprocessors often are required, even when standard data-compression schemes (for reducing the amount of data required to encode a specific image, sound, or specific text) are used.

MIDI

Although the world of multimedia MIDI is identical to the 1.0 spec used by every electronic musician, the need for conformity between patch types when standard MIDI files are played on various computers has brought about the need for a unified standard. This standard ensures that what should be a snare drum riff will be just that and not a Macedonian bleating sheep. As a result, the multimedia MIDI patch specification (known as *General MIDI* and provided in Table 14.1) was universally adopted.

Table 14.1. General MIDI sound patch groupings.

Program Number	Instrument Group
1–8	Pianos
9–16	Chromatic percussion
17–24	Organs
25–32	Guitars
33–40	Basses
41–48	Strings
49–56	Ensembles
57–64	Brass
65–72	Reeds
73–80	Pipes
81–88	Synth leads
89–96	Synth pads

continues

Table 14.1. continued

Program Number	Instrument Group
97–104	Synth effects
105–112	Ethnic
113–120	Percussive
121–128	Sound effects

Digital Audio

Digital audio is an integral part of interactive multimedia. Its processing can be handled by converters that have been factory designed into a computer, or they can be added to any popular PC simply by installing a hardware sound card. The ever-present SoundBlaster for IBM-compatible PCs and other similar hardware boards have brought digital audio into millions of homes with varying degrees of quality ranging from pitiful to professional.

Such digital audio cards can be used for a number of applications, including the following:

◆ Music and sound effects for games (*Blooop, Blop . . Zap! Ha Ha, you missed, loser!*)
◆ General computer-related effects (startup, error, and Arnold Schwarzenegger shutdown sounds)
◆ Sound for CD-ROMs
◆ Synthesized text readout
◆ Audio tracks that can be synched to MIDI tracks

A number of cards are designed to operate at the professional 16-bit stereo rates of 48, 44.1, and 32 kHz. At the time of this writing, however, much of the data used to encode multimedia audio is authored at 22 and 11 kHz and uses both 16- and 8-bit data formats. These sample and bit-rate combinations save on memory space and data processing overhead. In addition, much of the program material will be speech, so a wide bandwidth and increased bit rate aren't required. Again, no prescribed guidelines exist for sample-rate and bit-rate usage, so the choice is up to the project authors. Table 14.2 provides the available sample and bit-rate formats as well as their data requirements per minute.

Table 14.2. Sample-rate and bit-rate chart showing the amount of data required for one minute of audio.

Sample Rate	Bit Width	File Size / Minute
44.1 kHz	16-bit, stereo	10.5 MB
44.1 kHz	16-bit, mono	5.3 MB
44.1 kHz	8-bit, stereo	5.3 MB

Sample Rate	Bit Width	File Size / Minute
44.1 kHz	8-bit, mono	2.6 MB
22 kHz	16-bit, stereo	5.3 MB
22 kHz	16-bit, mono	2.6 MB
22 kHz	8-bit, stereo	2.6 MB
22 kHz	8-bit, mono	1.3 MB
11 kHz	16-bit, stereo	2.6 MB
11 kHz	16-bit, mono	1.3 MB
11 kHz	8-bit, stereo	1.3 MB
11 kHz	8-bit, mono	660 KB

CD Recorders

Before the development of CD-recording systems, the only way a test pressing, "one-off" disc (single disc created for test press or personal archive use), data backup, or other type of CD could be made was to go through the entire glass master, electroplating, and disc-pressing process. The process usually was considered too expensive and time-consuming, so these discs weren't normally made at all for personal purposes or until deemed absolutely necessary. In recent years, however, the CD-recorder (CD-R) has ushered in an era of affordable desktop CD mastering (see Figures 14.7 and 14.8) by enabling individual compact discs that adhere to both the Red Book (CD-Audio) and Orange Book (CD-ROM) standards to be produced directly from a PC-based or stand-alone system.

Figure 14.7. The PC-based *Personal Scribe 750 CD-R system. (Courtesy of Meridian Data, Inc.)*

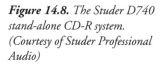

Figure 14.8. The Studer D740 stand-alone CD-R system. (Courtesy of Studer Professional Audio)

Currently, two types of CD-recording systems are available: the CD-WO (write once) and CD-Erasable. A write-once system uses a permanent encoding structure that enables data to be written to disc only once. In general, additional data can be written to the disc, but the previously existing data can't be recorded over. (It often is possible, however, to "erase" the existing data by permanently blanking out unwanted data).

Erasable technology, on the other hand, enables data to be written, erased, and rewritten any number of times. As a result, this technology is being used in systems with increasing frequency as a high-density, removable alternative to the hard disk.

High-Density CD

When discussing the compact disc, it would be an oversight to ignore an emerging technology that has begun to affect the multimedia and consumer production media: the high-density compact disc, or HD-CD.

Beyond the simple benefits gained from increasing the data density of a standard CD by factors of 2x, 4x, and even 8x (see Figure 14.9), HD-CD enables higher data transfer rates, making it easier for high-quality digital video images to be encoded to and reproduced from disc.

Using standard compression techniques, this CD technology raises high hopes for replacing standard VHS videotape technology with optical CD technology as the preferred media for home video use. Because the pit size is reduced (compared to the standard CD), a shorter-wavelength optical laser is required to record and playback these discs; beyond this difference, however, much of the production and replication process is similar to the standard CD.

Figure 14.9. *Detailed relief showing standard 1x CD and 8x HD-CD pit densities.*

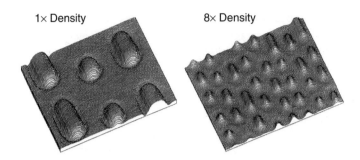

1× Density

8× Density

Cassette Duplication

On a worldwide basis, the prerecorded music cassette is still a strong, cost-effective medium for getting commercial music out to the masses. It's evident from the previous section on compact disc mastering and manufacture that a great deal of artistry and quality control goes into the manufacture of CDs. Contrary to public misconception, the same amount of care and quality control is given cassette duplication in order to produce a quality product. Currently, there are three basic methods of cassette duplication: real-time duplication, bin-loop high-speed duplication, and in-cassette high-speed duplication.

Real-Time Duplication

With *real-time duplication,* the actual production process occurs in real time. That is, both the master playback machine and the slave recording machines are operating at the normal cassette speed of 1 7/8 ips. Thus, a program lasting 30 minutes requires 30 minutes of duplication time, and as many programs can be produced as there are slave recording decks operating.

Most industry insiders agree that this format gives the highest quality of cassette reproduction because both the master and slave machines are operating at the optimum speed for the medium. The audio signal is kept within the audio bandwidth and is not shifted into a higher bandwidth as it is with the high-speed processes. This technique allows for maximum signal, transport, and tape optimization. Dual-cassette tape recorders use this straightforward method, which permits simple, high-quality duplication. Unfortunately, this method also permits the commercial pirating of mass-duplicated tapes, violating copyright laws and resulting in costly production and royalty losses.

The recorded tape, which is played on the master deck and from which the duplicated slave tapes are made, is called the *duplication,* or *dupe master.* When using the real-time method, it is possible for the dupe master to be in any audio format desired. It may take the form of a musically sequenced DAT copy; a 1/2-track, 15 or 30 ips copy of the master tape; a 1/4-track, 7 1/2 ips copy; or a cassette.

The dupe master is often recorded to an open-reel deck in 1/4-track stereo format onto 1/4" tape or 1/2" 4-track tape to ensure high quality. By recording in the 1/4 or 4-track stereo format, side A's stereo program information can be recorded in one direction and then, by turning the tape over, side B's stereo program can be recorded on the remaining tracks of the tape. Through the use of 4-track playback heads on the master machine and similar recording heads on the slave recorders, it's possible to copy both sides simultaneously—in a single tape pass—thereby cutting the duplication time in half (see Figure 14.10).

Figure 14.10. By recording side A in one direction onto 2-tracks of a 1/4-track stereo or 4-track tape, and side B on the remaining tracks in the opposite direction (side A's tracks play forward while side B's tracks play backward), it's possible to simultaneously duplicate both sides of a cassette tape in a single pass.

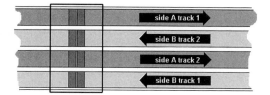

Bin-Loop High-Speed Duplication

With the bin-loop method, duplication takes place without the duplicated tape being housed in cassette shells. The duplicated tape is recorded on reel-to-reel machines, which give a higher quality and better tape handling at high speeds. In this method, a dupe master is recorded from the master tape at 3 3/4, 7 1/2, or 15 ips—depending on whether the program is speech or musical quality—and generally in the 1/4- or 4-track format previously mentioned. The dupe master is then wound onto a bin-loop master reproducer (see Figure 14.11). This is a playback machine that, instead of storing the tape on a reel, stores the tape in a special tape bin in a freestanding fashion. After the tape has been wound into the bin, it is spliced end to end to create an endless tape loop. At this point, a recorded tone signal, corresponding to between 5 and 15 Hz at 1 7/8 ips, is spliced between the program ends to mark the beginning and end of the program loop on the duplicated slave programs. The program outputs of the bin-loop master can then be fed to any number of slave recording machines. These slave machines are of the open-deck, reel-to-reel type, except that they are designed to record onto 1/8" cassette-grade tape, which is supplied on bulk 10 1/2" pancake reels.

During the duplication process, both the master and slave machines operate at high rates of speed: 60, 120, and up to 240 ips, with duplication ratios of 16:1, 32:1, and up to 64:1, respectively. After the duplication process has started, the bin-master machine plays back the looped dupe master repeatedly—at high speed—onto the recording slave machines (see Figure 14.12). Thus, each successive repeated performance is recorded on the bulk reels of cassette-grade tape.

Figure 14.11. *The Tapematic 5200 master reproducer and slave recorder bin-loop system. (Courtesy of Tapematic)*

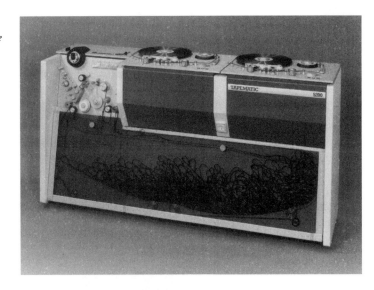

Figure 14.12. *The Versadyne bin-loop duplication system. (Courtesy of Versadyne)*

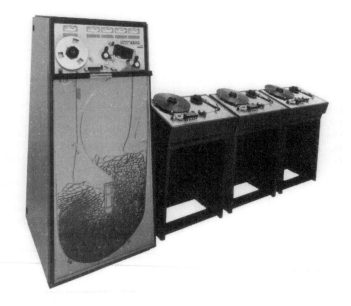

With the duplication process occurring at ratios that are many times the normal speed, the frequency spectrum is also shifted up into a high-frequency range well beyond the audio spectrum. The head, the frequency response, and the bias currents must be tailored for this demanding application. The bias current at these high speeds can reach 3.5 MHz.

The next stage in the duplication process is to load the prerecorded programs, which are now repeatedly recorded onto bulk tape, into cassette housings. This is accomplished using a machine known as a *self-feeding cassette loader* (see Figure 14.13). The duplicated bulk tape is loaded into

this device, and a cassette-feed magazine is then filled with C-0 cassettes (a cassette loaded only with a short section of leader tape). Next, a C-0 cassette is dropped into the loading section and the recorded tape is spliced automatically onto the leader in the empty cassette at the point at which the beginning sensing tone of the program appears. The loader then fast-forwards the tape, loading it into the cassette until the next tone is sensed. At that point, the loader splices the program's end onto the cassette's tail leader and ejects the cassette. The process then repeats. After loading has been accomplished, the final stage of the process is to label and package the cassette for sales and distribution. With a large-scale bin-loop method of production, it is possible to produce tens of thousands of cassettes each day.

Figure 14.13. *The Tapematic 2002 self-feeding cassette tape loader. (Courtesy of Tapematic)*

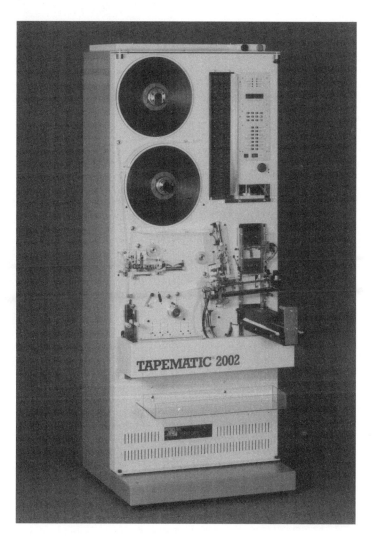

Product quality control is of great importance. The major emphasis in this process rests on the quality of the duplication master and the master-slave alignment. In the mastering process, it is usual for the master to be rerecorded in a stereo 1/4-track, 7 1/2 ips format. In this format, care must be taken to prevent distortion, high-frequency saturation, and a higher noise figure. Distortion and saturation can often be dealt with by using peak limiting and compression sparingly. If alteration of the sound occurs or EQ changes are required, the producer should be consulted. The problem of noise may be reduced with Dolby B noise-reduction procedures.

In-Cassette High-Speed Duplication

In-cassette high-speed duplication makes use of high-speed ratios (8:1 or 16:1) by reproducing from a dupe master to a set of cassette slave recorders (see Figure 14.14). These slave recorders are designed to handle the duplication of cassette tape which is already loaded into its shell.

Figure 14.14. Telex ACC 4000 XL 8:1 in-cassette sound duplicator. (Courtesy of Telex)

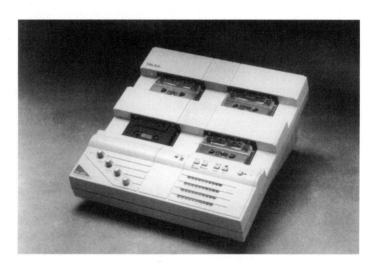

In-cassette units often are self-contained, with both a master and one or two slaves located in the same duplicating unit. Slave machines can often be added on to these devices, allowing for cost-effective expansion. With this method of duplication, the master is often a stereo cassette, although units are available in a 1/4-track, reel-to-reel master format. This method is cost-effective, but it may have trade-offs in frequency, distortion, and wow/flutter limitations.

Vinyl Disc Manufacture

Although the popularity of this medium has waned in recent years (due, of course, to the increased marketing, distribution and public acceptance of the CD), the vinyl record can't be declared dead quite yet. In fact, from Dance DJ hip-hopsters to die-hard classical buffs, the record is still a viable

reproduction and production medium. But the fact remains that many record pressing facilities have gone out of business in recent years, and there are far fewer mastering labs that cut master lacquers. It may take a bit longer to find a facility that fits your needs, budget, and quality standards, but it's definitely not a futile venture.

Disc Cutting

The first stage of production is the disc-cutting process. As the master tape is played on a specially designed tape playback machine, its signal output is fed through a disc-mastering console to a disc-cutting lathe. Here the electrical signals are converted into the mechanical motion of a stylus and are cut into the surface of a lacquer-coated recording disc.

Unlike the compact disc, a record rotates at a constant angular velocity, such as 33 1/3 or 45 revolutions per minute (rpm), and has a continuous spiral that gradually moves from the disc's outer edge to its center. The time relationship of the recorded material can be reconstructed by playing the disc on any turntable that has the same constant angular velocity as the original disc cutter.

The system of recording used for stereo discs is the 45/45 system. The recording stylus cuts a 90°-angle groove into the disc surface so that each wall of the groove forms a 45° angle with the vertical. Left-channel signals are cut into the inner wall of the groove and right-channel signals are cut into the outer wall, as shown in Figure 14.15. The stylus motion is phased so that a signal that is in-phase in both channels (a mono signal or a signal centered between the two channels) produces lateral motion of the groove (see Figure 14.16a); out-of-phase signals (channel difference information) produces vertical motion (see Figure 14.16b), that is, changes in groove depth. Because the system is compatible with mono disc systems, which use only lateral groove modulation, a mono disc can be accurately reproduced with a stereo playback cartridge.

Figure 14.15. *The 45/45 cutting system is used to encode stereo waveform signals into the grooves of a vinyl record.*

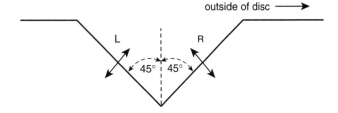

Disc-Cutting Lathe

The main components of the modern disc-cutting lathe are the turntable, the lathe bed and sled, the pitch/depth control computer, and the cutting head. Such a lathe, with its principal components indicated, is illustrated in Figure 14.17. Basically, the lathe consists of a heavy, shock-mounted steel base (A). A 65-pound turntable (B) is isolated from the base by an oil-filled coupling (C), which reduces wow and flutter to extremely low levels. The lathe bed (D), which moves perpendicular to the turntable in a sled fashion, is used to support and house the cutter suspension (E) and the cutter head (F).

Figure 14.16. *Groove motion in stereo recording. The solid line is the groove with no modulation.*
***a.** In-phase.*
***b.** Out-of-phase.*

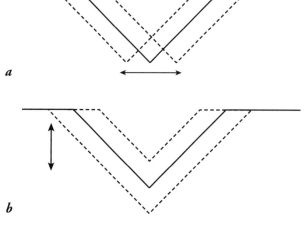

a

b

Figure 14.17. *A disc-cutting lathe with automatic pitch and depth control.*

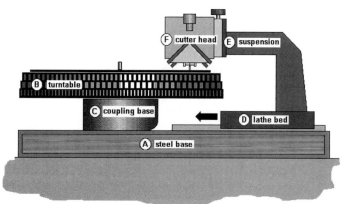

Cutting Head

The cutting head (refer again to letter F in Figure 14.17) translates the electrical signals applied to it into mechanical motion at the recording stylus. The stylus gradually moves in a straight line toward the center hole of the disc as the turntable rotates, creating a spiral groove on the surface of the record. This spiral motion is achieved by attaching the cutting head to a sled. A spiral gear known as the *lead screw* drives the sled in a straight track.

The stereo cutting head consists of a stylus that is mechanically connected to two drive coils and two feedback coils (mounted in a permanent magnetic field) and a stylus heating coil (wrapped around the tip of the stylus). This is illustrated in Figure 14.18. When a signal is applied to the drive coils, the alternating current flowing through them creates a changing magnetic field that alternately attracts and repels the permanent magnet. Because the position of the permanent magnet is fixed, the coils move in proportion to the strength of the field created and the attached stylus moves with them. The drive coils are wound and mounted so that energizing either one

causes the stylus to move in a plane that is 45° to the left or right of vertical, depending on which coil is energized. Feeding both coils an in-phase signal causes the stylus to move in the lateral plane; feeding the coils an out-of-phase signal causes the stylus to move in the vertical plane.

Figure 14.18. *Simplified drawing of a stereo cutting head.*

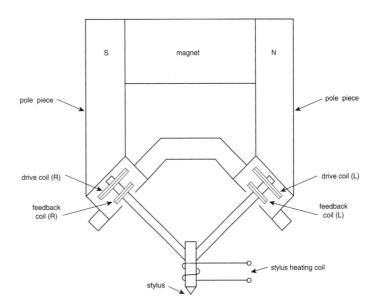

Pitch Control

The head speed is called the *pitch* of the recording and is measured by the number of grooves, or lines per inch (lpi), cut into the disc. As the head speed increases, the number of lpi decreases, so the pitch and playing time also decrease. There are several methods for changing pitch: the lead screw can be replaced by one with a finer or coarser spiral, the gears that turn the lead screw can be changed to alter the speed of the lead screw rotation, or the lead screw rotational speed can be varied directly by varying the speed of the motor driving it. The latter method is used in the Neumann lathe and provides a continuously variable pitch.

The space between grooves is called a *land*. Modulated grooves produce a lateral motion that is proportionate to the in-phase signals contained in the two channels being cut. If the cutting pitch is too high (too many lines per inch, making the grooves very closely spaced) and high-level signals are cut, it is possible for the groove to break through (*cutover*) the wall of an adjacent groove or for the grooves to overlap (*twinning*). The former is likely to cause the record to skip when played. The latter causes either distortion of the signal or an echo of a signal in the adjacent groove due to the deformation of one groove wall by the information cut in the next. Groove echo can occur even if the walls don't touch; it is a function of groove width, pitch, and level, and it decreases as the signal frequency increases. In addition, high-frequency echoes decrease in level as groove diameter decreases.

These cutting problems can be eliminated either by reducing the cutting level or by cutting fewer lines per inch. A conflict arises here because a louder record (compared to a softer record) sounds brighter, punchier, fuller, and more present. As a result, record companies and producers are concerned about the competitive level of their discs relative to those cut by others, so they don't want to reduce the cutting level.

The solution to these level problems is to vary the pitch—cutting more lines per inch during soft passages and fewer lines per inch during loud passages—by splitting the program material into two paths: undelayed and delayed. The *undelayed* signal is routed to the lathe's pitch/depth-control computer, which determines the pitch needed for each program portion and varies the speed of the bad screw motor. The *delayed* signal, which usually is achieved by using a high-quality digital delay line, is fed to the cutter head, thereby giving the pitch/depth control computer time to change the pitch.

Pitch is divided into two categories: coarse, which refers to between 96 and 150 lpi, and microgroove, which is between 200 and 300 (or more) lpi. Microgroove records have less surface noise, wider frequency range, less distortion, and greater dynamic range than coarse-pitch recordings. They can also be tracked with lower stylus pressure, which results in longer life. This lower tracking force, however, makes the stylus more likely to skate across the record if the turntable isn't level. The playback stylus for a stereo microgroove record must have a tip radius of 0.7 mil or less, compared to 2.5 mils + or -0.1 for coarse-groove records. The old 78 rpm and early 33 1/3 rpm records were recorded with a coarse pitch. Virtually all current records are microgroove, with an average pitch of 265 lpi. At maximum pitch, the playing time of one side of a 12-inch disc, with no modulation in the grooves, is 45 minutes. The duration of modulated 12-inch discs cut at average levels is 23 to 26 minutes per side when they are cut with a variable-pitch lathe.

Recording Discs

The recording discs used on the lathe are flat aluminum discs coated with a film of lacquer, dried under controlled temperatures, coated with a second film, and dried again. The quality of these discs, called *lacquers*, is determined by the flatness and smoothness of the aluminum base; any irregularities in the surface, such as holes or bumps, will cause similar defects in the lacquer coating. Lacquers are always larger in diameter than the final record, making it easy to handle the master without damaging the grooves. A 12-inch album is cut on a 14-inch lacquer and a 7-inch single is cut on a 10- or 12-inch lacquer. Producers often cut a reference lacquer to hear how the master tape will sound after being transferred to disc.

The Mastering Process

The mastering engineer sets a basic pitch on the lathe. A lacquer is placed on the lathe, and compressed air is used to blow any accumulated dust off the lacquer surface. Chip suction is started and a test cut is made on the outside of the disc to check groove depth and stylus heat. The start

button is pressed, the lathe moves into the starting diameter, the cutting head is lowered onto the disc, the starting spiral and lead-in are cut as preset, and the tape machine is started automatically. As the side is cut, the engineer changes the console settings as previously determined. A photocell mounted on the tape deck senses the white leader tape between the selections on the master tape and signals the lathe to automatically expand the grooves to produce bands. After the last selection on the side, the lathe cuts the lead-out groove and lifts the cutter head off the lacquer.

The master lacquer is never played because the pressure of the playback stylus would damage the recorded soundtrack. The damage consists of high-frequency losses and increased noise and is aggravated by each successive playing. Reference lacquers, also called *reference acetates* or simply *acetates*, are cut to hear how the master lacquer would sound. Thus, damage to the soundtrack is confined to the reference acetate.

After the reference acetate is approved, the record company assigns each side of the disc a master (or matrix) number that the cutting room engineer scribes between the grooves of the ending spiral on the lacquer. This number identifies the lacquer and any metal parts made from it and eliminates the need to play the record to identify it. If a disc is remastered for any reason, some record companies retain the same master numbers; others add a suffix to the new master to differentiate it from the previous one.

When the final master arrives at the plating plant, it is washed to remove any dust particles and is electroplated with nickel. When the electroplating is complete, the nickel plate is pulled away from the lacquer. This damages the master, so it can only be plated once. If something goes wrong at this point, the plating plant must order a new master from the cutting room.

Vinyl Disc Plating and Pressing

The nickel plate pulled off the master is called the *matrix* and is a negative image of the master lacquer (see Figure 14.19). This negative image is then electroplated to produce a nickel positive image called a *mother*. Because the nickel is stronger than the lacquer disc, several mothers can be made from one matrix. Because the mother is a positive image, it can be played to test it for noise, skips, and other defects. If it is acceptable, the mother is electroplated several times, producing the *stampers*, which are the negative images of the disc used to press the record.

The stampers for the two sides of the record are mounted on the top and bottom plates of a hydraulic press. A lump of vinylite record compound, called a *biscuit*, is placed in the press and sandwiched between the labels for the two sides. The press is closed and heated by steam to make the vinylite flow around the raised grooves of the stampers. The pressed record is too soft to handle when hot, so cold water is circulated through the press to cool it before the pressure is released. When the press opens, the operator pulls the record off the mold. The pressing process causes excess compound to flow to the outer edges of the disc, so the disc is oversize. This excess, called *flash*, is trimmed off after the disc is removed from the press. The edge of the disc is buffed smooth and then the product can be packaged for distribution and sales.

Figure 14.19. *The various stages in the plating and pressing process.*

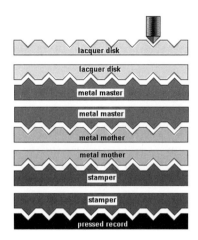

Marketing and Sales

Although this category is mentioned last, it is by far one of the most important areas to be dealt with when contemplating the time, talent, and financial endeavors involved in creating a recorded product. Questions such as: Who is my audience? How will the product be distributed? What should be the media format of the final product? How much will this project cost me? All of these questions should be answered long before the record button is pressed and the first downbeat is played.

In this short section, I won't even attempt to cover this extremely important and complex topic. It has been fully discussed in a number of well-crafted books, including *How to Make and Sell Your Own Recording*, Diane Sward Rapaport, Prentice Hall, 1992; *Releasing an Independent Record*, Gary Hustwit, Rock Press, 1993; and *This Business of Music*, Shemel & Krasilovsky, Watson Guptill, 1990. These books discuss in detail the three primary methods by which a finished recording can be distributed and sold:

◆ Major label record company
◆ Independent record label
◆ Selling the product yourself

Each of these marketing and sales approaches represent varying degrees of financial outlay, as well as artistic and distribution control. No matter which avenue you choose, every aspect of a deal should be fully and carefully investigated before a final commitment is made.

In addition to reading books that relate to this topic, it's often a wise decision to retain the counsel of a trusted music lawyer. The music industry is fraught with its own special legal and financial language. Having someone on your side who has insight into the language, quirks, and inner workings of this unique business can be an extremely valuable asset.

CHAPTER 15

Studio Session Procedures

As you learned earlier, the first rule of recording is simply this: There are no rules. This rule holds true insofar as inventiveness and freshness tend to play a key role in keeping the industry alive and new. In the recording process, however, there are guidelines and procedures which, when followed, can help you have a smooth, professional recording session. At the very least, these procedures can help you solve potential problems when you use two of the best tools you possess: creative insight and common sense.

Recording

Before beginning a recording session, it's always a good idea to mentally prepare yourself for what lies ahead. The best way to do this is to have all those involved in the recording process sit down with the engineer, well in advance of the session, and discuss instrumentation, studio layout, musical styles, and production techniques. This meeting lets all parties know just what to expect during the session and enables everyone to get to know one another.

Just before the session, it's always a good idea to check out the equipment. If analog tape machines are to be used, they should be cleaned, demagnetized, and (if necessary) aligned for the type of tape being used for the session—preferably using the actual session tape itself. Generally it's a good idea to record a 100 Hz,

1 kHz, and 10 kHz tone at 0 VU (–10 VU if 7 1/2 ips) on all tracks at the beginning of the tape to indicate the proper operating level. Then if it's necessary to overdub or mix the tape at another studio, you'll be able to calibrate the unknown tape machine to your reference set of tones. If digital tapes are to be used, it might be wise to ensure that the tapes have been formatted. Although certain recorders let you record onto unformatted tapes in a single pass or in a special update mode, it's never as easy as having preformatted tapes. Remember to make it as easy on yourself as possible.

At the time of the session, and after the number and types of instruments being recorded are known, a presession setup should take place. Instrument placement often varies from one studio to the next because of the acoustics of the room, the number of instruments, the isolation (or lack thereof) among instruments, and the degree of visual contact that's needed. Should further isolation beyond careful microphone placement be required, flats or baffles can be placed between the instruments in order to prevent loud sound sources from spilling over (in the form of leakage) into other open mikes. Alternatively, the instrument or instruments in question can be over-dubbed.

During an ensemble recording, the setup should permit the musicians to see each other as much as possible so that they can give and receive visual cues. The arrangement of baffles and mics will depend on your personal preferences, as well as on the type of sound the producer wants. If the mics are close to the instrument and the baffles are packed in close, a tight sound with good separation is achieved; a looser, more "live" sound, along with greater leakage, is achieved when the mics and baffles are farther away. An especially loud instrument can be isolated by putting it in an unused iso-room or vocal or instrument booth. Electronic amplifiers played at high volumes can also be recorded in such a room. Alternatively, the amps and mics can be surrounded on all four sides and the top in a box built from baffles. Another approach is to cover both the amplifier and the mic with a blanket or other flexible sound-absorbing material (see Figure 15.1), ensuring that it doesn't interrupt the path between the amplifier and the mic. Separation can also be improved by placing the softer instruments in an isolation booth or by plugging loud electronic instruments directly into the console through a D.I. (direct injection) box, thereby bypassing the miked amplifier. For a piano, leakage can be reduced by placing the mic inside it, putting the lid on its short support stick, and covering it with blankets (see Figure 15.2).

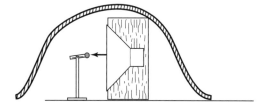

Figure 15.1. *Isolating an instrument amplifier by covering it with a sound-absorbing blanket.*

Figure 15.2. *Preventing leakage from getting into a piano mic.*

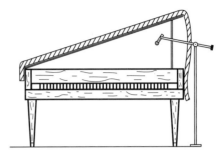

The microphones for each instrument are selected either by experience or by experimentation and are then connected to the desired console inputs. The input used for each mic should be noted on a piece of paper so that you can easily keep track of which instrument has been plugged into a specific input module at the console. Some engineers find it convenient to standardize a system by using the same mic input and tape track for the same instrument at every session. Thus, one engineer might consistently plug the kick drum mic into input 1 and record it onto track 1, the snare mic onto 2, and so on. That way, the engineer knows which fader controls what during both a record and a mix session without having to think much about it.

Electric and Electronic Instruments

Electric and electronic instruments, such as guitars, that have low-level, unbalanced, high-impedance outputs, can be recorded in the studio without using their amplifiers, if a D.I. box is used. As you have already learned, these devices convert high-level, high-impedance output signals into low-impedance, balanced signals that can be fed directly into a console's mic preamp. Instruments often are recorded "direct" in order to avoid instrument leakage problems in the studio and to avoid noise and distortion resulting from the instrument's amplifier or just to get the clean, tight "direct" sound. When using electronic instruments in the control room (such as electronic drum machines and synthesizers, which more closely match the impedance of studio equipment), they often can be plugged directly into the line-level input of the console. Both direct and line-level instruments can be played in the control room while you're listening over the main studio monitor speakers without fear of leakage. If you prefer, both the direct box and the instrument amplifier can be fed simultaneously from the instrument through the use of parallel jacks located on the direct box (see Figure 15.3). This technique enables the engineer to blend the direct and miked pickups in any way he or she wants. The sound of a direct box is very clean but is more susceptible to noise from strings than the rougher, gutsier sound of a miked amplifier.

Figure 15.3. *A schematic for a direct box.*

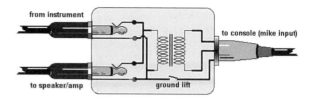

On most guitars, the lowest hum pickup and best tone for the direct connection occurs with the instrument volume control fully on. Because guitar tone controls consist of a variable treble rolloff, maximum control over the sound is achieved by leaving the tone controls on a treble setting and using a combination of console EQ and different guitar pickups to vary the tone. If the treble is rolled off on the guitar, boosting the highs with EQ increases noise that is being picked up because of the guitar's high impedance.

Drums

During the last decade, drums have undergone a substantial change with regard to playing technique, miking technique, and the choice of acoustic environment used for recording. The 1960s and 1970s saw the drum set normally placed in a small isolation room called a *drum booth*. This booth effectively isolated the instrument acoustically from the rest of the studio and had the effect of tightening the drum sound because of the limited space (and often, dead acoustics). The drum booth also isolated the musician from the studio, and this physical separation often caused the musician to feel a loss of involvement.

The conventional drum set has also undergone changes in miking techniques. Many engineers and producers have moved the drum set out of smaller iso-rooms and back into larger open studio areas where the sound can fully develop and mix with the studio's own acoustics. In many cases, this effect is exaggerated by placing a distant mic pair in the room—a technique that often produces a fuller, larger-than-life sound.

If drums are to be recorded, the drummer should tune them while the mics and baffles for the other instruments are being set up. Each drum head should be adjusted for the desired pitch and for constant tension around the rim by hitting the head at various points around its edge and adjusting the lugs for the same pitch all around the head. After the drums are tuned, the engineer should listen to each drum individually to make sure that there are no buzzes, rattles, or resonant after-rings when the heads are hit. Drums that sound great in live performance don't always sound that way when close miked. In a live performance, the rattles and rings are covered up by the other instruments and are lost before the sound reaches the listener. Close miking, on the other hand, picks up the noises just as well as it picks up the desired sound.

If tuning the drums doesn't bring the extraneous noises or rings under control, masking tape can be used to dampen them. Pieces of cloth, dampening rings, paper towels, or a wallet can also be taped to the head in various locations, determined by experimentation, to eliminate rings and buzzes. Although this method of head damping has been used extensively in the past, present methods use this damping technique with discretion and rely more on proper tuning.

For studio recording, it's best to remove the entire built-in damping mechanism from the drum set because it applies tension to only one spot on the head and, therefore, unbalances the head tension. The built-in dampers often vibrate when the head is hit and are chief sources of rattles. The kick drum can be damped by removing the front head and placing a blanket or some damping material inside the drum so that it's pressing against the head. By adjusting the pressure of the damping material against the head, the bass tone can be varied from a resonant boom to a dull thud. Kick drums are usually recorded with their front heads removed, while other drums are recorded with their bottom head either on or off. Tuning the drums is more difficult if two heads are used because the tensions of the heads interact in producing the pitch; however, two heads provide a more resonant tone than one head. After the drums are tuned, the mikes can be put in their appropriate positions. Just make sure that they don't get in the drummer's way. If mikes are in the way, they may be hit by a stick or moved out of position during a performance.

The drum machine often is used to lay down rhythm bed tracks. Sometimes, the "beat box" will eliminate the need for a physical drum set because the tracks have been previously sequenced into the drum machine's memory or a MIDI sequencer. Under certain circumstances, triggers can be built into a drum kit or controller surface that enable the drum machine's sampled sounds to be manually triggered in a live studio setting.

Set Up

After the instruments and rough mike, pickup, and baffle placements have been made, headphones equipped with enough extra cord to allow free movement should be distributed to each player. The engineer then confers with the producer to find out how many instruments are to be used on the song, including overdubs, to determine how many tracks must be left open. The number of tracks influences the number and way mics will be assigned to available tracks, especially for the drums. If many instruments are to be recorded, the number of drum tracks may be limited. If plenty of tracks are available, five or even seven or more tracks may be used for the drums: kick drum, snare, rack toms, floor toms, and cymbals.

When all the mics have been set up, the engineer labels each input fader with the name of the corresponding instrument, usually below the fader on the console or on a piece of masking tape or special console tape. The mics are then assigned to the desired tracks, and the assignments are noted on a track log that remains from then on in the tape box (see Figure 15.4). Master faders and group faders are set to their 0 dB design center (about 3/4 up). The engineer partly turns up the monitor send for the input signals or solos each mic. After all the labeling has been completed, the engineer begins to set the level of each instrument/mic input by asking each musician to play individually or by asking for a complete run-through of the song. Starting with the EQ settings at the flat position, the engineer checks each track's meter readings, listens for mic preamp overload, and adjusts mic preamp gain using the pad and/or mic preamp on the console or, if necessary, on the mic so as to eliminate any distortion.

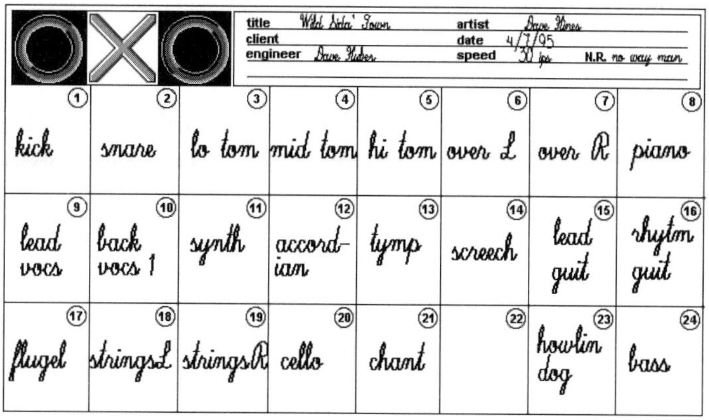

Figure 15.4. *A track log used for instrument/track assignments.*

After these levels have been set, a rough headphone mix can be set up so that the musicians can hear themselves. EQ can be adjusted (if necessary) to obtain the sound the producer wants on each instrument, and dynamic limiting or compression can be inserted into the chain if necessary. If the desired sound can't be achieved with a minimal amount of EQ, a different mic or mic placement should be tried until the instrument's sound is acceptable. The engineer and producer can then listen for any extraneous sounds—such as buzzes or hum from guitar amplifiers or squeaks from drum pedals—and try to eliminate them. This process of selective listening can be assisted by soloing the individual tracks as needed. If several mics are to be mixed onto one or more tracks, the balance between them can be set at this point. To do that, the engineer solos the group the mics are assigned to.

After this procedure has been followed for all the instruments, the musicians should do a couple of practice ("rundown") songs so that the engineer and producer can listen to how the instruments sound together before they are recorded to tape. First, they should listen to all the drums, next to the bass guitar with the drums, after that to the entire rhythm section, and finally, they should listen to all the instruments together. Changes in EQ can be made to compensate for one instrument's covering up another, thereby making them blend better. While the song is being run down, the engineer makes final adjustments to the recording levels and the headphone monitor mix. The engineer can check the headphone mix either by putting on a pair of headphones connected to the cue system or by routing the mix to the monitor loudspeakers to make sure that all the instruments can be clearly heard. If the musicians can't properly hear themselves, the mix can be varied to intensify the sound of particular instruments, regardless of their recorded levels. If several cue systems are available, several headphone mixes can be built up for those who need different balances. During loud sessions, the musicians often require high sound-pressure levels in their headphones in order to hear the mix above the ambient room leakage through the phones.

High sound-pressure levels can cause the pitch of instruments to sound flat, so musicians may have trouble tuning or even singing with headphones on. To avoid these problems, tuning shouldn't be done while listening through headphones. The musicians should play their instruments at levels to which they are accustomed and adjust their headphone levels accordingly. The importance of

proper headphone levels and a proper cue balance can't be stressed enough; these things can either help or hinder a musician's overall performance. The same situation exists in the control room with respect to high monitor-speaker levels: some instruments may sound out of tune even when they aren't, and ear fatigue may impair your ability to judge sounds and relative balance properly.

During the practice rundown, it's a good idea for the entire song to be performed so that the engineer knows where the loudest sections are. This helps the engineer ensure that the recorded level won't overload the tape and, if compression or limiting is used, to ensure that the instruments don't trigger an undue amount of gain reduction. Even though an engineer may ask the musicians to play their loudest when they are playing one at a time, they invariably play louder when performing together. This fact may require changes in the mic preamp gain, record level, and compression/limiting threshold. Separation between the instruments can be checked by soloing each mic and listening for leakage. The relative position of mics, instruments, and baffles can be changed at this time, if necessary.

At the beginning of each performance, the name of the song and a take number are recorded to tape for easy identification. A take sheet should be carefully kept to note the position of the take on a tape (see Figure 15.5). Comments are written on the take sheet to describe the producer's opinion of the performance, as well as whether it is a complete take, an incomplete take, or a false start.

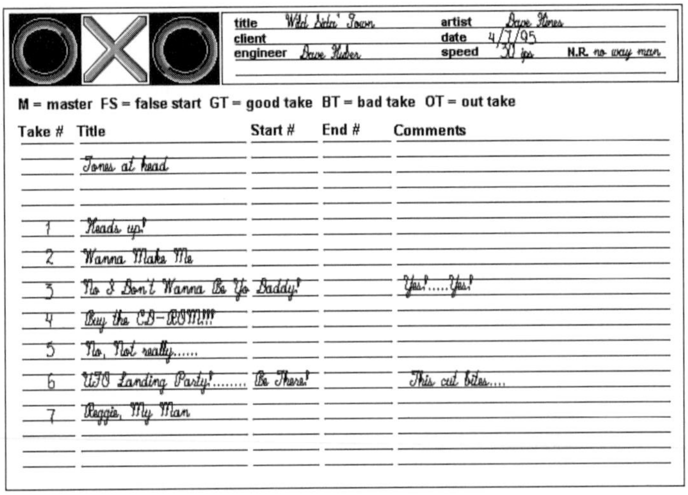

title Wild Sidin' Town   artist Dave Hines
client   date 4/7/95
engineer Dave Huber   speed 30 ips   N.R. no way man

M = master FS = false start GT = good take BT = bad take OT = out take

Take #	Title	Start #	End #	Comments
	Tones at head			
1	Heads up!			
2	Wanna Make Me			
3	No I Don't Wanna Be Yo Daddy!			Yes!.....Yes!
4	Buy the CD-ROM!!!			
5	No, Not really......			
6	UFO Landing Party!........	Be There!		This cut bites....
7	Reggie, My Man			

Figure 15.5. *An example of a studio take sheet.*

During the recording, the engineer watches the level indicators and—only if necessary—controls the faders to prevent overloading the tape. The engineer acts as another set of production ears by listening both for performance and quality factors. If the producer doesn't notice a mistake in the performance, the engineer just may catch it and point it out. The engineer should try to be helpful but also should remember that the producer has the final say and that his or her judgment of the quality of a performance or recording must be accepted.

When a take is played back, the tape is rewound and the monitor system is switched from the program to the tape playback mode. The musicians can then listen to the performance over the control room monitors, over their headphones, or through the studio speakers.

Overdubbing

Overdubbing, illustrated in Figure 15.6, is used to add more instruments to a performance after recording the basic tracks. These instruments are added by monitoring the previously recorded tape tracks while simultaneously recording new, doubled, or augmented instruments onto one or more available tracks.

Figure 15.6. Overdubbing enables additional instruments to be added to existing tracks on a multitrack recording medium.

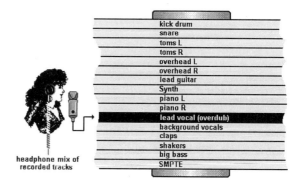

headphone mix of
recorded tracks

kick drum
snare
toms L
toms R
overhead L
overhead R
lead guitar
Synth
piano L
piano R
lead vocal (overdub)
background vocals
claps
shakers
big bass
SMPTE

In an overdub session, the same procedure is followed for mic selection, EQ, and levels as during the recording session. If only one instrument is being overdubbed at a time, the problem of leakage directly from other instruments doesn't exist; however, leakage can occur if the musician's headphones are too loud or not seated properly on his or her head. If the recorder to be used is analog, it should be placed in the master sync mode (thereby reproducing the previously recorded tracks from the record head in sync). Master sync mode is set either at the recorder or its autolocator/remote control. Monitor switching between source (monitoring signals being fed to the recorder or console) and tape/sync (monitor signals from the playback or record/sync heads) often are switched automatically by the tape machine. The control room monitor mix should make the instruments being recorded somewhat prominent so that any mistakes can be easily heard. The headphone mix can be adjusted to fit the musician's personal taste.

Should a mistake or bad take be recorded onto an overdub track, it's a simple matter to rewind the tape and re-record over the unwanted track. If only a small part of the take was bad, it's easy to *punch-in* (silently enter the record mode on that track while the tape is rolling in the record-ready mode) and record over the unwanted portion of the take (see Figure 15.7). After the section has been corrected, the track can be *punched-out* of record, thereby silently exiting record mode and returning to the originally recorded track signal.

Figure 15.7. *Punching-in allows an engineer to selectively replace material and correct mistakes.*

punch-in point		punch-out point
kick drum		
snare		
toms L		
toms R		
overhead L		
overhead R		
lead guitar		
Synth		
piano L		
piano R		
lead vocal (punch)		
background vocals		
claps		
shakers		
big bass		
SMPTE		

On the sections in which the performer isn't playing or singing, he or she should be careful not to make any undue noise. If noises are recorded, they may present weird problems during the punch that may have to be spot-erased before mixdown or be muted during the mix. In any of these cases, they can come back to haunt you and cause more trouble than they are worth.

An additional technique to be used in conjunction with or as an alternative to punching-in over a track is to overdub the instrument onto another available track or tracks. The advantage to recording onto and saving several tracks (if they are available) is that a good take can be saved and the musician can try to improve the performance, rather than having to erase the previous performance in order to improve it. When several tracks of the overdub have been saved, parts of each track may be acceptable and can then be combined to create a complete "composite" performance. This composite is created by playing the tracks back in the sync mode, mixing them together at the console, and recording them on another track of the tape. The overdubbed tracks are turned on or off as necessary to transfer only the best parts of each performance to the composite track (see Figure 15.8).

Figure 15.8. *A single composite track can be created from several, partially acceptable takes.*

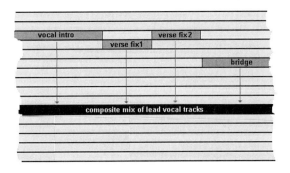

Another procedure that is an extension of the composite track process is known as track *bouncing* or *ping-ponging*. The bouncing tracks procedure is often used to mix entire performances onto one track, a stereo pair of tracks, or several tracks. Bouncing can be performed either to make the final mixdown easier (by grouping instruments together onto one or more tracks, as seen in Figure 15.9a) or to open up needed tape tracks by bouncing similar instrument groups down to one or to a stereo pair of tracks. This frees the originally recorded tape tracks for more overdubs. Suppose

that you have two ADAT digital recorders (which gives you 16 available tracks) with which to do a demo recording. During the basic tracks, suppose that you have decided to record the drums onto all 8 tracks of the first ADAT and then record a stereo piano, bass, and lead guitar (4 tracks) onto their own separate tracks of the second ADAT. After these 13 tracks have been recorded, you could go back and mix the drums down to 2 of the 4 available tracks on the second ADAT (see Figure 15.9b). As a result of the bounce, the 8 original drum tracks can now be freed up for added overdubs, giving you a total of 10 available tracks. Using the previous 16-track scenario, you would have:

12 recorded tracks = (16 tracks – 12 tracks = 4 available tracks)
2 bounced tracks = (4 tracks – 2 tracks = 2 available tracks)
8 tracks freed up = (2 tracks + 8 tracks = 10 available tracks)

Figure 15.9. Track bouncing is a common production technique used in multitrack recording.

a. Instruments can be grouped together onto one or more tracks.

b. Bouncing can be used to expand the number of available tracks and thus allow for additional overdubs.

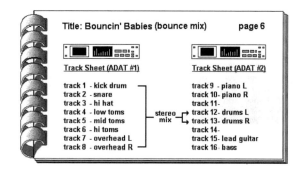

a

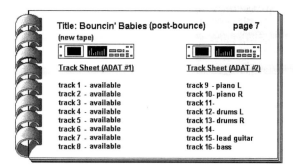

b

It's always a good idea, whenever possible, to keep the original, non-bounced tracks intact. In the example just described, it would be wise to record all eight of the drum tracks onto a single ADAT tape. The composite mix can then be bounced to tracks on the second recorder and the original drum tracks can be saved and safely stored. Another tape can then be formatted, and you're ready to add some fresh, new tracks to your demo. If there is a problem with the mix at a later time, or if the producer wants to use the demo's drum and basic tracks for the CD, you can transfer the original tracks to 24-track tape, and you're in business.

Mixdown

After all the tracks for a song have been recorded, the multitrack tape can be mixed down to either mono or stereo for subsequent duplication and distribution to the consumer. All analog multitrack and mixdown tape machines must first be demagnetized, cleaned, and aligned. After the alignment procedure has been completed, the engineer should record 0 VU level tones at the head of the mixdown reel at 1 kHz, 10 kHz, and 100 Hz. This procedure enables the mastering engineer to align the tape playback machine to play back these tones at 0 VU at these reference frequencies, resulting in a proper playback EQ. If Dolby noise reduction is used, the Dolby level tone should be added to these alignment tones. When mastering to a digital medium, such as DAT, it's only necessary to record a 1 kHz reference tone at the manufacturer's recommended reference level (usually 0 dB or –12 dB).

In mixdown, the console is placed into the mix mode (or each input module is switched to the line or tape position) and the fader's label strip is labeled with the respective instrument names. Master and group faders are set to 0-dB design center (about 3/4 up). The engineer then can monitor the stereo mix bus. The engineer sets up a rough mix of the song by adjusting the levels and the left-right pan position. The producer then listens to this mix and may ask the engineer to make specific changes. The instruments are often soloed one by one or in groups, and necessary EQ changes are made. The engineer and producer then begin the cooperative process of "building" the mix into a final form. Compression and limiting can be used on individual instruments as required, either to make them sound fuller and more consistent in level or to prevent them from overloading the mixdown tape when they're raised to the desired level in the mix. At this point, if it's available, the console's automation features can be used. After the mix begins to take shape, echo and effects are added to give close-miked sounds a more "live," spacious feeling and to help blend the instruments.

If the mix isn't automation-assisted and the fader settings have to be changed during the mix, the engineer memorizes the various fader moves or uses a grease pencil to mark the various levels on paper tape next to the fader scale. Then the engineer learns when to move each fader from one level mark to the next (often noting the tape-counter to help keep track of time). If more changes are needed than the engineer can handle alone, the producer or artist can help by controlling certain faders. It's best, however, if the producer doesn't have to handle any controls and can concentrate fully on the music rather than the mechanics of the mix. The engineer listens to the mix from a technical standpoint to detect any sounds or noises that shouldn't be present in the mix. If noises are recorded on tracks not in use during a section of a song, these tracks can be muted until needed. After the engineer practices the song enough to determine and learn all the changes, the mix can be recorded and the ending faded out with the master fader or faders. The engineer also may not want to fade the song at this time, saving the fade for the computer during the playlist sequencing phase because the computer can perform a fade much more smoothly than can even the smoothest hand. Songs with exceptionally difficult control changes can be mixed in sections that can be edited together afterwards.

The different takes of a mix should be slated as they are recorded, and a complete, detailed take sheet should be maintained with the differences between takes noted. When an acceptable take is recorded on an analog machine, the engineer may want to splice white paper leader tape at both the head and tail of the take so that it can be found easily. The leader can be inserted with either rough or tight accuracy depending on whether the producer is in a hurry to go on to the next mix. It should be noted that the practice of leadering tape has declined with the advent of DAT re-corders because they have built-in index search and absolute time features that make this process unnecessary (and impossible since they use rotary head technology).

Mixes should be made at consistent, moderate listening levels because the variation in the frequency response of the ear at various sound-pressure levels results in a mix sounding quite different at different monitoring levels.

Often it's a simple matter to smooth out differences in level when you use a hard-disk recorder for the final order sequencing. However, it is always wise to pay attention to the details of relative volume. Ideally, the level used should be the same as the level at which the listener will hear the record. Because most people listen to music at moderate volume, moderate monitoring levels between 70 and 90 dB SPL should be used.

The mix should be tested for mono/stereo compatibility to see what changes in instrumental balances will occur when the material is played in either format. If the changes are drastic, the original mix may have to be modified to make, for example, an acceptable mono video mix of a stereo LP mix. A mono mix also should be monitored over a car radio type speaker or boom box to see how it sounds on a system with limited frequency response. If the mix is for a single rather than for a CD, the entire mixdown session can be done at low volume over speakers that are used for the intended medium—such as car radio speakers.

Editing

After all the mixes for a recording are complete, a master mix can be assembled. The producer and artist decide on the sequence of the songs on the basis of their tempos, musical keys, how they flow into one another, and which songs will best attract the listener's attention.

Analog Sequence Editing

When using analog tape, the engineer edits the mixes out of their original reels and splices them together in sequence on the master reels, tightening up the leaders at the same time (if this wasn't already done). The level-set tones are included as the first band on the initial master reel.

The beginning of a mix can be listened to at high volume; the tape is moved back and forth over the heads by hand, and the tape machine is in the stop-edit mode (head lifters defeated) to a point just before the first sound of the performance begins. The tape over the playback head gap at this

point is marked with a grease pencil. If there is no noise directly in front of this spot, it is a good practice to cut the tape one-half inch before the grease pencil mark as a safety precaution against editing out part of the first sound. If there is noise ahead of the first sound, the tape should be cut at the mark and the leader inserted.

The tail of the song must be monitored at even higher volume because it usually is a fade-out or the overhang of the last note and, therefore, is much softer than the beginning of the song. The tape is marked and cut just after the last sound dies out to eliminate any low-level pops that may have been recorded when the bias fed to the record head was shut off and also to remove the tape hiss of the blank tape.

The length of time between the end of the song and the beginning of the next can be constant, or the timing can vary according to the musical relationship between the songs. Decreasing the time between songs can make a song seem to be a continuation of the previous one if they are similar in mood, or decreasing the time can make a sharp contrast with the preceding song if the moods are dissimilar. Longer times between songs enable listeners to get out of the mood of the previous song and lets them prepare to hear something that may be quite different without accentuating the contrast between the two.

In an analog edit, the length of leader tape used determines the time between songs. Paper, rather than plastic, leader tape is used because the plastic can cause static electricity pops. Blank tape can be used rather than leader, but it does not provide a visual division of the songs on the reel, which makes it more difficult to find a particular song. Blank tape also produces tape hiss between songs rather than silence.

When the sequencing is complete, one or two analog or DAT backup copies should be made of the final, sequenced master. The safety copies are made before the master leaves the studio. These copies serve as a backup in case the original mixes are lost or damaged. Several DAT, cassette, or one-off CD-R copies of the master reels can also be made for the producer, the artist, and the record company executives for final approval.

Digital Sequence Editing

With the advent of digital audio editing systems, the relatively cumbersome process of sequencing music tracks in the analog domain using magnetic tape has given way to the faster, easier, and more flexible process of editing the final masters on hard disk. Using this system, all the songs and initial test tones in a given project can be recorded to disk either as a single, continuous file or as a number of individually named soundfiles. The waveform of the song appears on-screen. The start and end points of each song are fine-tuned using a mouse. Each song can then be defined as a named region, which is then placed in the system's playlist source window. (For further information on hard disk editing and playlist-based editors, refer to Chapter 6, "Digital Audio Technology.") By dragging individual song regions into the system's playlist window (see Figure 15.10), the system plays each song directly from hard disk according to its sequential order in the list.

Figure 15.10. An example of a
source and playlist window
containing various song titles and
project-related instructions.

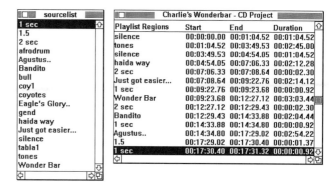

Silence can be inserted between each song by recording a 15-second soundfile that contains nothing but digital silence or by recording silence at the beginning of an overall soundfile. Defined regions from this section can be named accordingly, such as 1-sec silence, 1.5-sec silence, and 2-sec silence, and these regions can then be appropriately placed between each song. Fades, crossfades, volume changes, and other signal processing functions can also be performed at this time. The entire playlist should be saved to disk for possible future changes and as a precaution against a potential system crash.

After you have finally settled on the proper order, fades, and signal processing, the final step is to make several copies of the final, sequenced master on a master DAT tape. If your system is capable of backing up the playlist and soundfile data (either in its original data form or as a specially tagged archive file that later can be restored to disk), it is always a good idea to do so, just in case the record company, producer, or artist wants to make changes later. This simple precaution could save you a lot of time and frustration. The editlist would be restored automatically, with all the program data and playlist instructions being restored fully to disk.

CHAPTER 16

◆

Yesterday, Today, and Tomorrow

I'm sure you've heard the phrase, "Those were the good old days." I've usually found it to be a catch-all term that referred to a time in one's life that had a sense of great meaning, relevance, and all-around fun. Personally, I've never met a group of people who seem to carry that sense of relevance and fun with them more than music and audio students, enthusiasts, and professionals. Fortunately, I was born into that clan and have reaped the benefits all my life.

Music and audio industry professionals, by necessity, keep their noses to the grindstone. But market forces and personal visions cause them to keep one eye focused on future technologies—be they actual new developments, such as advances in digital processing or optical storage devices, or rediscovered ones that are decades old, such as the reemergence of tube technology and the reconditioning of older console designs that sound far too good to put out to pasture. Such is the paradox of music and audio technology, which leads me to the final task of addressing the people and technologies in the business of sound recording: yesterday, today, and tomorrow.

Yesterday

I have always looked at the history of music and sound technology and applied techniques with a sense of awe and wonder. I can't really explain it. Like so many in this industry, I have shivers run up my back when I see a wonderful, old mic

or an original tube compressor, or when I read about the Redd 37 (the console at Abbey Road that recorded much of the early Beatles albums, including *Sergeant Pepper*) or see an original Ampex 200 (the first commercially available professional tape machine). I experience the same sense of awe when I read about personal historical heroes, such as Mary C. Bell (see Figure 16.1), one of the first women sound engineers, and John T. Mullin (see Figure 16.2). John is the man who stumbled onto a couple of German Magnetophones at the end of WWII and brought them back to the United States. With the help of Alexander Poniatoff and Bing Crosby, John and his machines played a pivotal role in bringing the tape recorder into commercial existence (see Figure 16.3).

Figure 16.1. Mary C. Bell in NBC's dubbing room #1 (April, 1948)—working with three turntables to build a 15-minute program from pieces. (Courtesy of Mary C. Bell)

Fortunately, this new interest in technological history has manifested itself in the form of an almost frenzied lust for old gear or new toys based on decades-old technologies. At the forefront of this latest craze is the return to tube technology, particularly in the form of new and used tube condenser microphones, mic preamps, and signal processing gear (such as EQ, compressors, and limiters). Beyond the simple fact that many of the older mic designs were carefully crafted and have a particular sound that is very different from many of their modern counterparts, tube electronics have an inherently different sound than integrated-circuit-based systems. For starters, when overdriven to the point of clipping, instead of having sharply distorted waveform edges, a tube generally has a much smoother edge. This results in a greatly reduced odd-harmonic (square wave) distortion and generally yields a much less grating and "fatter" sound that is often sought after by musicians, producers, and engineers.

Figure 16.2. John T. Mullin (on the left) proudly displaying his two WWII vintage German Magnetophones, which were the first two tape-based recorders in the U.S. (Courtesy of John T. Mullin)

Figure 16.3. Early Ampex tape machines. (Courtesy of Mary C. Bell)

In addition to the resurgence of tubes, older recording and design techniques have slowly begun to come back into vogue. For example, stereo miking techniques aren't new; they were initially invented by Allen Dower Blumlein in the 1930s. The M/S (mid/side) stereo techniques, which spawned Ambisonic four-channel discrete/matrixed miking, are increasingly being used in studios, as well as for on-location audio-for-visual applications. Major music studio design has fortunately swung back towards the style of the '30s through the '50s of having a larger overall acoustic volume. This swing back in time is seen as a backlash against dry, lifeless acoustic recordings that often result from recording in a small, acoustically dead environment.

In short, we all benefit when those involved with music and recording technology can look to the past as well as to the future for their tools of the trade. A wealth of experience in design and application has been laid out for us. It's there for the taking; all we have to do is search it out and put it to good use.

Today

Every once in a while, major milestones in technological developments come along that affect almost every facet of technology. Such milestones have ushered us past the Edison and Berliner era of acoustic recordings, into the era of electrical recording and tape, on into the sonic environment of the multitrack recording studio (see Figure 16.4), and finally into the digital age and the age of the integrated circuit (IC).

Figure 16.4. Little Richard at the legendary L.A. Record Plant's Studio A (circa 1985) recording "It's a Matter of Time" for the Disney film Down and Out in Beverly Hills. *(Courtesy of the Record Plant Recording Studios, photo by Neil Ricklen)*

When you get right down to it, the IC has drastically changed the technology and techniques of present day recording by allowing electronic circuitry to be easily designed and mass produced at a fraction of what it would have cost a decade ago or even just a few years back. When combined

with advances in digital technology, we have seen the development of new technologies that have spawned new industries and ways by which music is produced. By combining cost-effective, yet powerful production consoles with modular digital multitracks, hard-disk digital workstations, MIDI synths/samplers, and other electronic instrument toys, music-related software, and digital signal processors (and that's just for starters), you've got the recipe for putting a project studio in your home, apartment, or place of business. (Presently, I'm writing this book from my studio in a mountainside log cabin.)

The project studio has made it possible for new types of musical forms (such as techno-dance, ambient, and a host of other styles) to emerge. The project studio also enables musicians and producers to record and produce their own projects at home and then take these masters into the studio where they can be professionally mixed down.

A concept envisioned by Peter Gotcher (President of Digidesign) is the concept of the "studio-in-a-box" (see Figure 16.5). This conceptual spark, which started a present-day Fortune 500 company, was to build a system that would offer the power of professional hard-disk-based audio at a low price that most could afford (especially when you compare the price to earlier existing systems that started at about $100,000). His goal was to create an integrated system that would link together the many facets that go into audio and audio-for-visual production, via a PC. Years later, this concept has transformed the very nature by which audio production is carried out.

Figure 16.5. ProStation furniture fitted with an example of Peter Gotcher's "studio-in-a-box" concept—the digital audio workstation. (Courtesy of Omnirax)

Another present-day mover and shaker is the CD-ROM. CD-ROM systems can be easily and affordably added to a PC platform to provide you with interactive graphics, audio, and text. Its proliferation has also brought about a new industry that continues to grow and mature.

One of the numerous occurrences that has worked to change and shape the audio industry is the proliferation and increased distribution of knowledge. Through the existence of quality books, trade magazines, university programs, workshops, and Internet nodes, a wide base of information in trade-related subjects is now being distributed to and understood by a greater number of professionals and aspiring individuals than ever before. These resources often provide a strong foundation for on-the-job training and general all-around experience. I urge you to jump in and read, read, read. I promise, the increased knowledge will be well worth the time and effort. This book's appendix includes a number of books, magazines, and educational resources to help you get started. So dive in and have fun!

Tomorrow

By and large, I'll leave tomorrow to follow its own eventuality. However, if you'll permit me to look into my crystal ball (I refuse to disclose its accuracy specs), there are a few points that I'd like to bring up. On the digital front, of course, there will be tons of new whiz bangs that will set all our hearts a flutter. However, I've noticed that the core of technology has begun to settle down. From here on out, we can expect "bigger, better, more" rather than the big-time shakeouts we saw during the '80s and up to the mid '90s. One tech thing is gnawing at me, however: HD-CD (high-density compact disc). The main thrust of this medium, beyond simply having more memory, is to increase data throughput so that we can attain full-motion, full-screen, high-definition video on a CD. Think of the implications: video stores can have more shelf space, production costs will go down (of course, that probably won't be passed on to us), and we'll have two CD-Video players in every garage. So what's the big deal? Well, where CD players go, CD-Video recorders can't be too far behind. Once the industry has gone through its copy code fits (again), we'll have high-quality optical recorders that might be able to be plugged into our computers or used as digital audio mastering media (I can see those DAT tapes rusting away already). Who knows what will really happen, but it's an interesting scenario to ponder.

And, finally, the information highway. Right now it seems to me to be the information dirt path. I won't even venture any guesses, except to say that I personally won't pay money to download a video or music project, unless I can hold the product in my hot little hands. Hey, maybe I could download it to the CD-Video recorder as a permanently protected file. Sending video faxes over the Net or downloading music tracks from my collaborators in Toronto or Nashville would be pretty cool, though. Let's see what's out there: Number One ... Engage!

Figure 16.6. *The record store of tomorrow? Fact or fiction?*

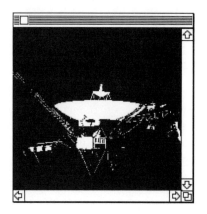

APPENDIX

◆

Continuing Education

As the recording and communication industries place an ever-increasing emphasis on technology, education must play a greater role in the acquiring of basic skills. Education can take many forms, ranging from classes taken in a formal educational setting to keeping abreast of industry directions by reading the many industry magazines. From my personal point of view, one of the most important things you can do to stay ahead in this fast-paced and ever-changing industry is to keep reading: it will definitely pay off for you in the long run.

The following resources are recommended for anyone wishing to further his or her education in recording technology.

Books

Many of these publications are available from *The Mix Bookshelf Catalog,* 6400 Hollis Street, Suite #12, Emeryville, CA 94608, (800) 233-9604 or (415) 653-3307. I urge you to call them to request this catalog. It contains the most complete listing of books and industry-related resources available.

Anderton, Craig. *Home Recording for Musicians.* New York: GPI Publications, Music Sales Corporation, 1978.

Anderton, Craig. *MIDI for Musicians.* New York: GPI Publications, Music Sales Corporation, 1986.

Balou, Glen, ed. *Handbook for Sound Engineers: The New Audio Cyclopedia,* 2nd ed. Indianapolis: Howard W. Sams & Company, 1991.

Bartlett, Bruce. *Practical Recording Techniques.* Indianapolis: Howard W. Sams & Company, 1992.

Borwick, John. *Sound Recording Practice.* 3d ed. London: Oxford University Press, 1988.

Davis, Don and Carolyn. *Sound System Engineering.* 2d ed. Indianapolis: Howard W. Sams & Company, 1987.

Eargle, John. *Sound Recording.* 2d ed. New York: Van Nostrand Reinhold Company, 1986.

Galluccio, Greg. *Project Studio Blueprint.* Indianapolis: Howard W. Sams & Company, 1992.

Giddings, Phillip. *Audio Systems: Design and Installation.* Indianapolis: Howard W. Sams & Company, 1990.

Huber, David. *Audio Production Techniques for Video.* Boston/London: Focal Press, 1987.

Huber, David. *Hard Disk Recording for Musicians.* New York: Music Sales Corporation, 1995.

Huber, David. *Microphone Manual: Design and Application.* Boston/London: Focal Press, 1988.

Huber, David. *MIDI Manual.* Indianapolis: Howard W. Sams & Company, 1991.

Martin, George. *All You Need Is Ears.* New York: St. Martin's Press, 1979.

Molenda, Michael. *Making the Ultimate Demo.* Emeryville, CA: Mix Books, 1993.

Petersen, George. *Modular Digital Multitracks: The Power User's Guide.* Emeryville, CA: Mix Books, 1994.

Pohlmann, Ken. *Principles of Digital Audio.* 2d ed. Indianapolis: Howard W. Sams & Company, 1989.

Rappaport, Diane Sward. *How to Make and Sell Your Own Record.* Englewood Cliffs, NJ: Prentice Hall, 1992.

Woram, John. *Recording Studio Technology.* Indianapolis: Howard W. Sams & Company, 1989.

Magazines

db, 203 Commack Road, Suite 1010, Commack, NY 11725.

Electronic Musician, 6400 Hollis Street, Suite 12, Emeryville, CA 94608; (800) 233-9604 or (510) 653-3307 (outside CA); fax (510) 653-5142.

EQ Magazine, 2 Park Avenue, Suite 1820, New York, NY 10016; (212) 213-3444; fax (212) 213-3484.

Keyboard, 411 Borel Avenue, San Mateo, CA 94402; (415) 358-9500; fax (415) 358-9527.

Mix Magazine, 6400 Hollis Street, Suite 12, Emeryville, CA 94608; (800) 233-9604 or (510) 653-3307 (outside CA); fax (510) 653-5142.

Post (Audio and Video Postproduction), 25 Willowdale Avenue, Port Washington, NY 11050.

Pro Sound News, 2 Park Avenue, Suite 1820, New York, NY 10016; (212) 213-3444; fax (212) 213-3484.

Pro Sound News (European Edition), Spotlight Publications Ltd., 8th Floor, Ludgate House, 245 Blackfriars Road, London SE1 9UR, England, (UK); 071-620-3636; fax 071-401-8036.

Studio Sound, Spotlight Publications Ltd., 8th Floor, Ludgate House, 245 Blackfriars Road, London SE1 9UR, England, (UK); 071-620-3636; fax 071-401-8036. In the US: *Studio Sound Magazine*, 2 Park Avenue, 18th Floor, New York, NY 10016.

Organizations

Hearing Education and Awareness for Rockers (HEAR)
P.O. Box 460847, San Francisco, CA 94146; (415) 441-9081; fax (415) 476-7113; TTY (415) 476-7600.

International MIDI Association
5316 West 57th Street, Los Angeles, CA 90056; (213)-649-MIDI.

National Association for Music Therapy, Inc.
8455 Colesville Road, Suite 930, Silver Springs, MD 20910; (301) 589-3300; fax (301) 589-5175.

National Academy of Recording Arts & Sciences, Inc. (NARAS)
3402 Pico Boulevard, Santa Monica, CA 90405; (310) 392-3777; fax (310) 392-2778.

Recording Industry Environmental Task Force
University of Massachusetts Lowell, Center for Recording Arts, Technology and Industry; College of Fine Arts, One University Avenue, Lowell, Massachusetts 01854.

Recording Schools

In the third edition of *Modern Recording Techniques*, I included a list of many of the top schools, universities, and workshops in recording and music technology. In recent years, this list has become too large to provide here. The list alone could constitute an entire book. The following resources can lead you to schools and workshops relevant to the recording and music industries:

◆ Each year, the July issue of *Mix Magazine* offers a comprehensive directory of schools, workshops, and universities throughout the US and the world.

◆ A complete reference guide to audio education can be found in the book *New Ears Productions: A Guide to Education in the Audio and Recording Sciences,* 2d ed. Syracuse, NY: New Ears Productions, 1992.

G

Glossary

absorption The dissipation of sound energy at a surface as the sound changes into heat. The absorption of acoustic energy effectively is the inverse of reflection. Whenever sound strikes a material, the amount of acoustic energy absorbed (often in the form of physical heat dissipation), relative to the amount reflected, can be expressed as a simple ratio known as the *absorption coefficient*.

AES/EBU (Audio Engineering Society/European Broadcast Union) protocol A professional transmission protocol that conveys two channels of interleaved digital audio data through a single, two-conductor XLR cable (for example, a standard microphone cable).

alignment The adjustment of an analog tape machine's tape head and electronic circuitry to standardize playback and record frequency response and signal levels within industry accepted standards for reasons of compatibility.

alignment tape A reference reproduction tape used for aligning analog tape machines.

amplification The process by which a signal level is increased by a device according to a specific input/output ratio.

amplitude The distance above or below the centerline of a signal's waveform. The greater the distance or displacement from the centerline, the more intense the pressure variation, electrical signal, or physical displacement within a medium.

analog-to-digital (A/D) converter   A device that converts analog signals into digital form.

assistant engineer   The person normally responsible for microphone and headphone setup, operation of the tape machines, session breakdowns, and, in certain cases, for positioning rough mixes on the console for the engineer. Often, larger studios train future staff engineers by having them work as assistant engineers.

attack   The initial transient or first part of the envelope of a signal. The beginning of a note.

attenuate   To reduce the signal level.

autolocator   A feature that enables a specific cue point location on a tape transport to be stored into and recalled from memory. The autolocator can then shuttle the tape to a time point entered by the operator.

automated mixdown   Enables the console to remember and re-create any settings or changes (regarding level and other mix-related functions) made by the engineer, while allowing continual improvements until the desired final mix quality is achieved.

auxiliary sends   Provide the overall effects or monitor sends of a console. These sections (of which up to eight or more sends can be provided on a single input module) are used to create separate and controlled submixes of any (or all) of the input signals to a mono or stereo output.

balance   The relative level of various instruments within a mix.

balanced line   A cable having two conductors and a ground connection and often surrounded by a shield. With respect to ground, the conductors are at equal potential but opposite polarity. These lines are often used in professional settings to reduce or eliminate induced noise and interference from external electromagnetic sources.

bass trap   Used to reduce low-frequency buildup at specific frequencies in a room. These low-frequency attenuation devices are available in a number of design types, such as the quarter-wavelength trap, the pressure zone trap, the functional trap, and the Helmholtz resonator trap.

bias   An ultrasonic signal mixed with the input signal at the record head of an analog tape recorder to reduce distortion.

bidirectional (figure-of-eight) microphone   A mic sensitive to sounds arriving from on-axis (front) and 180° off-axis (rear), with its maximum rejection occurring at both sides.

bin-loop high-speed duplication   High-speed duplication in a process in which the duplication takes place without the duplicated tape being housed in cassette shells. The tape is recorded on a reel-to-reel machine, which provides higher quality and better tape handling at high speeds.

black burst generator   A device that produces an extremely stable timing reference. The function of this signal is to precisely synchronize the video frames and time code addresses received or transmitted by every video related device in a production facility to a specific clocking frequency. This process ensures that the frame and address leading edges occur at exactly the same instant in time.

Blumlein pickup   The placement of two coincident bidirectional mics crossed at 90° and aiming 45° left and right of center.

board   American term for console. (*See also*: **recording console**)

bouncing tracks   Commonly used to mix entire performances onto one track, a stereo pair of tracks, or several tracks. This makes the final mixdown easier by grouping instruments together onto one or more tracks. It also opens up needed tape tracks by bouncing similar instrument groups down to one or a stereo pair of tracks, thereby freeing the originally recorded tape tracks for more overdubs.

bus   A common signal path that routes a signal, throughout a console or connected network, from one or more signal sources to one or more signal destinations.

cardioid microphone   A common mic pickup pattern designed to attenuate signals arriving 180° off-axis while picking up fully those sounds that arrive at the front (on-axis).

CD–ROM (compact disc–read-only memory)   A disc capable of holding as much as 680 MB (at the time of this writing) of any type of computer-based data, including graphics, digital audio, MIDI, text, and raw data. Unlike the CD-Audio disc, a CD-ROM isn't tied to a specific data format; the manufacturer or programmer can specify what is contained on the disc.

compansion   The process in which an incoming signal is compressed before it is recorded on tape; then, upon reproduction, the signal is reciprocally expanded back to its original dynamic range, with a resulting reduction in background tape noise.

composite tracks   The result of combining the best "takes" from a number of performances that exist on different tracks onto a final single or stereo pair of tracks. This is done to "open up" tracks for further overdubbing or to ease the number of mix movements that must be performed during mixdown.

compressor   In effect, an automatic fader. When the input signal exceeds a predetermined level, called the *threshold*, the gain is reduced by the compressor and the signal is attenuated.

compression ratio (slope)   The ratio of signal dynamic range between the compressor input and output (such as 2:1, 4:1, and 8:1) above the device's set threshold point.

condenser microphone   A microphone that operates on an electrostatic principle rather than on the electromagnetic principle used in dynamic and ribbon mics. The head, or capsule, of the mic consists of two very thin plates—one movable and one fixed. When the distance between these plates decreases, the capacitance increases; when the distance increases, the capacitance decreases. According to the equation:

$$Q = CV$$

where

Q is the charge, in coulombs,
C is the capacitance, in farads,
V is the voltage, in volts.

If the charge (Q) is constant and sound pressure moves the diaphragm-changing capacitance (C), the voltage (V) across these plates must change in direct proportion to the acoustic signal.

control room   In a recording studio, this room serves several purposes: it is acoustically isolated from the sounds produced in the studio and surrounding vicinities, it is optimized to act as a critical listening environment using critically balanced and placed monitor speakers, and it houses the majority of the studio's recording, control, and effects equipment.

crosstalk   The unwanted leakage of a signal from one channel or track onto another.

cue send   The auxiliary send used for the musicians' headphone mix.

cycle   The period in which an acoustic or electrical signal varies over one completed excursion of a wave, which is plotted over the 360° axis of a circle.

digital audio tape (DAT)   A compact, dedicated PCM digital audio recorder that combines rotary head technology and PCM digital technology to create a professional recorder with a wide dynamic range, low distortion, and immeasurable wow and flutter.

decibel (dB)   A unit of audio measurement of sound-pressure level (SPL), signal level, and changes or differences in signal level. The decibel is a logarithmic (log) mathematical function that reduces large numeric values into smaller, more manageable numbers. Decibel is calculated as being 10 times the log of the ratio of two powers and 20 times the log of the ratio of two voltages.

> **dBm**: Decibels referenced to 1 milliwatt.
> **dBu** or **dBv**: Decibels referenced to 0.775 volt. (dBu is preferred.)
> **dBV**: Decibels referenced to 1 volt.

de-esser   A frequency-dependent compressor used to reduce excessive sibilance ("sss," "sh," and "ch" sounds).

degaussing (demagnetizing)   The process by which small amounts of residual magnetism are eradicated from an analog magnetic tape head. It's often wise to degauss a magnetic tape head after 10 hours of continuous operation.

destructive editing   When the audio data recorded on a hard disk is altered and rewritten to disk in such a way that it can't be recovered in its original form.

diffraction of sound   Sound inherently has the capability to bend around a physical acoustic barrier or go through a hole in the barrier. It bends around an object in a manner that will reconstruct the original waveform in both frequency and amplitude.

digital-to-analog (D/A) converter   A device that converts digital signals into analog form.

digital console   A console in which analog input signals are converted directly into digital data (or are directly inserted into the console's chain as digital data) and are distributed and processed entirely in the digital domain.

digitally controlled analog consoles   A console that distributes and processes the signal path in analog form with control over all console parameters being carried out in the digital domain.

In most cases, this means that the console's control surface (containing all the knobs, faders, assignment buttons, and so on) will output its control parameters as digital signals.

digital signal processing (DSP)   The processing of a signal in the digital domain in such a way as to follow basic binary computational rules according to a specialized program algorithm. This algorithm is used to alter the numeric values of sampled audio in a predictable way.

direct injection (D.I.) box   A box for converting high-level, high-impedance instrument signals to low-impedance microphone-level signals for direct injection into a console mic input.

directional (polar) response   The variations in microphone sensitivity versus the angle of sound incidence, plotted on a polar graph. The sensitivity on-axis is called 0 dB, and the sensitivities at other angles are relative to that. This chart, known as the *polar response* or *polar pattern* of a microphone, shows microphone output with respect to direction and frequency over 360°.

dither   Adding to a signal small amounts of white noise that are less than the least-significant bit (that is, less than a single quantization step), thereby increasing signal-to-error and reducing distortion.

dry signal   An unprocessed signal that doesn't contain reverb or echoes.

dynamic microphone   A microphone that operates by electromagnetic induction to generate an output signal. When an electrically conductive coil of wire or ribbon is made to cut across the flux lines of a magnetic field, a current of specific magnitude and direction is generated within that coil or ribbon.

edit decision list (EDL)   A sequential editlist containing permanent SMPTE time code and related edit information.

effects send   An auxiliary send feeding an effects device.

electret-condenser microphone   A condenser microphone that has the polarizing charge stored permanently in the diaphragm or on the backplate. Because of this electrostatic charge, no external powering is required to charge the diaphragm or backplate.

engineer   The person responsible for expressing the artist's music and the producer's concepts through the medium of recording. This job is actually an art form because both music and recording are subjective in nature and rely on the tastes and experiences of those involved.

equalizer   A frequency-dependent amplifier that enables a recording or mix engineer to control the relative amplitude of various frequencies in the audible bandwidth. Put another way, the equalizer lets you exercise tonal control over the harmonic content or timbre of a recorded sound.

expander   A device that increases the dynamic range of a signal.

fade   A slow change in volume—up from silence or down to silence—accomplished manually or by calculation in a DAW or hard-disk recorder. The fading in or fading out of a region is a DSP function carried out by calculating the soundfile's relative amplitude over a defined duration.

fader   A linear attenuation device or linear volume control.

feedback   The returning of a loudspeaker signal back into a microphone feeding that loudspeaker. Excessive feedback results in unpleasant, screaming buildups at particular frequencies, which the English call *howlround*.

flanging   A process whereby a delayed signal is combined with itself undelayed. The delay is varied to create continual changes in timbre. Also called a *comb-filter effect*.

Fletcher-Munson curves   A group of curves that plot the subjective sensitivity of humans to various frequencies at different sound-pressure levels.

flutter   A fast, periodic variation in a tape transport's speed.

frequency   The rate at which an acoustic generator, electrical signal, or vibrating mass repeats a cycle of positive- and negative-going amplitude. The number of cycles that occurs over the period of one second is measured in hertz (Hz). Often, the perceived range of human hearing is from 20 Hz to 18,000 Hz.

gain   Amount of amplification in dB.

gate   A device that fully attenuates a signal which falls below a predetermined threshold level. Often used to reduce noise or extraneous pickup leakage.

ground loop   A condition that exists in an improper grounding situation, whereby a DC current differential exists between one signal path and another, resulting in 60 Hz or 50 Hz (European) hum.

hard-disk recorder   A system that uses a computer hard disk to record, edit, and reproduce digital data.

harmonic content   A factor that allows us to differentiate between instruments. The presence of several different frequencies within a complex sound wave, in addition to its fundamental note. The frequencies present in a sound, other than the fundamental, are called *partials*. Partials higher than the fundamental frequency are called *upper partials* or *overtones*. These overtones play an important part in determining the sonic character of an instrument. Harmonics are integral multiples of the fundamental frequency.

hertz   Frequency measurement unit (cycles/second).

hiss   Broadband tape or amplifier noise.

impedance   The opposition of a circuit to the flow of an alternating current.

input (I/O) module   The vertical array of controls on a console that relates to a specific input signal.

iso-room/iso-booth   Isolation rooms and the smaller iso-booth are acoustically sealed areas built into and easily accessible from the main studio area. These areas provide improved separation between loud and soft instruments, vocals, and so on.

jitter   A time-base error caused by varying time delays in a digital audio circuit path.

leader tape   A paper tape that can be spliced into analog audio tape for the purpose of inserting silent spaces, visual separation, and identification for various songs or selections.

leakage   The "spilling" or "bleeding" of sound from one instrument onto another instrument's microphone.

limiter   A device used to keep signal peaks from exceeding a certain level in order to prevent the clipping or distortion of amplifier signals, recorded signals on tape or disc, broadcast transmission signals, and so on.

line level   A signal level that is referenced to either +4 dBm (pro) or −10 dBV (semi-pro/consumer). Devices other than mics, speakers, and power-amplifier outputs operate at these levels.

maintenance engineer   The person who ensures that the equipment in the studio is maintained in top condition, regularly aligned, and repaired when necessary.

masking   The phenomenon by which loud sounds prevent the ear from hearing softer sounds. The greatest masking effect occurs when the frequency of the sound and the frequency of the masking noise are close to each other.

mastering   The processing and transferring of a final, sequenced audio tape to a medium for duplication.

MIDI (Musical Instrument Digital Interface)   A digital communications language and compatible hardware specification that allows multiple electronic instruments, performance controllers, computers, and other related devices to communicate with one another within a connected network.

MIDI interface   A digital hardware device used to translate MIDI's serial message data into a structure that can be understood by and communicated to a personal computer's internal operating system.

MIDI machine control (MMC)   A standardized series of transport-related commands that are transmitted over standard MIDI lines from one controller to one or more other MMC-capable devices within a connected system.

MIDI sample dump standard (SDS)   A protocol developed and ratified by the MIDI Manufacturers Association that enables the transmission of sampled digital audio and loop information from one sampling device to another.

MIDI time code (MTC)   Provides a cost-effective and easily implemented means for translating SMPTE time code throughout a MIDI chain as a stream of MIDI messages.

mixdown   The process in which the separate audio tracks of a multitrack tape machine are combined, balanced, and routed through the recording console. At this point, volume, tone, special effects, and spatial positioning can be artistically set by the engineer to create a stereo or four-channel mix that is then recorded to a master recording device such as a DAT recorder.

modular digital multitrack recorders (MDMS)   Small, cost-effective multitrack digital audio recorders capable of recording eight tracks of digital audio onto readily available videotape cassettes. They are called "modular" because they can be linked together in a proprietary sync fashion, with a theoretical maximum limit of up to 128 tracks!

moving-coil microphone   Generally consists of a Mylar diaphragm of roughly 0.35 mil thickness attached to a finely wrapped coil of wire that is precisely suspended within a high-level magnetic field. When an acoustic pressure wave hits the face of this diaphragm, the attached voice coil is displaced in proportion to the amplitude and frequency of the wave, causing the coil to cut across the lines of magnetic flux supplied by a permanent magnet. In so doing, an analogous electrical signal (of a specific magnitude and direction) is generated across the voice coil leads.

multimedia   A field that encompasses the mixed media of text, graphics, MIDI, and digital audio sound for the personal computer. The production and distribution of educational, entertainment, and data storage for PC users has quickly established itself as an emerging market with an enormous potential for growth.

multitimbral   The capability of an electronic musical instrument to respond to and output multiple voice patches at one time.

multitrack recording   A process that provides an added degree of production flexibility to the recording process by enabling multiple sound sources to be recorded to and played back from isolated tracks that are synchronously locked in time. The recorded tracks are isolated from one another, so any number of sound sources can be recorded and rerecorded without affecting other tracks.

mute   To turn off or silence an input signal, tape track, and so on.

nearfield monitoring   Monitoring with a small "bookshelf" style speaker on or slightly behind the meter bridge of a console, close to the engineer and producer. This technique ensures that a greater portion of the direct sound mix is heard relative to the room's acoustics.

noise gate   A device that acts as an infinite expander, allowing a signal above the selected threshold to be passed through to the output at unity gain and without dynamic processing. When the input signal falls below this threshold level, the device effectively shuts down the signal by applying full attenuation to the output.

nondestructive editing   Editing a hard-disk soundfile by moving pointers, without altering in any way the digital audio data originally recorded to disk.

normalizing   A specialized gain-related process that makes the best use of a digital system's dynamic range by automatically determining the amount of gain required to increase the level of the highest amplitude signal to its full-scale amplitude value, and then increasing the level of the selected region or entire file by this gain ratio.

Nyquist Theorem   A theory which states that in order to digitally encode the entire frequency bandwidth, the selected sample rate must be at least twice as high as the highest desired recorded frequency (sample rate $2 \times$ highest frequency).

omnidirectional microphone   A mic that outputs signals received from all incident angles at the same (or nearly equal) level.

open tracks   Available tracks on a multitrack recorder.

operational amplifier (*op amp*)   A stable high-gain, high-bandwidth amplifier with a high input impedance and a low output impedance. These qualities enable it to be used as a basic building block for a wide variety of audio and video applications.

outboard equipment   Signal processing and other devices external to the mixing console.

overdubbing   Enables one or more of the previously recorded tracks to be monitored while simultaneously recording one or more signals onto other tracks. This process can be repeated until the song or soundtrack has been built up. If a mistake is made, it generally is a simple matter to recue the tape to the desired starting point and repeat the process until you have the best take on tape.

overload   The distortion that occurs when an applied signal exceeds a system's maximum input level.

oversampling   A process commonly used in professional and consumer digital audio systems to improve Nyquist filter characteristics, thereby reducing intermodulation and other forms of distortion. This process effectively multiplies the sampling rate by a specified factor—commonly ranging from between 12 to 128 times the original rate. This increased sample rate likewise results in a much wider frequency bandwidth, so a simple, single-order filter can be used to cut off the frequencies above the Nyquist limit.

pan pot   A dual-potentiometer that can place a single signal source at any point between the left and right channels of a stereo image, or the left/right, front/back quadrants of a surround sound image.

patch bay   A panel that, under the best of conditions, contains a jack corresponding to the input and output of every discrete component or group of wires in the control room. It acts as a central point where console signal paths, pieces of audio gear, and other devices can be connected.

patch chords   Short cables used for routing signals through a patch bay.

peak amplitude   The maximum instantaneous amplitude of a signal.

peaking filter   Used to create a peak- or bell-shaped equalization curve in the frequency response that can be either boosted or cut at a selected center frequency.

phantom power   Power for a condenser mic that comes directly from the console through balanced mic cables by supplying a positive DC supply voltage of +48V (usually) to both conductors (pins 2 & 3) with respect to pin 1. This voltage is distributed through identical value resistors so that no differential exists between the two leads; therefore, the voltage is electrically invisible to the alternating audio signal. The DC circuit is completed by connecting the negative side of the supply to the cable's grounding shield.

phase   The degree of progression in the cycle of a wave, where one complete cycle is 360°. Waveforms can be added by summing their signed amplitudes at each instant of time. A cycle can begin at any point on a waveform, so it's possible for two waveforms having the same or different frequency and peak levels to have different amplitudes at any one point in time. These waves are said to be "out of phase" with respect to each other. Phase is measured in degrees of a cycle (divided into 360°) and will result in audible variations of a combined signal's amplitude and overall frequency response.

phase shift   The difference in degrees of phase angle between corresponding points on two waves.

ping-ponging   See **bouncing tracks**.

pitch control   A control that varies the speed of a tape transport or the sample rate of a digital audio device, changing the pitch of the reproduced signal.

pitch shifting   Used to vary the pitch of a program either upward or downward so as to transpose the relative pitch of an audio source.

playlist   A sequential list of soundfile regions that can be played as a single, continuous program or sequentially triggered at specific time code addresses.

polar pattern   A polar graph of the sensitivity of a microphone at all angles of sound incidence relative to the sensitivity on-axis.

polyphonic   The capability of an electronic musical instrument to output multiple notes at one time.

pop filter   A foam or wire screen placed between the mic and the instrument or performer to reduce wind and breath blasts.

potentiometer (pot)   A rotary gain, pan, or other type of continuously variable signal control.

pre-fade listen (PFL)   See **solo**.

print-through   The transfer of a recorded signal from one layer of magnetic tape to an adjacent layer by means of magnetic induction, which gives rise to an audible false signal (pre-echo or post-echo) on playback.

producer   The person who handles the scheduling, budgetary, and coordination aspects of a recording project. It is the producer's responsibility to create the best recorded performance and final product possible. In effect, a producer is often chosen for his or her ability to understand the many phases of the overall process of creating a final recorded project, from the standpoints of business, musical performance, and creative insight into recording technology.

project studio   A high-quality MIDI and/or recording facility in a home, apartment, or personal place of business that is used to record the owner's own projects rather than outside projects.

proximity effect   A bass boost that occurs with single-D directional mics at close working distances.

pulse code modulation (PCM)   The most common encoding scheme for storing digital data onto a medium with a maximum degree of data density.

punch-in/punch-out   The entering into and out of record mode on a track that contains existing program material for the purpose of correcting or erasing an unwanted segment.

quantization   The amplitude component of the digital sampling process. In an A/D converter, the process of generating a binary number (made of 1s and 0s) that represents the voltage of the analog waveform at the instant it is measured or sampled.

recording console   A device that enables the engineer to mix and control most (if not all) of the device input and output signals that can be found in the studio. The console's basic function is to allow for any combination of mixing (control over relative amplitude and signal blending between channels), spatial positioning (left/right, as well as possibly front/back), routing (the capability to send any input from a source to a signal destination), and switching for the multitude of audio input/output signals that are commonly encountered within an audio production facility. The recording console can be thought of as the recording engineer/producer's color palette.

recording studio   One or more acoustic environments specially designed and tuned for the specific purpose of getting the best sound possible onto tape when using a microphone pickup.

release   The final portion of a note's envelope, which falls from the sustain signal level to silence.

release time   Once dynamic processing has begun, the time taken for a dynamic range changer (such as a compressor, limiter, or expander) to return the signal to 63% of its original (unprocessed) level.

resistance   The opposition to the flow of DC current in a wire or circuit.

reverberation (reverb)   The persistence of a signal, in the form of reflected waves in an acoustic space, after the original sound has ceased. These closely spaced and random multiple echoes result in perceptible cues as to size and surface materials of a space and add to the perceived warmth and depth of recorded sound. Reverb plays an extremely important role, both in the enhancement of our perception of music and in proper studio design. The reverberated signal itself can be broken into three components: direct signal, early reflections, and reverberation.

reverb time (RT_{60})   Measurement unit of a room's reverberation. The time taken for a reverberated signal (once the initial signal has stopped) to reduce in level by 60 dB.

ribbon microphone   A microphone that uses a diaphragm of extremely thin, aluminum ribbon suspended in a strong field of magnetic flux. As sound-pressure variations displace the metal diaphragm in accordance with air-particle velocity, the ribbon cuts across the magnetic lines of flux. This induces a current in the ribbon of proportional amplitude and frequency to the acoustic waveform.

S/PDIF (Sony/Phillips Digital Interface)   This digital protocol was adopted for the purpose of transmitting digital audio between consumer digital devices in a manner that is similar to but not identical in data structure to its professional AES/EBU counterpart.

safety copy   A high-quality analog or digital copy of a production tape or final master recording. *Safetys* should be carefully stored under moderate temperature and humidity conditions.

sampler   A device capable of recording, musically transposing, processing, and reproducing segments of digitized audio to and from RAM.

samplefile   A computer file that contains sampled audio data.

sampling   The process of taking periodic samples of an audio waveform and transforming these sampled signal levels into a representative stream of binary words that can be manipulated or stored for later reproduction.

scratch vocal   A rough vocal track recorded live along with the initial rhythm instruments to help the basic tracks keep in the "groove" of the song. Final vocals can be rerecorded later during overdubs.

SCMS (Serial Copy Management System—pronounced *scums*)   A system implemented in many consumer digital devices in order to prohibit the unauthorized copying of digital audio at 44.1 kHz (standard CD sample rate). With the SCMS, you can often make a digital copy of a commercial DAT or CD, but you cannot make a copy from that copy.

SCSI (small computer system interface—pronounced *scuzzy*) sample dump format   A bidirectional communications bus used by many PCs and digital devices to exchange data between systems at high speeds.

sensitivity rating   The output level (in volts) that a microphone will produce given a specific and standardized input signal (rated in dB SPL). This specification implies the degree of amplification required to raise the microphone signal to line level (−10 dBV or +4 dBm).

sequence editing   The process of editing songs into a final order (either in the analog or digital domain). The end result of this process is a completed project ready for mastering into a finished, saleable product.

sequencer   A digitally based device used to record, edit, and output performance-related MIDI data.

servo-driven fader   A resistive attenuator that is driven automatically by a servo motor interface. During the playback of an automated mix, the faders will move on their own, in accordance with the requirements of the mix.

shelving filter   A rise or drop in frequency response at a selected frequency that tapers off to a preset level and continues at this level to the end of the audio spectrum.

shock mount   A suspension system that isolates a microphone from stand- and floor-borne noises. A shock mount built into a mic reduces handling noise.

slate   A verbal identification of the song, take, and other identification on the original master tape tracks.

SMPTE time code A standard method for synchronously interlocking audio, video, and film transports. The use of time code allows identification of an exact position on a magnetic tape by assigning a digital address to each specified position. This address code cannot slip and always retains its original location, allowing continuous monitoring of tape position to an accuracy of roughly 1/30th of a second.

SMPTE-to-MIDI converter Used to read SMPTE time code and convert it into such MIDI-based sync protocols as MIDI time code, Direct Time Lock, or song position pointer.

solo A monitor function that lets the engineer hear a single instrument or group of instruments without affecting the studio's headphone monitor mix, recorded tracks, or mixdown signal.

sound-pressure waves Sound waves generated by a vibrating body in contact with the air, such as an instrument or loudspeaker. Sound arrives at the ear in the form of a periodic variation in atmospheric pressure. The atmospheric pressure is proportional to the number of air molecules in the area being measured.

soundfile A computer file that contains audio data. Rather than being reproduced from a sampler, these files are often played back from a hard-disk-based system.

spatial perception of direction The capability of two ears to localize a sound source in an acoustic space. Although one ear is not able to discern the direction from which a sound originates, two ears can. This is called spatial or binaural localization.

speaker polarity (phase) Speakers are said to be electrically in-phase whenever the same signal applied to both speakers will cause their cones to move in the same direction (either positively or negatively). If they are wired out-of-phase, one speaker cone will move in one direction while the other will move in the opposite direction.

splice To join two pieces of analog magnetic tape using a special adhesive, nonbleeding tape called *splicing tape*.

standing wave An apparently stationary waveform created by multiple reflections between opposing room surfaces. At certain points along the standing wave, the direct and reflected waves cancel each other, and at other points, the waves add together or reinforce each other. Standing waves cause boomy sounding peaks in a room's low-frequency response.

studio management Businesspeople who are knowledgeable about the inner workings of the music studio, music business, and people. Running a studio requires the constant attention of a studio manager, booking department (to keep track of most of the details relating to studio use, billing, and possibly marketing), and competent secretarial staff.

submix A grouped set of signals that can be varied in overall level from a single control or set of controls.

sweetening Overdubbing strings, horns, chorus, and sometimes percussion to give added impact to a recorded production.

synthesizer   An electronic musical instrument that uses multiple sound generators to create complex waveforms that synthesize a unique sound character.

synchronization (sync)   The locking of relative transport or playback speeds of various devices to allow them to work together as a single, integrated system.

sync (repro)   The use of an analog tape machine's record head to play back tracks during the overdub process in order to synchronize with the current tracks being recorded.

system-exclusive (Sys-Ex) messages   Messages that enable MIDI manufacturers, programmers, and designers to communicate customized information between MIDI devices. These messages communicate device-specific data of an unrestricted length.

take sheet   A written sheet that notes the position of each take on a tape. Comments are written on the take sheet to describe the producer's opinion of the performance, as well as whether it is a complete take, an incomplete take, or a false start.

time code   A standard encoding scheme (hours:minutes:seconds:frames) for encoding time-stamped address information. Time code is used for address location, triggering, and synchronization between various analog, video, digital audio, and other time-based media.

track sheet   A sheet that indicates what instrument (or instruments) is on each track of a multitrack tape. The track sheet should always be stored in the box with the reel.

transducer   Any device that changes one form of energy into another, corresponding form of energy. For example, a microphone is an example of a transducer because it converts sound waves into an electrical signal.

three-to-one (3:1) rule   A miking guideline that states that leakage and phase cancellations can be reduced by keeping the distance between mics at least three times the distance the mikes are placed from their respective sound sources.

threshold of feeling   An SPL rating that will cause discomfort in a listener 50% of the time and which occurs at a level of about 118 dB SPL between 200 Hz and 10 kHz.

threshold of hearing   The quietest sound humans can hear: 0 dB SPL. A convenient pressure-level reference that constitutes the minimum sound pressure that produces the phenomenon of hearing in most people. It is equal to 0.0002 microbar. One microbar is equal to one-millionth of normal atmospheric pressure, so it's apparent that the ear is extremely sensitive. The threshold of hearing is defined as the SPL for a specific frequency at which the average person can hear only 50% of the time.

threshold of pain   An SPL rating that causes pain in a listener 50% of the time. It corresponds to an SPL of 140 dB in the range between 200 Hz and 10 kHz.

transient response   The measure of how quickly a mic diaphragm, speaker, or physical mass reacts to an input waveform.

triggered event digital audio   The automated search and firing on command of a specific file or segment of digital audio data.

unbalanced line   A cable having only one conductor plus a surrounding shield, in which the shield is at ground potential. The conductor and the shield carry the signal.

velocity of sound   The speed at which sound waves travel through a medium. At 70°F, the speed of sound waves in air is approximately 1130 feet per second (ft/sec) or 344 meters per second (m/sec).

wow   A slow, periodic variation in a tape transport's speed.

voltage controlled amplifier (VCA)   An amplifier in which audio level is a function of a DC voltage (generally ranging from 0 to 5 V) applied to the control input of the device. As the control voltage is increased, the analog signal is proportionately attenuated. Thus, an external voltage is used to change the audio signal level. Console automation and automated analog signal processors often make extensive use of VCA technology.

waveform   A graph of a signal's sound pressure or voltage level versus time. The waveform of a pure tone is a sine wave.

wavelength   The distance in a medium between corresponding points on adjacent waveform cycles.

I

Index

V

W

X–Y–Z

 **Focal Press**

Focal Encyclopedia of Electronic Media (CD-ROM)

Editorial Board Members:
Christopher H. Sterling, Editor-in-Chief
Stanley R. Alten
Lynne S. Gross
Skip Pizzi
Joan M. Van Tassel

- Over 4,000 articles, from Acoustics to Zoom lens
- Contributions from over 50 leading experts in their fields
- Powerful search and retrieval functions link topics for quick reference
- Video and audio clips bring information to life
- Extensive bibliographies make this an excellent research tool
- Contains important industry and academic Website links for users with web access
- Network version available for multiple users

This is an up-to-date, authoritative, and comprehensive reference source. Directed to applications of electronic media technology, it clearly explains technologies and methods. Entry emphasis is on clear language, illustrations, organization, and ease of use. Applications, processes, definitions and use of equipment are described in detail. This information is important for students, libraries, creative personnel, technical professionals, management decision-makers, corporate media users and directors, and anyone seeking background on applications of and trends in media equipment and processes.

 June 1997   *ISBN 0-240-80133-4, $149*
Network Version (Up to 10) - ISBN 0-240-80310-8, $250

Available from all good book stores or in case of difficulty call:
1-800-366-2665 in the U.S. or +44 1865 310366 in Europe.